T0201933

THE THERMOPHYSICAL PROPERTIES
OF METALLIC LIQUIDS

The Thermophysical Properties of Metallic Liquids

Volume 1: Fundamentals

Takamichi Iida

Professor Emeritus of Materials Science and Processing
Osaka University, Japan

Roderick I.L. Guthrie

Macdonald Professor of Metallurgy at McGill University, and
Director and co-founder of the McGill Metals Processing Centre
McGill University, Montreal, Canada

OXFORD

UNIVERSITY PRESS

OXFORD
UNIVERSITY PRESS

Great Clarendon Street, Oxford, OX2 6DP,
United Kingdom

Oxford University Press is a department of the University of Oxford.
It furthers the University's objective of excellence in research, scholarship,
and education by publishing worldwide. Oxford is a registered trade mark of
Oxford University Press in the UK and in certain other countries

© Takamichi Iida and Roderick I. L. Guthrie 2015

The moral rights of the authors have been asserted

First Edition published in 2015
Reprinted 2015

Impression: 2

Published in the United States of America by Oxford University Press
198 Madison Avenue, New York, NY 10016, United States of America

British Library Cataloguing in Publication Data
Data available

Library of Congress Control Number: 2015930577

ISBN 978–0–19–872983–9

Printed in Great Britain by
Clays Ltd, St Ives plc

Preface

High temperature materials processing, such as smelting, refining, casting, welding, crystal growth, melt spinning, zone melting, and spraying, is an extremely complex area in which many thermophysical properties of metallic liquids are intimately involved. Numerous studies have been made to unravel this complexity on the basis of thermodynamics, reaction kinetics, and hydrodynamic analyses.

Between the 1950s and 1980s, great progress has been made in the industrial technology and science of process metallurgy, particularly for iron- and steelmaking processes. The advances in the science of extractive metallurgy have been described in numerous books, such as *Physical Chemistry of Metals* by Darken and Gurry (1953) and *Physical Chemistry of Melts in Metallurgy* by Richardson (1974). Advances in science and technology relating to liquid metallic processing operations are covered by books such as *Physical Chemistry of High Temperature Technology* by Turkdogan (1980) and *Engineering in Process Metallurgy* by Guthrie (1989). These books cover a wide range of topics relating to the physical chemistry of melts (i.e. liquid metals, salts, and slags): structure, physical properties, thermodynamic properties, reaction kinetics, interfacial phenomena, related mass/heat transport phenomena, and attendant fluid flows. As such, they are of great significance for process metallurgists. However, the former books place an emphasis on chemical, and reaction kinetics, while the latter emphasize process kinetics in metallurgical systems, related mainly to iron- and steelmaking. For such macroscopic or continuum treatments, the values of thermophysical properties are generally taken as empirical constants.

The advent of semiconductors, metallic glasses, and other functional materials since the late 1950s, and ever-increasing demands for the manufacture of high quality metallic materials have broadened the scope of high temperature science and technology of metallic liquids. At the same time, metallurgy has expanded into materials science and materials engineering. As such, materials process science has come to be based not only on chemistry but also on physics. This is especially true for the efficient manufacture of high quality metallic materials, where a detailed knowledge of the thermophysical properties of metallic liquids in question, based on the theory of liquids, is required. This atomistic, or microscopic, approach to materials processing, in contrast to equilibrium chemical thermodynamics, and process kinetics, provides an essential underpinning to understanding and interpretation of various phenomena in liquid metallic processing operations. From this viewpoint, the present authors published a technical book *The Physical Properties of Liquid Metals* (1988). The purpose of that book was to introduce theoretical equations based on structure, semi-empirical (semi-theoretical) equations, empirical equations, together with methods of experimental measurement and experimental data, for the physical properties of liquid metals. This book was written for metallurgical and materials research workers.

Since the 1970s, there has been renewed interest in the thermophysical properties of almost all liquid metallic elements or simple metallic substances (i.e. liquid metals, semimetals, and semiconductors), following the advent of mathematical modelling techniques supported by powerful computers. Nowadays, computer simulation studies of materials processing operations, based on mathematical models, are widely used as a very useful tool for improving liquid metal and liquid metallic processing operations and product quality. Naturally, accurate and reliable data for the thermophysical properties of metallic liquids are indispensable, not only for the execution of computer simulations and for the development of mathematical models, but also for the direct solution of industrial high temperature processing operations. In particular, accurate data for almost all liquid metallic elements are first needed. To elaborate further on this matter, while multicomponent alloys are typically treated in liquid metallic processing operations, accurate and reliable data for the respective pure components of an alloy system are first needed as a starting point. In addition, we must go deeply into a study of an element's (i.e. simple substance) properties, in order to clearly understand the essence of a metallic liquid's thermophysical properties. For example, in the liquid state, the motions of atoms through the liquid are impeded by frictional forces set up by their nearest neighbours. Thus, the atomic transport coefficients, i.e. molecular viscosity and diffusivity, of metallic liquids are dominated by the frictional forces among atoms. Furthermore, the atomic transport coefficients cannot be entirely formulated in terms of thermodynamic properties alone, because the coefficients involve the movement of atoms. Owing to this, model theories, such as the hole theory or the rate process (activated state) theory, are unlikely to provide fruitful predictions of transport coefficients of metallic liquids.

A large number of research articles on the thermophysical properties of metallic liquids have been published in the last quarter century or so. Their results or main points have been organized in numerous review articles and books such as *Measurement and Estimation of Physical Properties of Metals at High Temperatures* by Mills in *Fundamentals of Metallurgy* edited by Seetharaman (2005). This book emphasizes the need for reliable data for the thermophysical properties of liquid materials (e.g. liquid metals) involved in high temperature processes. Mills states that thermophysical property data are beneficial in two ways: (1) in the direct solution of industrial problems, and (2) as input data for the mathematical modelling of processes. Unfortunately, the explanations for the various thermophysical properties of melts are condensed into 70 pages in that book, so details for each property are not described. (However, over 170 important articles are referred.)

As mentioned previously, there is an urgent need for reliable data for the thermophysical properties of metallic liquids. Indeed, accurate and reliable data for the thermophysical properties, even of liquid metallic elements, are still not necessarily plentiful. The absence of accurate data for metallic liquids is mainly due to the experimental difficulties in obtaining accurate values for these thermophysical properties at high temperatures. The measurements of the thermophysical properties of metallic liquids at high temperatures are time-consuming, expensive, and in some cases, impossible. Further, they require considerable experimental and metrological expertise. Thus, much effort has been directed towards the development of reliable models for accurate predictions of the thermophysical properties of metallic liquids. Even so, for many years, little progress has been made on reliable models for accurate predictions. Early in this century, a new approach to making such predictions was presented for several thermophysical properties of all liquid metallic elements.

As seen from this brief historical outline (presented above), describing scientific approaches to the manufacture of metallic materials, the purposes of materials process science in this area are (1) to provide a clear understanding of the structure and thermophysical properties of metallic liquids based on the liquid state physics, and (2) to develop models that can be used for accurate predictions of the thermophysical properties of metallic liquids, particularly for all, or almost all, liquid metallic elements. In the area of materials process science, both accuracy and universality are required of any model for predicting the thermophysical properties of metallic liquids. Other purposes of materials process science are (3) to provide appropriate evaluation of the thermophysical property data for metallic liquids, and (4) to act as a guide to creation of new types of materials. In order to achieve these four main purposes, materials process science must incorporate a blending in of knowledge in many subject area, including process metallurgy, materials science, condensed matter physics (particularly, liquid state physics), molecular physics, chemistry (particularly, theoretical chemistry, and inorganic chemistry), and finally metrology (particularly for high temperature experiments on the thermophysical properties of metallic liquids).

The present book is divided into two volumes. This book of two volumes is a completely revised version of the authors' previous book *The Physical Properties of Liquid Metals* (1988): the present book lays emphasis on both the *Fundamentals* (*Volume 1*) and on *Predictive Models* (*Volume 2*) for accurate predictions of the thermophysical properties of metallic liquids. The performances of models/equations for several thermophysical properties (e.g. sound velocity, surface tension, viscosity) of liquid metallic elements are quantitatively assessed by determining relative differences between the calculated and experimental property values.

Volume 1 is intended as an introductory text explaining the structure and thermophysical properties of metallic liquids for students of materials science and engineering, and also for research scientists and engineers who have an interest in liquid metal and liquid metallic processing. The authors have tried to give simple explanations. However, the level of its contents is necessarily relatively advanced, since knowledge of the various scientific fields, mentioned above, is blended into one.

It comprises nine chapters, i.e. Chapters 1–9. Several basic matters for understanding the thermophysical properties of metallic liquids and for developing reliable models to accurately predict their thermophysical properties, together with methods for assessment of models/equations, are briefly described in Chapter 1. Chapter 2 is an introduction to the structure of metallic liquids. Two fundamental quantities in the theory of liquids, i.e. the pair distribution function and the pair potential are described. In particular, a reasonably detailed description of the distribution function is presented, because an understanding of the thermophysical properties of liquids must be based on a fundamental understanding of a liquid's atomic arrangement. Chapter 3 is concerned with the density of metallic liquids. Although density, or number density (i.e. number of atoms per unit volume), is an indispensable and basic quantity, accurate data are not plentiful. A new model for the temperature dependence of liquid metallic element density (or volume expansivities) is introduced. Thermodynamic properties of a metallic liquid, i.e. evaporation enthalpy, vapour pressure, heat capacity, are outlined in Chapter 4. The velocity of sound in a liquid metallic element is described in Chapter 5. Dimensionless new common parameters for better predictions of several important thermophysical properties of metallic liquids can be extracted from their sound velocity data. The new common parameters give an indication of an atom's hardness or softness, as well as a piece of useful information about the structure of metallic liquids. The new parameters are also useful in discussions of anharmonic effects of atomic motions in metallic liquids. This theme is further developed in Chapter 6, dealing with surface tension, and Chapter 7, which covers the dynamic property, or transport property, viscosity, (also considered in volume 2). In Chapter 6, characteristic features of experimental data for metallic liquid surface tensions are identified. Reasons for large discrepancies among experimental data for metallic liquid viscosities are clarified in Chapter 7. In Chapters 6 and 7, the relationship between surface tension and viscosity for liquid metallic elements is discussed in some detail. Chapter 8 is concerned with diffusion in metallic liquids. Knowledge of diffusion is needed for many fields of engineering. However, even self-diffusivity data are extremely scarce; a predictive model for metallic liquid self-diffusivity, expressed in terms of well-known physical quantities, is presented. Metallic liquids, like solid metals, are characterized by high electrical and thermal conductivities. Fundamentals of electronic transport properties, i.e. electrical conductivity, or electrical resistivity, and thermal conductivity, of metallic liquids are discussed in Chapter 9, the final chapter. The essential points of methods for measuring density, surface tension, viscosity, diffusivity, electrical resistivity, and thermal conductivity are also described.

Volume 2 is designed for research scientists and engineers engaged in liquid metallic processing. In Volume 2, using as a basis the fundamental issues presented in Volume 1, we discuss models used for predicting accurate values of metallic liquid thermophysical properties.

It contains eight chapters, a glossary, and ten appendices. Essential points in building reliable models for accurate predictions of the thermophysical properties of liquid metallic elements are outlined in Chapter 10, from the standpoint of materials process science. Chapter 11 is devoted to the velocity of sound in liquid metallic elements. Useful dimensionless common parameters (or dimensionless numbers), which characterize the metallic liquid state, can be revealed through data for the velocity of sound. These common parameters allow for better predictions of several thermophysical properties of liquid metallic elements. Models, in terms of the common parameters, are discussed for the volume expansivity, evaporation enthalpy, surface tension, viscosity, and self-diffusivity of liquid metallic elements, in Chapters 12–16. The performances of the various models are evaluated by comparing them against experimental values (provided experimental data are available). Predicted, or calculated, data for sound velocity, volume expansivity, evaporation enthalpy, surface tension, viscosity, and self-diffusivity of liquid metallic elements are given in Chapters 11–16. In Chapters 10, 11, and 13–16, atomic periodicity in values of each thermophysical property, discussed in this book, is illustrated for a large number of liquid metallic elements. In Chapter 17, the final chapter, a large number of experimental data for the physical quantities and the thermophysical properties of liquid metallic elements are compiled, although we emphasize that this book is not primarily a reference data book.

Prior to the Appendices, a glossary and/or supplementary explanations are provided. The periodic table contains many pieces of useful information about the thermophysical properties of elements; one form of periodic table is given in Appendix 1. Appendices 2 and 3 provide numerical expressions for determining the minimum values of relative standard deviation. SI units, unit conversions, fundamental physical constants in SI units, and the Greek alphabet, are given in Appendices 4–7. Occam's razor is an essential guide to the development of any model for accurate predictions of the properties of materials; Appendix 8 cites Occam's razor. In Appendices 9 and 10, calculated values of isothermal compressibility, structure factor, and the ratio of heat capacity of some liquid metallic elements at their melting point temperatures, are all listed.

Over the last half century, a huge number of research articles and review articles on the thermophysical properties of metallic liquids have been reported. Even so, our present knowledge of the thermophysical properties of metallic liquids is still lacking from the materials process science and engineering points of view. Accurate and reliable data for the thermophysical properties of metallic liquids are still not plentiful. Systematic investigations based on theory and experiment are greatly needed from the standpoint of materials process science and engineering. As such, we hope that this two volume book will not only be used for obtaining the relevant constants for the properties of specific liquid metallic elements, but will also help the user recognize the continuum between the microscopic and macroscopic approaches to liquid metallic processing operations. For the future, we hope this book will help in enabling great progress in materials process science.

Takamichi Iida
Roderick I.L. Guthrie
August, 2015

Acknowledgements

We wish to express our gratitude towards the offices of the McGill Metals Processing Centre for infrastructure support for this long running endeavour.

We thank Professor S. Iitaka of Gakushuin University, and Professor Emeritus H. Ishigaki of Waseda University for their helpful comments on the mathematical treatment given in Appendices 2 and 3. Similarly, we wish to thank Dr I. Yamauchi for his patient and expert work in producing the final typescript of the present, two-volume text book.

The first author sincerely thanks his wife, the late Sugaho, for her devoted assistance in typing the draft manuscripts and numerical calculations. He is also grateful to his daughter, Eri, for her continued cooperation.

Finally, we would like to express our sincere gratitude towards the Canadian Immigration Authorities and the Natural Science and Engineering Research Council of Canada for their support in allowing this extended period of collaboration in Canada.

Contents

Volume 1 Fundamentals

Volume 2 Predictive Models

Principal Symbols

Numbers in parentheses refer to equations.

Capital Italic

A	area (molar surface, oscillating plate, etc.)
A	parameter (2.11)
A, B, C, D	constants
A_i	area of component i
A_r	relative atomic mass
A^*	reduced area
B_s	isentropic, or adiabatic, bulk modulus
C_A	Andrade coefficient
$C_{Aw}(\eta)$	correction factor (8.13)
C_{el}	heat capacity of electron gas
C_L	constant (1.16)
C_P, C_V	heat capacity at constant pressure, at constant volume
D	self-diffusivity
D_{HS}	self-diffusivity in the hard-sphere fluid
$D_i, D_{S,M}$	solute, solvent (or base metal) diffusivity
D_0	frequency factor (8.21)
D^*	reduced rectilinear diameter (3.4), parameter (corresponding -state principle (8.17))
$E, \boldsymbol{E}$	kinetic energy; electric field
E, E_a	resonant amplitude of plate in liquid, in air
E_C^0	molar cohesive energy (solid) at 0 K at 1 atm (101.325 kPa)
E_F	Fermi energy
E_V^*	height of potential barrier
F	Helmholtz free energy
G	Gibbs free energy
G_s	total molar surface energy
$\tilde{G}(Q)$	Fourier transform of $\{g(r) - 1\}$
H	enthalpy; height of liquid sample
H^E	enthalpy of mixing
H_μ, H_D	constants or parameters (apparent activation energy for viscous flows or for diffusion)
$1/H$	shape factor of liquid drop

$\Delta_s^l H_m$	enthalpy of melting
$\Delta_l^g H_b$	enthalpy of evaporation at the boiling point temperature
$\Delta_s^g H_0$	enthalpy of sublimation at 0 K
I	intensity (of X-ray beam); moment of inertia of suspended system
$\mathcal{J}$	flux of matter
K	coverage independent adsorption coefficient
$K, L, M, \ldots$	shells (energy level)
K, K_0	apparatus constants
K_f	force constant
M	molar mass
N	number of atoms; number of samples
N_A	Avogadro constant
P	pressure
ΔP	pressure difference
$P(T)$	probability function
$P(A), P(B)$	proportion of vibrator A, B (7.38)
P_1	Legendre polynomial
P_m	maximum bubble (gas) pressure
Q	parameter (6.11)
Q_E	activation energy (8.21)
R	molar gas constant; radius
S	entropy
$S, S(N)$	relative standard deviation
S_A	surface area
$S(Q)$	structure factor, or interference function
$S_{\alpha,\beta}(Q)$	partial structure factor
S_S	molar excess surface entropy
$\Delta_s^l S_m$	entropy of melting
$\Delta_l^g S_b$	entropy of evaporation at the boiling point temperature
T	absolute temperature; period of oscillation
T^*	reduced temperature
U	sound velocity; total internal energy
U_b	mobility
U_l	total internal energy (liquid)
$U(Q)$	pseudopotential
V	volume; molar volume (e.g. (7.14))
V_{mol}	molar volume
V_A	molar volume for a binary system
V^E	excess volume (3.23)
ΔV_m	volume change on melting
$W(\phi)$	probability of a thermal fluctuation
X, X', Y, Z, Z'	parameters (measurements of surface tension)
Z	first coordination number; partition function (2.15)
Z_i, Z_s	first coordination number within bulk, at surface

Lower Case Italic

a	average interatomic distance
a, b, c, d, q	constants
a_i, a_s	activity of component i or s
c	concentration
d	diameter of an atom (or ion); distance between solute and solvent ions
dv	volume element
e	electron (or electronic) charge
f	atomic scattering factor
f	activity coefficient; surface-packing, (or configuration), factor
f, f_a	resonant frequency in liquid, in air
$f(s), f_z$	interatomic forces
g	acceleration of gravity
$g(r)$	pair distribution function
$g_{\alpha,\beta}(r)$	partial pair distribution function
g_s	surface energy per unit area
h	depth of immersion
$h, \hbar$	Planck, Dirac constants ($\hbar = h/2\pi$)
$\bar{h}$	difference in height
i	chemical constant
j	electric current density
k	Boltzmann constant
k_F	radius of the Fermi sphere, or wave (number) vector
k_f	force constant
k_0	dimensionless numerical factor (6.37)
l	orbital quantum number
l	effective thickness of interface; length; mean free path (electron)
m	atomic mass (or mass of an atom); mass of a liquid drop
m	magnetic quantum number
m_e	electron mass
n	principal quantum number; repulsive exponent
n	amount of substance (mole in SI units)
n	number density; electron number density (9.3)
n_0	average number density
Δn	fluctuation in number density
$\dot{q}$	power input (9.15)
r	radial distance; radius
r_0	value of r at the left-hand edge of the first peak in $g(r)$ curve
r_m	value of r at the first peak, or the main peak, in $g(r)$ curve
r.d.f.	radial distribution function
s	spin quantum number
s	surface tension correction (3.16); distance of displacement
s, d, p, f	block (cf. Table 1.7)

$s, p, d, f \ldots$	subshells (energy level)
t	time
u_v	internal energy (during vibration) (1.14)
u, v	pairwise interaction energy among atoms within bulk, on surface
v	volume of immersed suspension wire (3.16)
v	average electron velocity (9.11)
v_0	close-packed molecular volume
v_F	Fermi velocity
v_f	average free volume per molecule
w	work, or energy, necessary to separate an atom
Δw	apparent loss of weight
x	molar, (or mole), fraction; length
x, y, z	coordinate axes
z	number of valence electrons
z^E	excess valence

Greek

α	volume expansivity
α_a	absorption coefficient
$\underline{\alpha}$	parameter related to the distance over which the interatomic force extends
α, β	phases
β	correction factor (for Lindemann's equation)
β_S	coefficient of sliding friction
Γ	damping constant
Γ_s	excess surface concentration of solute s
Γ_s^0	saturation coverage by solute s
γ	surface tension
$\gamma_G, \gamma_{G,E}, \gamma_{G,T}$	Grüneisen constants (cf. Section 5.5)
γ_h	$\equiv C_P / C_V$ (ratio of the isobaric and isochoric heat capacities)
γ_M	surface tension of mixtures
γ_0	surface tension of pure solvent (6.58)
γ_0	$\equiv k_0 C_A$
Δ	dimensionless variable (3.6)
$\Delta, \Delta(N)$	global delta (1.21)
δ	amplitude of vibration of each atom (1.12)
δ, δ_0	logarithmic decrements
δ_i	relative difference between experimental and calculated values for χ_i
ζ_A	parameter (8.20)
ζ_f	friction coefficient
$\zeta_H, \zeta_S, \zeta_{SH}$	friction coefficients due to hard-core collision, to soft interaction, to cross effect
η	packing fraction (or packing density)
η_l	phase shift

θ	Einstein characteristic frequency
θ	contact angle; angle
2θ	scattering angle
θ_S	fractional coverage
κ	transmission coefficient
κ_S	isentropic, or adiabatic, compressibility
κ_T	isothermal compressibility
Λ	temperature dependence of the density of liquid metallic elements
λ	thermal conductivity
λ	wavelength; screening radius (8.36)
μ	viscosity; constant ($\mu \equiv bT_c / \rho_c$, (3.4))
μ_A	viscosity of binary liquid mixtures (or alloys)
$\mu_\kappa, \mu_{\phi(\sigma)}$	viscosity due to kinetic contribution, to a hard-sphere collision
μ_ϕ	soft attractions
μ^E	excess viscosity
ν	kinematic viscosity ($\nu \equiv \mu / \rho$)
ν	atomic frequency
ξ_E, ξ_T, ξ	dimensionless common parameters ($\xi \equiv \xi_T / \xi_E$)
π	pi (3.141592 . . .)
ρ	density
ρ_A	alloy density
ρ_e	electrical resistivity
σ	effective hard-sphere diameter; diameter of molecule (7.11)
σ_e	electrical conductivity
τ	mean free time
Φ	total potential energy
ϕ	excess binding energy; fluctuation in kinetic energy (7.21)
ϕ_H, ϕ_S	hard-sphere potential, soft potential
$\phi(r)$	pair potential (energy)
$\tilde{\phi}_S(Q)$	Fourier transform of the long-range part of potential
χ_i	thermophysical property (e.g. sound velocity, surface tension, viscosity, etc.)
ω_c	characteristic oscillation frequency

Subscripts

b	at boiling point
c	at critical point
m	at melting point; value of r at the first peak, or the main peak, in $g(r)$ curve
g	gas
l	liquid
s	solid

Selected General References

1. D. Tabor, *Gases, Liquids and Solids and Other States of Matter*, 3rd ed., Cambridge University Press, 1991.
2. G.S. Rushbrooke, *Statistical Mechanics*, Oxford University Press, 1949.
3. C. Kittel, *Introduction to Solid State Physics*, 8th ed., John Wiley & Sons, 2005. (See also 2nd ed., 1956; 7th ed., 1996).
4. E.A. Moelwyn-Hughes, *Physical Chemistry*, Pergamon Press, 1961.
5. D.H. Trevena, *The Liquid Phase*, Wykeham Publications, 1975.
6. M.A. White, *Properties of Materials*, Oxford University Press, 1999.
7. W. Benenson, J.W. Harris, H. Stocker, and H. Lutz (eds.), *Handbook of Physics*, Springer-Verlag, 2002.
8. A. Stwertka, *A Guide to the Elements*, 2nd ed., Oxford University Press, 2002.

1

An Introductory Description of Metallic Liquids

1.1   Introduction

Solid, liquid, and gas are known as the three phases of matter or the three states of matter.[1] It is well known that matter can exist in one of these three phases depending on the conditions of temperature and pressure, and that there are, to all appearances, obvious differences between these three phases. Without exception, simple metallic substances, i.e. pure metals, semimetals (antimony and bismuth), and semiconductors (e.g. silicon and germanium), to be described in this book, and their alloys exist in one of these three phases. Incidentally, in the liquid state, the simple metallic substances (including the semimetals and semiconductors mentioned earlier) are good conductors of electricity and heat; metals are characterized by high electrical and thermal conductivities.

A liquid is a phase of matter intermediate between solid and gaseous phases. Nevertheless, values for the thermophysical properties of a liquid are never equal, or even approximately equal, to the mean determined from those of a solid and a gas. Thus, in some properties, a liquid resembles a solid, while in other properties, a gas. In this volume we try to explain, in a simple way, the structure and thermophysical properties of metallic liquids taking an atomistic, or microscopic, point of view. This is one of the most important purposes of materials process science in liquid metallic processing. However, of these phases of matter, it is the most difficult to seize the essence of the intermediate liquid phase (whose behaviour is the result of the cooperative interaction within a system of random order, or, at best, local order) on the basis of a rigorous atomistic analysis. Indeed, from a historical viewpoint, the theory of gases was first formulated in the latter half of the nineteenth century. Then, the theory of solids was begun in the early twentieth century, and has rapidly progressed since the introduction of quantum mechanics, i.e. since the 1920s. By contrast, the theory of liquids begun, at long last, in the 1930s, even

---

[1] Solid, liquid, gas, and plasma are called the four states of aggregation. The plasma (i.e. a gaseous mixture in which electrons and atomic nuclei, or ions, move freely) is regarded as the fourth state of matter. The plasma state occurs at very high temperatures, as in stars; the atoms are ionized and decomposed into charged constituents.

The Thermophysical Properties of Metallic Liquids: Volume 1 – Fundamentals. First Edition.
Takamichi Iida and Roderick I. L. Guthrie. © Takamichi Iida and Roderick I. L. Guthrie 2015.
Published in 2015 by Oxford University Press.

though for chemists, physical chemists, biologists, and metallurgists (extractive, casting), the liquid state is of great importance; since the late 1940s considerable progress has been made on characterizing the liquid state.

This chapter provides several basic concepts necessary for advancing from the fundamentals of knowledge of metallic liquids' structure and thermophysical properties to their applications.

1.2 Preliminary to Studies of Metallic Liquids

1.2.1 Three Phases of Matter

The three phases of matter can be shown using a phase diagram. This is a graph showing the relationship between solid, liquid, and gaseous phases over a range of conditions, e.g. temperature (T), pressure (P), and volume (V). The $P–V–T$, $P–T$, and $P–V$ phase diagrams for a simple substance are shown in Figure 1.1(a–c), respectively.

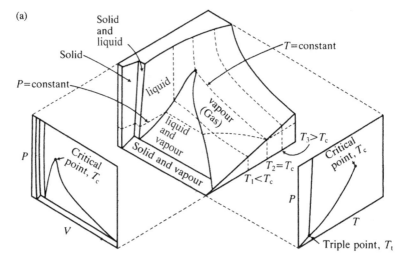

Figure 1.1 *(a) P–V–T phase diagram for a simple substance (after Giedt [1]). (b) P–T phase diagram for a simple substance. The temperatures of glass transition T_{gt}, triple T_t, melting T_m, boiling T_b, and critical points T_c, are indicated. The pressures of triple point P_t, atmosphere P_{at}, and critical point P_c are also indicated. The arrowed broken path represents a change from liquid to vapour (or from vapour to liquid) without phase transition. In the supercooled state, the liquid is metastable. (c) P–V phase diagram for a simple substance. V_l and V_g are the volumes of the liquid and vapour coexisting in equilibrium at temperature T_1 and pressure P_1. V_c is the volume of the critical point.*

(b)

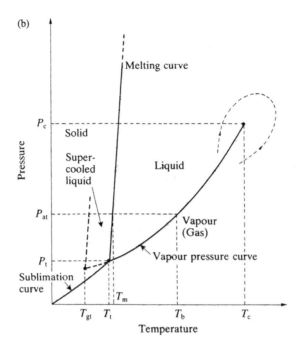

(c)

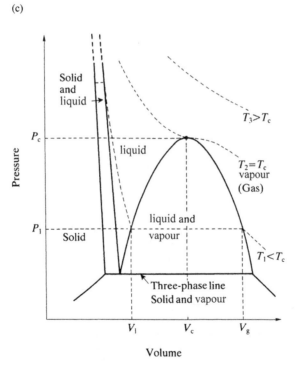

Figure 1.1 *(continued)*

Let us outline the main points of the phase diagram by referring to Figure 1.1(a–c). At the triple point, the three phases of matter, i.e. the solid, liquid, and gaseous phases, are in equilibrium. The critical point represents the upper limit of the matter's possible existence as a liquid. Above the critical point, no distinction between the liquid and the gas (vapour[2]) can be made, and a single, undifferentiated, liquid state of uniform density exists.[3] By contrast, however, there is no experimental or theoretical evidence that there is an end-point to the melting curve: there is no upper critical point for melting. The critical point is a material constant which characterizes the specific properties of the material. Unfortunately, few experimental data for the critical point of metallic elements[4] are available because of the experimental difficulties in the critical region. Between the critical and triple points, two phase changes can be observed: melting and evaporation (or condensation and solidification). This is shown in Figure 1.2. The variation of the melting or boiling point temperature with applied pressure is given

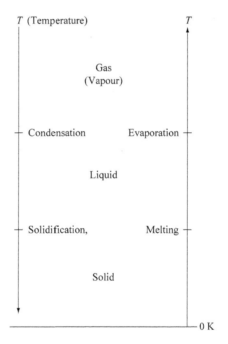

Figure 1.2 *Phase changes with increasing (or decreasing) temperature for a simple substance.*

---

[2] The vapour is a special case of a gas and not a distinct form. If the gas is below the critical temperature, it is usually called a vapour.

[3] For liquids at temperatures and pressures slightly above the critical conditions (or critical point), they are termed supercritical fluids. Supercritical fluids are neither liquid nor gas, and they have most unusual physical properties such as possessing a much lower viscosity than the liquid and a much higher density than the gas.

[4] 'Metallic elements' mean 'simple metallic substances'.

thermodynamically by the Clausius–Clapeyron equation (see Subsection 1.2.2.). The lower limit for the matter's existence as a liquid is generally taken to be its triple point temperature T_t. However, the phenomenon of supercooling is well known and the liquid can exist in the supercooled state at $T < T_t$. Incidentally, both liquids and gases are often called fluids, or the fluid phases, as a collective term embracing them; similarly, the liquid and solid are also both referred to as condensed matter, or the condensed phases.

1.2.2 Phase Transitions

If a solid such as a simple metallic substance is heated,[5] it starts to melt when the melting point temperature is reached. It then changes into a liquid, which, in turn, boils and changes into a vapour or a gas. An additional supply of heat at the melting (or boiling) temperature does not lead to an increase of temperature, but only to further melting (or evaporation) of the solid (or the liquid), as shown in Figure 1.3. The quantity of heat absorbed or released in a change of phase, i.e. a phase transition, is called the latent heat.[6] For the description of processes proceeding at constant pressure, enthalpy H (a

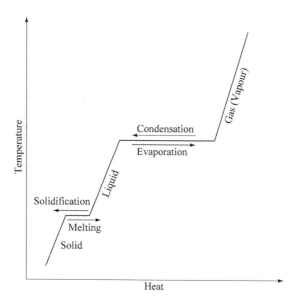

Figure 1.3 *Temperature increase (or decrease) by supply (or release) of heat.*

---

[5] From a microscopic point of view, if a solid is heated, the atoms acquire additional thermal energy; as a result, if the temperature is steadily raised, the atoms vibrate about their lattice points with amplitudes which will continually increase, depending on the strength of the bonding, or the interactions among the atoms.

[6] The heat is released when a gas condenses or a liquid solidifies.

thermodynamic function of a system) is of importance;[7] in phase transitions at constant pressure and at constant temperature, the change of enthalpy ΔH of the substance is equal to the latent heat absorbed (in melting, and boiling) or released (in solidifying, and condensation).

1.2.2.1   The Clausius–Clapeyron Equation

Let us consider two single-component phases α and β in equilibrium with each other, as shown in Figure 1.4. At the point A, corresponding to a pressure P and temperature T, the Gibbs energies of the two phases α and β that are in equilibrium are identical (i.e. this is the condition for equilibrium between phase α and phase β). Therefore, we can write

$$G^\alpha = G^\beta \tag{1.1}$$

in which G is the Gibbs (free) energy, or Gibbs function. At a neighbouring point B on the equilibrium curve where the pressure and the temperature are changed by dP and dT, respectively (see Figure 1.4), we then have

$$G^\alpha + dG^\alpha = G^\beta + dG^\beta \tag{1.2}$$

From Eqs. (1.1) and (1.2), we obtain

$$dG^\alpha = dG^\beta \tag{1.3}$$

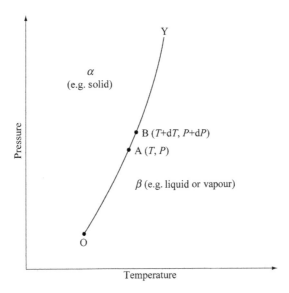

Figure 1.4 *OY is the $\alpha - \beta$ phase equilibrium curve, while A, B are two neighbouring points on this curve.*

---

[7] In general, liquid metallic processing operations are carried out at the atmospheric pressure, i.e. at about constant pressure. The atmospheric pressure fluctuates by about 10%, depending on the weather and temperature.

When we use a fundamental equation for dG $(=V dP - S dT)$, Eq. (1.3) becomes

$$V^{\alpha} dP - S^{\alpha} dT = V^{\beta} dP - S^{\beta} dT \tag{1.4}$$

where V is the volume, and S the entropy.

Thus,

$$\frac{dP}{dT} = \frac{S^{\beta} - S^{\alpha}}{V^{\beta} - V^{\alpha}} = \frac{\Delta_{\alpha}^{\beta} S}{\Delta_{\alpha}^{\beta} V} = \frac{\Delta_{\alpha}^{\beta} H}{T \Delta_{\alpha}^{\beta} V} \tag{1.5}$$

This is well known as the Clausius–Clapeyron equation. Here we give a short explanation of Eq. (1.5). At constant temperature, the change in Gibbs free energy ΔG is given by $\Delta G = \Delta H - T \Delta S$. The Gibbs free energies of two phases α and β in equilibrium are identical, so $\Delta_{\alpha}^{\beta} G$ due to phase transition is equal to zero: $\Delta_{\alpha}^{\beta} G = \Delta_{\alpha}^{\beta} H - T \Delta_{\alpha}^{\beta} S = 0$. Thus, $\Delta_{\alpha}^{\beta} H = T \Delta_{\alpha}^{\beta} S$, (or $\Delta_{\alpha}^{\beta} S = \Delta_{\alpha}^{\beta} H / T$), in Eq. (1.5), is the entropy change for a phase transition from α to β.

In liquid metallic processing operations, the phenomena of melting (or solidifying) and evaporation (or condensation) are of great importance.

For the transition from solid to liquid, the Clausius–Clapeyron equation can be expressed as

$$\frac{dP}{dT} = \frac{\Delta_{s}^{l} H_{m}}{T(V_{l} - V_{s})} \tag{1.6}$$

where $\Delta_{s}^{l} H_{m}$ is the molar melting enthalpy,[8] or the molar latent heat of fusion, T is the absolute temperature $(T > 0)$, V is the molar volume (the subscripts l and s denote liquid and solid, respectively). $\Delta_{s}^{l} H_{m}$ is positive (i.e. $\Delta_{s}^{l} H_{m} > 0$) and $V_{l} - V_{s}$ is positive for most metallic elements; dP/dT along the solid–liquid equilibrium curve is positive for most metallic elements, so that increasing the pressure raises the melting point temperature. However, for some metallic elements (e.g. gallium, bismuth, silicon, plutonium, etc.; see Table 3.1) $V_{l} - V_{s} < 0$, dP/dT is negative; an increase in pressure causes a drop in the melting point temperature (like the phase transition from ice to water).

For the transition from liquid to gas, the Clausius–Clapeyron equations is

$$\frac{dP}{dT} = \frac{\Delta_{l}^{g} H_{b}}{T(V_{g} - V_{l})} \tag{1.7}$$

where $\Delta_{l}^{g} H_{b}$ is the molar evaporation enthalpy,[9] or the molar latent heat of evaporation, at the boiling point. $\Delta_{l}^{g} H_{b}$ is positive, and the molar volume of gas V_{g} is always greater than that of liquid V_{l}. As such, dP/dT is always positive, i.e. the boiling point temperature always rises with increasing the pressure.

---

[8] This amount of heat must be absorbed during the phase transition from solid (one mole) to liquid at the melting point temperature. Instead of the molar quantities, we may use specific quantities (e.g. specific melting enthalpy).

[9] This amount of heat is needed to evaporate one mole of liquid at the boiling point temperature.

Table 1.1 *Values of relative atomic mass, melting point, boiling point, and liquid range of elements (metals, semimetals, and semiconductors) at standard pressure.*[a]

Element		Relative atomic mass	Melting point[a] °C	Boiling point[a] °C	Liquid range[b] °C
Actinium	Ac	227.0278*	1050[c]	3200 ± 300[c]	2150 ± 300
Aluminium	Al	26.981539	660.323[†]	2519	1858.677
Americium	Am	243.0614*	994 ± 4[c]	2607[c]	1613 ± 4
Antimony	Sb	121.757	630.63[†]	1587	956.37
Arsenic	As	74.92159	817 (at 2.8 MPa)[c]	613 (sublimation)[c]	—
Astatine	At	209.9871*	302[d]	~337[d]	~35
Barium	Ba	137.327	727	1897	1170
Beryllium	Be	9.012182	1287	2471	1184
Bismuth	Bi	208.98037	271.4	1564	1292.60
Boron	B	10.811	2300[c]	2550[c]	250
Cadmium	Cd	112.411	321.069[†]	767	445.931
Caesium	Cs	132.90543	28.44	671	642.56
Calcium	Ca	40.078	842	1484	642
Carbon	C	12.011	sublimation at 3652[c]		—
Cerium	Ce	140.115	798	3433	2635
Chromium	Cr	51.9961	1907	2671	764
Cobalt	Co	58.93320	1495[†]	2927	1432
Copper	Cu	63.546	1084.62[†]	2562	1477.38
Dysprosium	Dy	162.50	1412	2567	1155
Erbium	Er	167.26	1529	2868	1339
Europium	Eu	151.965	822[c]	1527[c]	705
Francium	Fr	223.0197*	27[d]	677[d]	650
Gadolinium	Gd	157.25	1313	3273	1960
Gallium	Ga	69.723	29.7646[†]	2204	2174.2354
Germanium	Ge	72.61	937	2830	1893

Table 1.1 *(continued)*

Element		Relative atomic mass	Melting point[a] °C	Boiling point[a] °C	Liquid range[b] °C
Gold	Au	196.96654	1064.18[†]	2856	1791.82
Hafnium	Hf	178.49	2233	4603	2370
Holmium	Ho	164.93032	1474	2700	1226
Indium	In	114.818	156.5985[†]	2072	1915.4015
Iridium	Ir	192.22	2446[†]	4428	1982
Iron	Fe	55.847	1538	2861	1323
Lanthanum	La	138.9055	918	3464	2546
Lead	Pb	207.2	327.462	1749	1421.538
Lithium	Li	6.941	180.5	1342	1161.5
Lutetium	Lu	174.967	1663	3402	1739
Magnesium	Mg	24.3050	650	1090	440
Manganese	Mn	54.93805	1246	2061	815
Mercury	Hg	200.59	−38.83	356.73	395.56
Molybdenum	Mo	95.94	2623	4639	2016
Neodymium	Nd	144.24	1021	3074	2053
Neptunium	Np	237.0482*	630 ±1[c]	—	—
Nickel	Ni	58.6934	1455[†]	2913	1458
Niobium	Nb	92.90638	2477	4744	2267
Osmium	Os	190.23	3033	5012	1979
Palladium	Pd	106.42	1554.9	2963	1408.1
Phosphorus (yellow)	P	30.973762	44.1[c]	280[c]	235.9
			590 at 4.3 MPa (red)[c] —		—
Platinum	Pt	195.08	1768.4	3825	2056.6
Plutonium	Pu	244.0642*	640	3228	2588
Polonium	Po	208.9824*	254	962	708
Potassium	K	39.0983	63.38	759	695.62

continued

Table 1.1 *(continued)*

Element		Relative atomic mass	Melting point[a] °C	Boiling point[a] °C	Liquid range[b] °C
Praseodymium	Pr	140.90765	931	3520	2589
Promethium	Pm	146.9151*	1170[d]	2460[d]	1290
Protactinium	Pa	231.0358*	1840[d]	—	—
Radium	Ra	226.0254*	∼700[d]	1140[d]	∼440
Rhenium	Re	186.207	3186	5596	2410
Rhodium	Rh	102.9055	1964	3695	1731
Rubidium	Rb	85.4678	39.30	688	648.70
Ruthenium	Ru	101.07	2334	4150	1816
Samarium	Sm	150.36	1074	1794	720
Scandium	Sc	44.955910	1541	2836	1295
Selenium	Se	78.96	220.5	685	464.5
Silicon	Si	28.0855	1412	3270	1858
Silver	Ag	107.8682	961.78[†]	2162	1200.22
Sodium	Na	22.989768	97.72	883	785.28
Strontium	Sr	87.62	777	1382	605
Sulphur	S	32.066	119.0[c] (monoclinic) —		—
(rhombohedral)			112.8[c]	444.674[c]	331.874
Tantalum	Ta	180.9479	3017	5458	2441
Technetium	Tc	96.9063*	2170[d]	—	—
Tellurium	Te	127.60	450	988	538
Terbium	Tb	158.92534	1356	3230	1874
Thallium	Tl	204.3833	304	1473	1169
Thorium	Th	232.0381*	1750	4788	3038
Thulium	Tm	168.93421	1545	1950	405
Tin	Sn	118.710	231.928[†]	2602	2370.072
Titanium	Ti	47.88	1668	3287	1619

Table 1.1 *(continued)*

Element		Relative atomic mass	Melting point[a] °C	Boiling point[a] °C	Liquid range[b] °C
Tungsten	W	183.84	3422	5555	2133
Uranium	U	238.0289*	1135	4131	2996
Vanadium	V	50.9415	1910	3407	1497
Ytterbium	Yb	173.04	819	1196	377
Yttrium	Y	88.90585	1522	3345	1823
Zinc	Zn	65.39	419.527[†]	907	487.473
Zirconium	Zr	91.224	1855	4409	2554

Data for relative atomic mass ($^{12}C = 12.000$) are taken from Benenson et al [2].
Data for melting point and boiling point, except for those bearing the superscripts c or d, are taken from Gale and Tolemeier [3].
[a] The standard pressure is one standard atmosphere

$$P = 101{,}325\ \text{Pa} = 1\ \text{atm} = 760\ \text{Torr}.$$

In general, melting and boiling points are given for standard pressure.
[b] Difference in temperature between melting and boiling points $(T_b - T_m)$ at 101.325 kPa.
[c] Data from Benenson et al. [2].
[d] Data from Nagakura et al. [4].
* Radioactive (stable isotopes not known).
[†] Defined fixed point of ITS-90 (The International Temperature Scale of 1990).

The two phase transitions of melting (or solidification) and evaporation (or condensation) are each called first-order phase transitions. During each phase transition, the pressure, temperature, and Gibbs free energy per mole remain unchanged;[10] however, values of the molar volume and molar entropy jump by finite amounts of $\Delta_\alpha^\beta V$ and $\Delta_\alpha^\beta S$, respectively, which are the characteristics of a first-order phase transition.

1.2.2.2 Richards' Rule

Richards' rule states that the molar entropy of melting $\Delta_s^l S_m$ for metals has a mean value of 8.96 J mol^{-1} K^{-1}:

$$\Delta_s^l S_m = \frac{\Delta_s^l H_m}{T_m} = 8.96\ \text{J}\,\text{mol}^{-1}\text{K}^{-1} \tag{1.8}$$

or

$$\frac{\Delta_s^l H_m}{RT_m} = 1.08 \tag{1.9}$$

where R is the molar gas constant (8.314 J mol^{-1} K^{-1}) and T_m is the absolute melting point temperature. Data for melting point temperature and molar melting enthalpy $\Delta_s^l H_m$ are listed in Tables 1.1, 17.1, and 17.3, respectively.

---

[10] We take one mole.

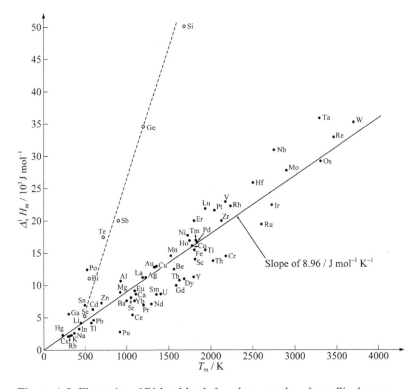

Figure 1.5 *Illustration of Richards' rule for a large number of metallic elements.*

$\Delta_s^l S_m$ is positive: the entropy of a liquid is greater than that of a solid. Figure 1.5 shows the correlation between $\Delta_s^l H_m$ and T_m, i.e. Richards' rule, for a large number of metallic elements. This figure indicates that most metals lie in the vicinity of 8.96 J mol^{-1} K^{-1}; however, semimetals (antimony and bismuth) and semiconductors (germanium, selenium, silicon, and tellurium) lie on their own straight line. For the melting entropy of 62 metals, Richards' rule represented by Eq. (1.8) works reasonably with $\Delta(62)$ and $S(62)$ values of 21.4 per cent and 0.329, respectively (see Section 1.5).

It is known that the values of entropy and volume changes that take place on melting depend somewhat on the crystal structure in the solid state.

1.2.2.3 Trouton's Rule

Trouton's rule states that the molar entropy of evaporation $\Delta_l^g S_b$ has a mean value of 102 J mol^{-1} K^{-1}:

$$\Delta_l^g S_b = \frac{\Delta_l^g H_b}{T_b} = 102 \text{ J mol}^{-1}\text{K}^{-1} \tag{1.10}$$

or

$$\frac{\Delta_l^g H_b}{RT_b} = 12.3 \qquad (1.11)$$

where T_b is the absolute boiling point temperature. Data for boiling point temperature and molar evaporation enthalpy $\Delta_l^g H_b$ are also listed in Tables 1.1, 17.1, and 17.3, respectively.

Figure 1.6 indicates the correlation of $\Delta_l^g H_b$ with T_b for a large number of metallic elements. Trouton's rule represented by Eq. (1.10) performs well with the molar entropy of evaporation for 62 metallic elements (including semimetals and semiconductors), giving $\Delta(62)$ and $S(62)$ values of 13.5 per cent and 0.161, respectively. As is clear from Figure 1.6, however, the metals with higher boiling point temperatures are apt to show larger positive deviations from Trouton's rule. The enthalpy of evaporation is a very useful thermophysical quantity. However, $\Delta_l^g H_b$ is not known for some ten metallic elements; reliable data for the enthalpy of evaporation of all metallic elements are needed. Trouton's rule is often used to estimate the values of $\Delta_l^g H_b$ for a liquid, knowing the liquid's boiling point temperature.

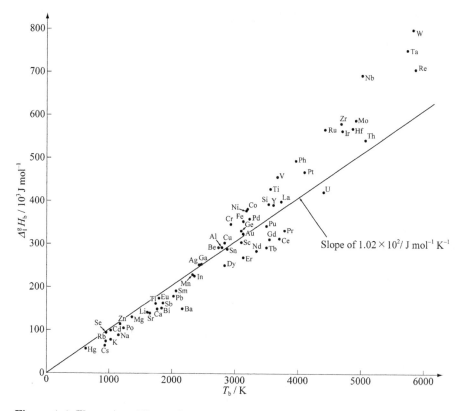

Figure 1.6 *Illustration of Trouton's rule for a large number of metallic elements.*

1.2.2.4 Lindemann's Melting Model

The phenomenon of melting and the sharpness of a metallic element's melting point are known to all materials scientists and engineers. However, in spite of many attempts, there is no quantitative theory based on quantum statistical mechanics. Even now, Lindemann's melting model [5], proposed about one century ago, is probably the most famous, and also most useful for researchers in the field of liquid metallic processing.

Lindemann considered the melting process from the standpoint of the solid, and suggested that a solid shakes itself to pieces and changes into a liquid once the amplitude δ of vibration of each atom due to thermal vibration becomes a certain critical fraction (about 1/10) of the average interatomic distance a, namely

$$\delta \approx \frac{1}{10}a, \quad (T = T_m) \tag{1.12}$$

Since melting point temperatures T_m are generally high, we will discuss the behaviour of classical particles.

Let us now consider that the solid consists of atoms of mass m, vibrating harmonically and isotropically, independent of each other, with the same frequency about their equilibrium positions (i.e. the Einstein model). The frequency ν as a harmonic oscillator and its internal energy (total energy during vibration) u_v are, respectively, given by

$$\nu = \frac{1}{2\pi}\left(\frac{k_f}{m}\right)^{1/2} \tag{1.13}$$

and

$$u_v = \frac{1}{2}k_f\delta^2 \tag{1.14}$$

where k_f is the force constant between one atom and its neighbour ($k_f = 4\pi^2 m\nu^2$). From the equipartition law of energy, the vibrational energy, or the internal energy, is kT (where k is the Boltzmann constant) per atom for each principal direction of vibration. Thus at the melting point, we have

$$\frac{1}{2}k_f\delta^2 = 2\pi^2 m\nu^2\delta^2 = kT_m \tag{1.15}$$

Since the average interatomic distance a is approximately equal to $(V/N_A)^{1/3}$, and the atomic mass m equals M/N_A (where N_A is Avogadro constant, and M is the molar mass), we have the following equation for the atomic frequency ν in terms of macroscopic physical quantities.

$$\nu = C_L\left(\frac{T_m}{MV^{2/3}}\right)^{1/2}\Bigg|_{T=T_m} \tag{1.16}$$

where $C_L \approx 10N_A^{1/3}R^{1/2}/\sqrt{2}\pi$. Equation (1.16) is called Lindemann's equation.

Lindemann's equation is regarded as a dimensional-analytical approach, and therefore the value of C_L requires experimental determination; the experimentally determined

value of C_L is roughly 8.9×10^8 in SI units. The results of examination of the Lindemann model using the theory of lattice dynamics indicate that the value of C_L depends
on the type of crystal lattice in the solid state.

The Lindemann model also suggests that the surface layer of a metal may melt at
a temperature lower than the melting point temperature of the bulk. We give several
references [6–10] on the surface melting temperatures of metals.

1.2.3 Liquid Ranges of Metallic Elements

At standard pressure,[11] in general, the ranges of liquids (i.e. $T_b - T_m$) are not wide (e.g.
everyone knows, in the case of water, its range is 100 K or $^\circ$C); however, those of metallic
elements are comparatively wide.

Metallic liquids are ordinarily treated and examined from the scientific and technological point of view at (or in the vicinity of) one standard atmosphere. Table 1.1 lists
values of liquid ranges for a large number of metallic elements. Incidentally, as can be

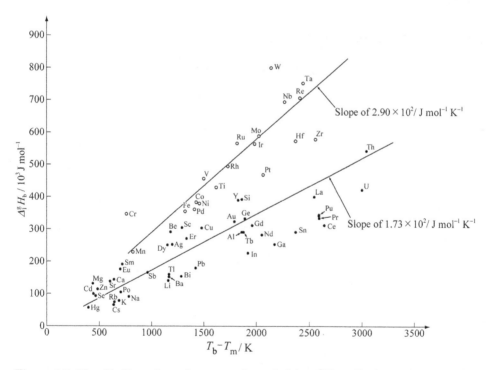

Figure 1.7 *Plot of boiling point molar evaporation enthalpies $\Delta_l^g H_b$ vs. liquid ranges $(T_b - T_m)$
for a large number of metallic elements. ○, d-block main transition elements, excluding Group IB
elements (or copper group metals); •, other metallic elements, cf. Table 1.7(a).*

[11] As already mentioned, the standard pressure, or normal pressure, is one standard atmosphere:
$P = 1.01325 \times 10^5$ Pa = 1 atm. Previously, 760 Torr (1 Torr = 1 mmHg), 1 atm = physical atmosphere.

seen from Table 1.1, the data for melting and boiling point temperatures at one standard atmosphere, and also relative atomic mass (or molar mass) are given.[12] All of these are very important physical quantities, because of material constants characterizing the properties and behaviour of materials. Thus, models or equations, particularly predictive models, for thermophysical properties of metallic liquids are often expressed in terms of these reliable, well-known physical quantities, or material constants.

Figure 1.7 shows a relationship between liquid ranges and molar evaporation enthalpies for 62 metallic elements. Although the overall scatter is rather high, a linear relationship between the two variables holds roughly. The linear relationship may be divided into two groups: *d*-block transition elements from Groups 4 to 10, i.e. three main transition elements, excluding Group 11 elements (which are sometimes called the copper group), (see Table 1.7(a)), and the other metallic elements; the slopes of 2.90×10^2 and 1.73×10^2 in J mol^{-1} K^{-1} refer to the mean values determined from those of the respective metallic elements of each group. For the *d*-block transition elements, $\Delta(19)$ and $S(19)$ values are 11.8 per cent and 0.176, respectively; for the other elements, $\Delta(43)$ and $S(43)$ values are 25.5 per cent and 0.292, respectively.

1.3   Approaches to the Liquid State

1.3.1   Characteristics of Solids, Liquids, and Gases

It is first appropriate to consider the characteristic features of the liquid state and to compare its structure, from the microscopic standpoint, with those of the solid and gaseous states. As such, our discussion is general and not limited specifically to metallic liquids. During the past 12–15 decades, numerous studies have been carried out on the properties and characteristics of the solid and gaseous states; both experimental and theoretical methods have been employed, and a large amount of knowledge has been accumulated. In contrast to that of the solid and gaseous states, our understanding of liquid structures and properties is still relatively imperfect despite the great efforts that have been devoted to such investigations. A major reason for this difficulty is that the liquid state lacks any 'idealized model' on which to establish a base.

The concept of an idealized model is extremely important in studying the properties of matter. Such a model can be defined as a hypothetical substance which has the characteristic features of real matter. Preferably, it should be relatively simple to treat mathematically, and its concept should be clear. Such idealized models exist for the solid and gaseous states.

Thus, in an ideal solid, or a perfect crystal, atoms are regularly arranged at the lattice points. This regular arrangement of atoms is long-range and three-dimensional, undisturbed by thermal agitation. The ideal solid forms a crystal of invariant shape. By contrast, in the ideal gas, each atom can freely translate throughout the volume in which it is contained.

---

[12] Formerly called atomic weight. Molar mass M is given by: $M = 10^{-3}$ kg (i.e. 1 g) $\times A_r$ mol$^{-1} = 10^{-3} A_r$ kg mol^{-1}, or $M = N_A$ mol$^{-1} \times m$ kg $= N_A m$ kg mol^{-1}, where A_r is the relative atomic mass and m is the atomic mass (i.e. the mass of one atom).

These simple models for the solid and gaseous states lead to useful results for real solids and gases (e.g. the Einstein model for the lattice heat capacity, the Brillouin zone, the Boyle–Charles law). Extensions and corrections to these idealized models for describing real substances can provide close agreement with experimental results.

Unfortunately, it is considerably more difficult to explain the behaviour of a liquid. From a macroscopic point of view, the most characteristic feature of a liquid is its inability to support any shearing; it does not take a specific shape. This is manifest in its capacity to flow. Thus, the viscosities of small-molecule liquids are very low; diffusivities are very high, compared with those in solids. On an atomic scale, this information suggests that an atom in the liquid state can easily migrate through fluctuations in density arising from the thermal motion of surrounding atoms. If we consider an atom in the liquid state at any moment in time, it will be interacting strongly with the atoms which surround it, vibrating as though it were an atom in the solid state because of the liquid's high packing density, like a solid (and unlike a gas). The atom repeats such motions in changing from place to place. We see that both atomic distance and time scales are of critical importance for a clear understanding of the structure and properties of liquids.

Figure 1.8 shows the trajectories of molecules in the solid, liquid, and gaseous phases, which were simulated with the aid of a computer. This figure indicates the characteristic features of the three phases of matter. Although it is often said that the liquid state exhibits intermediate properties between those of a solid and a gas, as seen from Figure 1.8 it should be noted that liquid behaviour and properties cannot merely be averaged between those of their respective solid and gaseous states. Thus, while matter is solid at low temperatures, gaseous at high temperatures, and liquid at intermediate values, we find that a liquid will resemble a solid, in some ways, but in other ways, it will more resemble a gas. In Figure 1.8, the differences and similarities among the molecular motions in the solid, liquid, and gaseous phases are shown clearly.

Liquids and solids exhibit similar densities. For example, in the case of metals, their densities in the solid and liquid states differ by less than approximately 2 to 6 per cent near their melting points. Both have similar number densities (i.e. number of atoms per unit volume) and average interatomic distances. Furthermore, the ratio of the molar enthalpy of melting $\Delta_s^l H_m$ to the molar enthalpy of evaporation $\Delta_l^g H_b$ for the majority of metallic elements is in the order of 2 to 6 per cent with the exceptions of semimetals, germanium, and silicon (semiconductors) (see Table 4.3). In addition, the ratio of the molar enthalpy of evaporation $\Delta_l^g H_b$ to the molar cohesive energy,[13] E_0^c, for the metallic elements is approximately 74 (polonium) to 99 (plutonium) per cent (the mean value for 59 metallic elements is 88.5 per cent) excluding semimetals, boron, selenium, and sulphur (see also Subsection 4.1.3.1 and Figure 4.3). These factual results suggest that liquids and solids resemble each other in their properties of cohesion. Atomic interactions in the liquid state must be quite strong like solids (unlike gases). Consequently, we can see that liquids and solids take fixed volumes, which change only under high pressure. 'Condensed matter' is a generic term for both solids and liquids.

---

[13] The cohesive energy E_0^c is defined as the energy required to form separated neutral atoms (in their ground electronic state) from the solid at 0 K at 1 atm (i.e. 101.325 kPa).

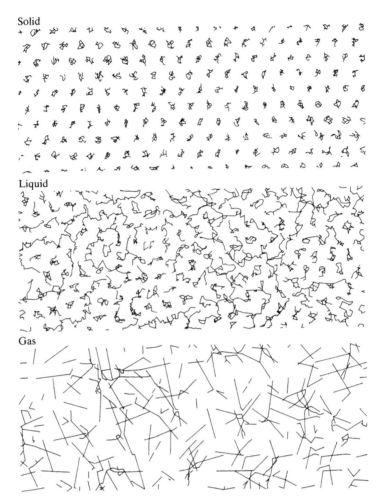

Figure 1.8 *Trajectories of molecules in a solid, a liquid, and a gas. These were simulated with the aid of a computer. The two-dimensional system of molecules has the same phases and phase transitions as a real substance, but the molecular positions and motions are more easily displayed. In the solid, the molecules are constrained to vibrate about fixed lattice sites, whereas molecules in the liquid and gaseous phases are free to wander. The only substantial differences between the two fluid states are those of density and of frequency of collision. The computer program calculates the trajectories by solving the equations of motion for some 500 two-dimensional molecules. The simulation was done by Farid Abraham of International Business Machines Corporation Research Laboratory in San Jose, California (after Barker and Henderson [11]).*

At temperatures exceeding the critical point of substance, as already mentioned, there is no discontinuous change from liquid to gaseous form. In other words, the liquid and gas can no longer be distinguished at the critical point: they have no permanent structure. From this fact, a liquid may then be considered as a dense gas. In fact, the atoms in the liquid state have so much freedom that a liquid, unlike a solid, flows very readily and easily takes up the shape of the vessel it occupies. Both gases and liquids are often referred to as 'fluids'. Most models of liquids are extensions of solid-like or gas-like models.

As yet, we have no idealized liquid, largely because of the fundamental difficulty of describing the liquid state. However, the structure and some properties of liquid metals have been explained quite skilfully and successfully on the basis of a relatively simple model, in which the atoms are treated as hard, inert spheres, i.e. the hard-sphere model. The hard-sphere model is very useful for an understanding of thermophysical properties of metallic liquids. Nevertheless, at present, it would be difficult to accurately predict the thermophysical properties of all liquid metallic elements including semimetals and semi-conductors (e.g. germanium, silicon, tellurium), using the hard-sphere model. In many high temperature processes, accurate predictions for the thermophysical properties of all liquid metallic elements are required; uncertainties in the calculations and predictions should be equal, or nearly equal, to the uncertainties associated with experimental meas-urements. In the field of materials process science, both 'accuracy' and 'universality' are required of any model for predicting the thermophysical properties of liquid metallic elements.

1.3.2 Approaches to the Liquid State Based on Model Theories

The spatial arrangement of the atoms in a liquid, i.e. the structure of a liquid, is best described by an important function referred to as the pair distribution function. This can be obtained from experiments using X-ray or neutron diffraction techniques. The approach is called the pair theory of liquids. A reasonably detailed explanation for the pair theory of liquid is given in Chapter 2. Before explaining the pair theory, however, we must first mention two approaches to the liquid state: the liquid as a modified solid, and the liquid as a modified gas.

1.3.2.1 *The Liquid as a Modified Solid*

Simplistically speaking, all crystalline solids consist of regular, periodic, three-dimensional arrangements of atoms, molecules, or ions, and they oscillate about their equilibrium positions or their lattice points. Actually, all crystalline solids (e.g. metallic elements) contain an equilibrium number of empty lattice points, or sites, above absolute zero (i.e. 0 K). These empty lattice points are called vacancies or holes. The equilibrium number of vacancies increases exponentially as the temperature rises. The existence of vacancies in crystalline solids has a great effect on various properties of the solids.

When metallic elements melt, most of them expand by 3 to 5 per cent in volume, and the volume change would lead to a mean increase in interatomic distance of only approximately 1.5 per cent. Thus, at the melting point, from the microscopic point of view, the spatial arrangement of the atoms in metallic liquids (in general, the molecules in a liquid) will be a fairly regular one which closely resembles the corresponding solids. The way of describing this can be classified broadly into the following two. One way is that the greater volume of liquid could be attributed to a greater number of sites, or lattices, most of which are occupied by atoms and the remaining few by holes, or from a slightly different viewpoint we can suppose that the melting process consists mainly in producing of vacancies, or holes, in a solid; as a result the solid melts and changes into a liquid. (The latent heat of melting might be spent mainly to produce vacancies in a solid.) Another way of describing the volume expansion on melting is that there is an increase in volume occupied by each atom, i.e. an increase in free volume. Each of the atoms can move around almost freely within the free volume.

The former is called the hole theory, and the latter the free volume theory or the cell theory. These model theories are essentially the same with minor differences; thus they are, all together, called the lattice theory of the liquid state, which are based on the solid-like structure.

1.3.2.2   *The Liquid as a Modified Gas*

We have already mentioned that above the critical point, no distinction between the liquid and the gas can be made, and a single undifferentiated, fluid state of uniform density exists. To put it another way, it is possible to change continuously from gas (vapour) to liquid (or from liquid to gas (vapour)), in the critical region, without phase transition. The liquid as a modified gas, i.e. a gas-like model for a liquid, is based on this fact.

First, we consider an ideal gas, or a perfect gas, which is made up of massive point particles that exert no forces on each other. This is an idealized, simple model for a gas. The ideal gas obeys exactly the Boyle–Charles law, (or the combined gas law):

$$PV = nRT \qquad (1.17)[14]$$

This equation shows the relationship between the pressure P, volume V, and thermodynamic temperature T of an amount of substance n (mole in SI units), which is called the equation of state for an ideal gas. The law represented by Eq. (1.17), to a good approximation, holds for a real gas at high temperatures and low pressures, or at low number densities.

A more accurate equation of state for one mole of the real gas would be

$$\left(P + \frac{a}{V^2}\right)(V - b) = RT \qquad (1.18)$$

where a and b are constants which depend on the type of gas. The quantity a/V^2 allows for the short-range attractive forces between molecules,[15] and b allows for the volume effectively occupied by molecules because of the very short-range repulsive forces.

---

[14] The defining equation for an ideal gas.
[15] An atom is a monatomic molecule.

Equation (1.18) is well known as the van der Waals equation of state (proposed semi-empirically by van der Waals in 1873). For a real gas, we must take account of repulsions when two molecules are extremely close together,[16] and take account of attractions when they are somewhat further apart. The van der Waals equation, in which the intermolecular forces (i.e. interactions between molecules) are taken into consideration, gives a more accurate explanation for the properties of gases, and further, can describe the liquid phase and liquid–gas equilibrium, in the critical region, although it is less accurate for the liquid phase.

Since the 1950s, studies on the thermophysical properties of liquid metals have been actively made on the basis of the hard-sphere model. A great many research articles on them have been published in the last half century or more. In the hard-sphere model, we treat the atoms as perfect, hard, non-attracting spheres of effective diameter σ (see Figure 2.5(a)). However, the fact that a liquid as well as a solid has a fixed volume, unlike a gas, suggests that there are attractive forces, or cohesive forces, to hold the molecules together. Thus, in the hard-sphere model, to deal with the attractive forces causing the hard-spheres to crowd together, the volume in which they move around is fixed, and further, the packing fraction is adopted as a convenient and yet important parameter. The packing fraction η is defined as the fraction of the total volume which is occupied by the hard-spheres themselves;

$$\eta = \frac{4}{3}\pi\left(\frac{\sigma}{2}\right)^3\frac{N}{V} = \frac{\pi}{6}\sigma^3 n_0 \tag{1.19}$$

where N is the number of (hard-sphere) atoms in volume V, and $n_0 (\equiv N/V)$ is the average number density. The packing fraction is determined by fitting structure factor, which is directly observed in X-ray or neutron experiments. The hard-sphere models give reasonable or great agreement with experimental data for several thermophysical properties of liquid metals.

Several equations based on the model theories and their performances are discussed in Chapters 5 to 8. Broadly speaking, the model theories provide semi-quantitative explanations for various thermophysical properties of metallic liquids. As we have no idealized model for a liquid, full considerations to the liquid from many different angles (e.g. molecular theory, pair theory, various model theories) are needed for a clear understanding of its structure and properties.

In general, matter consists of an extremely large number of molecules, and the three states of matter are the result of a competition between kinetic energy (which is positive: $3NkT/2$) and potential energy of intermolecular interactions (which is negative on average). In a gas, the total energy of the molecules is positive, whereas in condensed phases, the total energy is negative. Thus, in the liquid state, we see the importance of the mutual potential energy. If the mutual potential energy of a pair of molecules (i.e. the pair potential) is known, then the thermodynamic properties of a liquid can be calculated at once, using statistical mechanical theory.

---

[16] As the two molecules are brought together, their electronic charge distributions gradually overlap. At sufficiently close separations the overlap energy is repulsive, in large part because of Pauli exclusion principle.

Incidentally, building more detailed, complex (mathematical) models would not necessarily lead to success, especially as predictive models, because the number of parameters probably increase numerously; only a few accurate and reliable values of parameters, or physical quantities, are presently available for all, or almost all, liquid metallic elements (in other words, the more parameters used, the more difficult it is to find their accurate and reliable value).

1.4   Well-Known, Representative Models for the Thermophysical Properties of Liquid Metals

One of the most important purposes of materials science and engineering is to make the best use of each metallic element in the periodic table, namely to bring out its strength, or its unique properties. For this purpose, accurate and reliable data for the thermophysical properties of all metallic elements, or simple metallic substance, are indispensable. Unfortunately, accurate experimental determinations of the thermophysical properties of all metallic elements (i.e. metals, semimetals, and semiconductors) are actually impossible, because of difficult problems attendant on high temperature experiments, especially for the high melting point, toxic, reactive, and/or radioactive elements, as has already been mentioned in the preface of this book. Thus, great efforts have been poured into the development of reliable models for accurate predictions for thermophysical properties of metallic liquids. Even so, for many years, little progress has been made on such predictive models. Table 1.2 shows well-known, representative models/equations for the thermophysical properties of liquid or solid elements, which are presently referred to as valuable models. Most of these models for the thermophysical properties of liquid elements were presented more than a half century ago. Moreover, these are neither rigorous nor established models, as some assumptions and/or approximations are involved in all of them. The theoretical physics of solids has advanced remarkably in the past 100 years. Nevertheless, even today, it is impossible to accurately predict the thermophysical properties of liquid metallic elements based on the theory of liquids. This is mainly because a rigorous mathematical treatment of many-body correlations for the liquid state has yet to be established: the theory of liquid is based on the pair approximation (called the pair theory), in which three-body and more correlations are neglected. Even in the pair theory, however, it is extremely difficult to turn the complex intermolecular interactions, or anharmonic motions of molecules (atoms, ions), in the liquid state, into a rigorous quantitative model for accurate property predictions.

In broad terms, the agreement of the property values calculated using the models given in Table 1.2 with experimental data is fairly good. For example, the Andrade formula (see Table 1.2) for the melting point viscosity is rated highly as the most successful quantitative model (proposed on the basis of a quasi-crystalline model in 1934). The Andrade formula performs well with $\Delta(48)$ and $S(48)$ values of 14.9 per cent and 0.182, respectively (see Sections 1.5. and 7.6.). However, a better viscosity model is required for both a clearer understanding and, where experimental data are lacking, a more accurate

Table 1.2 *Representative models for the thermophysical properties of solid and liquid metallic elements, and published years of the models.*

Model (in SI units)	Researcher	Published year
$\Delta_l^g S_b = \dfrac{\Delta_l^g H_b}{T_b}$	Trouton (Trouton's rule)	1884
$\nu_m = 8.9 \times 10^8 \left(\dfrac{T_m}{M V_m^{2/3}} \right)^{1/2}$	Lindemann (Lindemann's equation)	1910
$\alpha_m = \dfrac{0.23}{T_b - 0.23\,T_m}$	Steinberg (Steinberg model)	1974
$U = \left(\dfrac{2E_c}{M} \right)^{1/2}$	Gitis and Mikhailov (Gitis–Mikhailov model)	1968
$\gamma_m = q\,\dfrac{T_m}{V_m^{2/3}}$	Schytil (Schytil model)	1949
$\mu_m = 1.80 \times 10^{-7}\,\dfrac{(MT_m)^{1/2}}{V_m^{2/3}}$	Andrade (Andrade formula)	1934
$D = \dfrac{kT}{\varsigma\,(V/N_A)^{1/3}\,\mu}$	Eyring and Ree (Modified Stokes–Einstein formula)	1961
$\dfrac{\lambda}{\sigma_e T} = \dfrac{\pi^2 k^2}{3e^2} = 2.45 \times 10^{-8}$	Wiedemann, Franz, and Lorenz (Wiedemann–Franz–Lorenz law)	1872

List of symbols

$\Delta_l^g S_b$	molar evaporation entropy at T_b
$\Delta_l^g H_b$	molar evaporation enthalpy at T_b
T	absolute temperature, or thermodynamic temperature
ν	mean atomic (or molecular) frequency
M	molar mass
V	molar volume
α	volume expansivity
U	sound velocity
E_c	molar cohesive energy
γ	surface tension
q	numerical factor
μ	viscosity
D	self-diffusivity
k	Boltzmann constant
ς	constant (5 to 6)
N_A	Avogadro constant
λ	thermal conductivity
σ_e	electrical conductivity
e	electron charge

Subscripts

m	at melting point
b	at boiling point

prediction of a liquid metal's viscosity. As for values of parameters used (i.e. M, T_m, V_m), these are known for all, or almost all, metallic elements. Thus, it is necessary to take a new approach to accurately predict the thermophysical properties of liquid metallic elements from the standpoint of materials process science. This subject is discussed further in Volume 2.

In this book, the results of calculations based on the representative models for the thermophysical properties of liquid metallic elements are compared against experimental data using relative standard deviations as a yardstick. The models' performances are thereby evaluated: most importantly, the judgment on a model should not only be based on its accuracy, but also on its universality.

1.5 Methods for Assessment of Models/Equations

Both accuracy and universality are required of any model for predicting the thermophysical properties of liquid metallic elements in the area of materials process science. In this book, the term 'accuracy' indicates that uncertainties in calculations are equal, or nearly equal, to uncertainties in experimental measurements; in other words, if calculated values fall, or almost fall, within the range of the uncertainties associated with experimental measurements, the model is deemed 'accurate'. As for 'universality', this term denotes that the model is applicable to all liquid metallic elements, i.e. to liquid metals, semi-metals, and semiconductors. Needless to say, it is no easy task to satisfy both accuracy and universality. Nevertheless, the compatibility with both accuracy and universality is indispensable in view of the purposes of materials process science. Properly, the performance of a model and/or an equation, if we can, should be quantitatively assessed by comparing calculated values with accurate, reliable experimental values.

The performance of models and/or equations is assessed by determining values for the parameters such as a relative percentage error or uncertainty δ_i, a global Δ, and a relative standard deviation S. The parameter δ_i is defined as

$$\delta_i = \frac{(\chi_i)_{exp} - (\chi_i)_{cal}}{(\chi_i)_{cal}} \times 100, \quad (\%) \tag{1.20}$$

where $(\chi_i)_{exp}$ and $(\chi_i)_{cal}$ are the experimental and calculated values for a liquid metallic element i, respectively, of thermophysical property χ. Equation (1.20) is set on the basis of the calculated value using a model or an equation. This parameter is useful for the values of $|\delta_i|$ within the range of approximately 30 to 40 per cent ($|\delta_i| < 30 - 40$).

Equation (1.20) indicates that an increase (or a decrease) in δ_i per cent in the calculated value is needed to exactly square, or reconcile, the calculated value with the experimental value. Here an example is given for the velocity of sound in liquid copper at its melting point temperature:

$$(\chi_{Cu})_{exp} = 3440 \text{ m s}^{-1}, \quad (\chi_{Cu})_{cal}{}^{17} = 3093 \text{ m s}^{-1}$$

---

[17] Calculated from the Gitis–Mikhailov model (see Subsection 5.3.3).

$$\delta_{Cu} = \frac{3440 - 3093}{3093} \times 100 = 11.2_2, \ (\%)$$

$$3093 + 3093 \times 0.1122 = 3440 \ (m\,s^{-1})$$

$$(\chi_{Cu})_{cal} + (\chi_{Cu})_{cal} \times 11.2_2 \ (\%) \ = (\chi_{Cu})_{exp}$$

The global $\Delta(N)$ is defined as

$$\Delta(N) = \frac{1}{N} \sum_{i=1}^{N} |\delta_i|, \ (\%) \tag{1.21}$$

where N is the number of samples.

The relative standard deviation $S(N)$ is defined as follows:

$$S(N) = \frac{1}{100} \sqrt{\overline{\delta_i^2}} = \sqrt{\frac{1}{N} \sum_{i=1}^{N} \left\{ \frac{(\chi_i)_{exp} - (\chi_i)_{cal}}{(\chi_i)_{cal}} \right\}^2} \tag{1.22}$$

In order to obtain a higher accuracy in calculations, the absolute values of the above parameters need to be lower, if the model is to be deemed 'more accurate'. By using these parameters, therefore, it is easy for us to objectively evaluate the model's performance.

1.6 Electron Configuration and the Periodic Table of the Elements

In order to build accurate and reliable models for the thermophysical properties of metallic liquids, knowledge of the ground-state electron configurations of the atom, based on quantum mechanics, would be of great importance. Atoms with similar electron configurations in the outermost shell have similar chemical and physical properties: the number of electrons in the outermost shell, which is known as the valence electrons, controls the chemical and physical properties of elements. The electron configurations underlie the periodic law and the periodic table of the elements. Incidentally, the periodic law states that the chemical and physical properties of elements are a periodic function of their proton number, or their atomic number.

Let us now give an example of the importance of the periodic law, i.e. electron configuration, in the field of materials process science. The Andrade formula is regarded as an ingenious dimensional consideration, so that the constant of proportionality, called the Andrade coefficient (denoted by C_A), must be determined experimentally. The value of C_A that gives the best agreement with experimental viscosity data for liquid metals is 1.80×10^{-7} $(J/K\,mol^{1/3})^{1/2}$. As a matter of fact, the C_A values determined for respective liquid metallic elements vary periodically with atomic number. In a word, the Andrade coefficient obeys the periodic law (see Subsection 7.3.1). Further discussions on this subject and some similar examples are given in Volume 2.

Atomic periodicity in the values of the Andrade coefficient has not at present received a theoretical explanation.

1.6.1 Energy Levels and Electron Configurations of Atoms

The term 'energy level' refers to a definite fixed energy that a system described by quantum mechanics, e.g. an electron, can have. The energy levels in an atom correspond to the orbitals in which its electrons move around the nucleus.[18] Traditionally, the energy levels, or the (electron) orbits corresponding to them, are often referred to as shells, and given letters K, L, M, and so on. Each energy level (K, L, M, . . .) can be divided into sublevels; in the other words, each energy level in the atom consists of one or more sublevels. The sublevels are represented by the letters s, p, d, and so on. The possible atomic orbitals correspond to subshells of the atom. Let us now touch briefly on the electronic structure of an atom. An electron in an atom is characterized by four quantum numbers, n, l, m, and s, where n is the principal quantum number, l the orbital quantum number,[19] m the magnetic quantum number, and s the spin quantum number. Table 1.3 shows the scientific names for the principal number (K, L, M, . . .) and the orbital quantum number (s, p, d, . . .), and the subshells, or the sublevels, made up of the combinations of n and l.

The four quantum numbers define the quantum state of the electron, and explain how the electronic structures of atoms occur. Thus, the numbers lead to an explanation of the periodic table of the elements.

The way the electrons fill the various energy levels of the many-electron atoms in their ground state (i.e. their lowest energy level) is termed the electron configuration. The electron configurations of the atoms obey the Pauli exclusion principle. According to the principle, a particular atomic orbital, which has fixed values of n, l, and m, can contain a maximum of two electrons, since the spin quantum number s can only be $+1/2$ or $-1/2$ ($\uparrow$ or $\downarrow$). Table 1.4 shows how many electrons in the atom can occupy atomic

Table 1.3 *Scientific names for n and l, and energy subshells, or energy sublevels, based on the combinations of n and l*

	s ($l=0$)	p ($l=1$)	d ($l=2$)	f ($l=3$)	g ($l=4$)	h ($l=5$)	i ($l=6$)
K shell ($n=1$)	$1s$	-	-	-	-	-	-
L shell ($n=2$)	$2s$	$2p$	-	-	-	-	-
M shell ($n=3$)	$3s$	$3p$	$3d$	-	-	-	-
N shell ($n=4$)	$4s$	$4p$	$4d$	$4f$	-	-	-
O shell ($n=5$)	$5s$	$5p$	$5d$	$5f$	$5g$	-	-
P shell ($n=6$)	$6s$	$6p$	$6d$	$6f$	$6g$	$6h$	-
Q shell ($n=7$)	$7s$	$7p$	$7d$	$7f$	$7g$	$7h$	$7i$

[18] In the quantum-mechanical model of the atom, the region of space in which an electron can be found with 95 per cent probability is called an orbital, or electron orbit.

[19] The orbital quantum number is sometimes called the azimuthal quantum number.

shells, (or orbitals); each sublevel, (or subshell), can hold a certain maximum number of electrons, and the total number of electrons that a main energy level can hold is the sum of these maxima for all its sublevels (i.e. $2n^2$).[20] The sublevels of some energy levels overlap; for example, as shown in Figure 1.9, the 4s sublevel has a lower energy than the $3d$ sublevel. Thus the 4s sublevel fills with electrons before the $3d$ sublevel does, even though the 4s sublevel is farther away from the nucleus than the $3d$ sublevel. The energy levels of the electrons in many-electron atoms, and also the order (with arrows) in which the electrons fill energy sublevels are shown in Figure 1.9.

To describe the electron configuration of atoms (and ions), the following notation is also used. An example is given for the electron configuration of hydrogen (a neutral atom in its ground state):

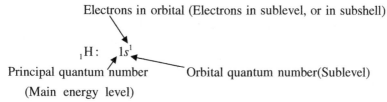

Electrons in orbital (Electrons in sublevel, or in subshell)

$_1$H: 1s^1

Principal quantum number   Orbital quantum number(Sublevel)

(Main energy level)

Table 1.4 *The quantum numbers of subshells and the maximum electrons in atomic shells, or how electrons occupy atomic shells*

Subshell (sublevel)	Principal quantum number n	Orbital quantum number[†] l	Magnetic quantum number[‡] m	Maximum electrons in subshell	Maximum electrons in shell[††]
1s	1	0	0	2	K[#] can have 2
2s	2	0	0	2	L can have 8
2p	2	1	$-1, 0, 1$	6	
3s	3	0	0	2	M can have 18
3p	3	1	$-1, 0, 1$	6	
3d	3	2	$-2, -1, 0, 1, 2$	10	
4s	4	0	0	2	N can have 32
4p	4	1	$-1, 0, 1$	6	
4d	4	2	$-2, -1, 0, 1, 2$	10	
4f	4	3	$-3, -2, -1, 0, 1, 2, 3$	14	

[†] $l = 0, 1, 2, \ldots, n-1$. [‡] $m = -l, -l+1, \ldots, l-1, l$.

[††] $2n^2$: The theoretical maximum number of electrons for the energy levels of O ($n = 5$), P ($n = 6$), and Q ($n = 7$) shells are 50, 72, and 98, respectively.

[#] Shell closest to nucleus.

[20] The number of orbitals is given by $n^2 \left(= \sum_{l=0}^{n-1} (2l + 1) \right)$.

We have already mentioned that each sublevel, represented by the letter s, (p, d, f, . . .) having orbital quantum number 0, (1, 2, 3, . . .), can hold a certain maximum number of electrons;[21] the superscript to the right of the letter s, (p, d, f, . . .) denotes the number of electrons in the sublevel, (or orbital). Here a few more examples:

$_3$Li   : $1s^2\,2s^1$

$_4$Be   : $1s^2\,2s^2$

$_5$B   : $1s^2\,2s^2\,2p^1$

$_{20}$Ca   : $1s^2\,2s^2\,2p^6\,3s^2\,3p^6\,4s^2$

$_{26}$Fe   : $1s^2\,2s^2\,2p^6\,3s^2\,3p^6\,4s^2\,3d^6$

Table 1.5 gives the electron configurations for all the elements.

We repeat here the importance of electron configurations of elements. The number of electrons in the outermost shell, or the valence electrons, has control of the chemical and physical (thermophysical) properties of an element.

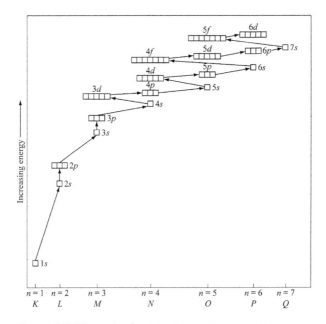

Figure 1.9 *Energy levels of the electrons in many-electron atoms, and the order in which the electrons fill energy sublevels (follow the arrows). Each box can hold a maximum of two electrons.*

[21] The maximum number of electrons in a given shell (i.e. a main energy level) is $2n^2$ (cf. also Tables 1.3 and 1.4).

Table 1.5 *Electron configuration of the elements.*

Period	Element		K	L		M			N				O				P			Q
			1s	2s	2p	3s	3p	3d	4s	4p	4d	4f	5s	5p	5d	5f	6s	6p	6d	7s
1	1	H	1																	
	2	He	2																	
2	3	Li	2	1																
	4	Be	2	2																
	5	B	2	2	1															
	6	C	2	2	2															
	7	N	2	2	3															
	8	O	2	2	4															
	9	F	2	2	5															
	10	Ne	2	2	6															
3	11	Na	2	2	6	1														
	12	Mg	2	2	6	2														
	13	Al	2	2	6	2	1													
	14	Si	2	2	6	2	2													
	15	P	2	2	6	2	3													
	16	S	2	2	6	2	4													
	17	Cl	2	2	6	2	5													
	18	Ar	2	2	6	2	6													
4	19	K	2	2	6	2	6		1											
	20	Ca	2	2	6	2	6		2											
	21	Sc	2	2	6	2	6	1	2											
	22	Ti	2	2	6	2	6	2	2											
	23	V	2	2	6	2	6	3	2											
	24	Cr	2	2	6	2	6	5	1											
	25	Mn	2	2	6	2	6	5	2											

continued

Table 1.5 *(continued)*

Period	Element	K	L		M			N				O				P			Q
		1s	2s	2p	3s	3p	3d	4s	4p	4d	4f	5s	5p	5d	5f	6s	6p	6d	7s
	26 Fe	2	2	6	2	6	6	2											
	27 Co	2	2	6	2	6	7	2											
	28 Ni	2	2	6	2	6	8	2											
	29 Cu	2	2	6	2	6	10	1											
	30 Zn	2	2	6	2	6	10	2											
	31 Ga	2	2	6	2	6	10	2	1										
	32 Ge	2	2	6	2	6	10	2	2										
	33 As	2	2	6	2	6	10	2	3										
	34 Se	2	2	6	2	6	10	2	4										
	35 Br	2	2	6	2	6	10	2	5										
	36 Kr	2	2	6	2	6	10	2	6										
5	37 Rb	2	2	6	2	6	10	2	6			1							
	38 Sr	2	2	6	2	6	10	2	6			2							
	39 Y	2	2	6	2	6	10	2	6	1		2							
	40 Zr	2	2	6	2	6	10	2	6	2		2							
	41 Nb	2	2	6	2	6	10	2	6	4		1							
	42 Mo	2	2	6	2	6	10	2	6	5		1							
	43 Tc*	2	2	6	2	6	10	2	6	6		1							
	44 Ru	2	2	6	2	6	10	2	6	7		1							
	45 Rh	2	2	6	2	6	10	2	6	8		1							
	46 Pd	2	2	6	2	6	10	2	6	10									
	47 Ag	2	2	6	2	6	10	2	6	10		1							
	48 Cd	2	2	6	2	6	10	2	6	10		2							
	49 In	2	2	6	2	6	10	2	6	10		2	1						
	50 Sn	2	2	6	2	6	10	2	6	10		2	2						
	51 Sb	2	2	6	2	6	10	2	6	10		2	3						

Table 1.5 *(continued)*

Period	Element	K	L		M			N				O				P			Q
		1s	2s	2p	3s	3p	3d	4s	4p	4d	4f	5s	5p	5d	5f	6s	6p	6d	7s
	52 Te	2	2	6	2	6	10	2	6	10		2	4						
	53 I	2	2	6	2	6	10	2	6	10		2	5						
	54 Xe	2	2	6	2	6	10	2	6	10		2	6						
6	55 Cs	2	2	6	2	6	10	2	6	10		2	6			1			
	56 Ba	2	2	6	2	6	10	2	6	10		2	6			2			
	57 La	2	2	6	2	6	10	2	6	10		2	6	1		2			
	58 Ce**	2	2	6	2	6	10	2	6	10	1	2	6	1		2			
	59 Pr	2	2	6	2	6	10	2	6	10	3	2	6			2			
	60 Nd	2	2	6	2	6	10	2	6	10	4	2	6			2			
	61 Pm	2	2	6	2	6	10	2	6	10	5	2	6			2			
	62 Sm	2	2	6	2	6	10	2	6	10	6	2	6			2			
	63 Eu	2	2	6	2	6	10	2	6	10	7	2	6			2			
	64 Gd	2	2	6	2	6	10	2	6	10	7	2	6	1		2			
	65 Tb	2	2	6	2	6	10	2	6	10	9	2	6			2			
	66 Dy	2	2	6	2	6	10	2	6	10	10	2	6			2			
	67 Ho	2	2	6	2	6	10	2	6	10	11	2	6			2			
	68 Er	2	2	6	2	6	10	2	6	10	12	2	6			2			
	69 Tm	2	2	6	2	6	10	2	6	10	13	2	6			2			
	70 Yb	2	2	6	2	6	10	2	6	10	14	2	6			2			
	71 Lu	2	2	6	2	6	10	2	6	10	14	2	6	1		2			
	72 Hf	2	2	6	2	6	10	2	6	10	14	2	6	2		2			
	73 Ta	2	2	6	2	6	10	2	6	10	14	2	6	3		2			
	74 W	2	2	6	2	6	10	2	6	10	14	2	6	4		2			
	75 Re	2	2	6	2	6	10	2	6	10	14	2	6	5		2			
	76 Os	2	2	6	2	6	10	2	6	10	14	2	6	6		2			
	77 Ir	2	2	6	2	6	10	2	6	10	14	2	6	7		2			

continued

Table 1.5 *(continued)*

Period	Element		K	L		M			N				O				P			Q
			1s	2s	2p	3s	3p	3d	4s	4p	4d	4f	5s	5p	5d	5f	6s	6p	6d	7s
	78	Pt	2	2	6	2	6	10	2	6	10	14	2	6	9		1			
	79	Au	2	2	6	2	6	10	2	6	10	14	2	6	10		1			
	80	Hg	2	2	6	2	6	10	2	6	10	14	2	6	10		2			
	81	Tl	2	2	6	2	6	10	2	6	10	14	2	6	10		2	1		
	82	Pb	2	2	6	2	6	10	2	6	10	14	2	6	10		2	2		
	83	Bi	2	2	6	2	6	10	2	6	10	14	2	6	10		2	3		
	84	Po	2	2	6	2	6	10	2	6	10	14	2	6	10		2	4		
	85	At	2	2	6	2	6	10	2	6	10	14	2	6	10		2	5		
	86	Rn	2	2	6	2	6	10	2	6	10	14	2	6	10		2	6		
7	87	Fr	2	2	6	2	6	10	2	6	10	14	2	6	10		2	6		1
	88	Ra	2	2	6	2	6	10	2	6	10	14	2	6	10		2	6		2
	89	Ac	2	2	6	2	6	10	2	6	10	14	2	6	10		2	6	1	2
	90	Th	2	2	6	2	6	10	2	6	10	14	2	6	10		2	6	2	2
	91	Pa	2	2	6	2	6	10	2	6	10	14	2	6	10	2	2	6	1	2
	92	U	2	2	6	2	6	10	2	6	10	14	2	6	10	3	2	6	1	2
	93	Np[†]	2	2	6	2	6	10	2	6	10	14	2	6	10	4	2	6	1	2
	94	Pu	2	2	6	2	6	10	2	6	10	14	2	6	10	6	2	6		2
	95	Am	2	2	6	2	6	10	2	6	10	14	2	6	10	7	2	6		2
	96	Cm	2	2	6	2	6	10	2	6	10	14	2	6	10	7	2	6	1	2
	97	Bk[‡]	2	2	6	2	6	10	2	6	10	14	2	6	10	8	2	6	1	2
	98	Cf	2	2	6	2	6	10	2	6	10	14	2	6	10	10	2	6		2
	99	Es	2	2	6	2	6	10	2	6	10	14	2	6	10	11	2	6		2
	100	Fm	2	2	6	2	6	10	2	6	10	14	2	6	10	12	2	6		2
	101	Md	2	2	6	2	6	10	2	6	10	14	2	6	10	13	2	6		2
	102	No	2	2	6	2	6	10	2	6	10	14	2	6	10	14	2	6		2
	103	Lr	2	2	6	2	6	10	2	6	10	14	2	6	10	14	2	6	1	2

* or [Kr] $4d^5 5s^2$; ** or [Xe] $4f^2 6s^2$; † or [Rn] $5f^5 7s^2$, ‡ or [Rn] $5f^9 7s^2$.

1.6.2   Periodic Table of the Elements

Although there is a large variety of materials around us, only 92 chemical elements exist on the Earth.[22] They range from hydrogen, the lightest element, to uranium, the heaviest. The classification of the (chemical) elements in tabular form in the order of their atomic numbers, or their proton numbers, is well known as the periodic table of the elements. Table 1.6 shows one common form of the periodic table of the elements (see also Appendix 1). As seen, the elements are arranged in order of increasing atomic number, in order to show the similarities of chemical elements with related electron configurations.

The elements are arranged in horizontal rows called periods. The periods are numbered from 1 to 7; the first three are called short periods, while the next four are termed long periods. In particular, periods 6 and 7 each have a long row of 14 additional elements; they are known as the lanthanides (or lanthanoids) (from cerium $_{58}$Ce to lutetium $_{71}$Lu) and the actinides (or actinoids) (from thorium $_{90}$Th to lawrencium $_{103}$Lr). The periods 6 and 7 are so long that we often write the lanthanides and actinides below the rest of table. Within a period, the atoms of all elements have the same number of shells $(K, \ldots, Q)$, or the same principal quantum number $(1, \ldots, 7)$, but with a steadily increasing number of electrons, or protons. The periodic table can also be divided into four blocks, depending on the type of shell being filled: the s-block, the p-block, the d-block, and the f-block. Table 1.7 gives the division of the periodic table into (a) the four blocks, and (b) their electron configurations.

The 18 vertical columns of the periodic table are called groups or families. Traditionally, the alkali metals are positioned on the left of the table and the groups are simply numbered in sequence from left to right, using Arabic numerals from 1 to 18 (according to the suggestion of the IUPAC[23]). However, there has been some disagreement about the numbering of the groups. In one commonly used system the groups are labelled with Roman numerals, and divided into A groups and B groups, as shown in Table 1.6. The A groups are called the representative elements; the B groups the transition elements (or metals). However, Group IIB metals are usually regarded as non-transition elements (they have filled d-orbitals). They are sometimes called the zinc group. Originally, the classification of chemical elements is valuable only in as far as it illustrates chemical and physical behaviour, and it is conventional to use the term 'transition elements' in a more restricted sense, particularly in the field of materials process science.

Basically, the electron configuration of the outermost energy level determines the chemical and physical behaviour of an element. Broadly speaking, elements in same group in the periodic table have the same number of electrons (or the same electron configuration) in their outer-shell. As a result, all of the elements within the same group behave in a similar way; they are sometimes referred to as families of elements. On the other hand, in general, the elements in any period have different outer-shell electron configurations, and therefore these elements have different chemical and physical properties. For example, in period 3, aluminium ($[Ne]3s^2\,3p^1$) is a metal, whereas chlorine ($[Ne]3s^2\,3p^5$) is a non-metal. However, this is not true for the transition metals. As we

---

[22] Several more elements exist, but they are artificially created in laboratories.
[23] The International Union of Pure and Applied Chemistry.

Table 1.6 *Periodic table of the elements.*

Phase (under standard conditions)
- g   gas
- l   liquid
- m   metal
- sm   semimetal
- sc   semiconductor

g H 1
Chemical symbol — g H 1 — Atomic number

Group	IA 1	IIA 2	IIIB 3	IVB 4	VB 5	VIB 6	VIIB 7	VIIIB 8	VIIIB 9	VIIIB 10	IB 11	IIB 12	IIIA 13	IVA 14	VA 15	VIA 16	VIIA 17	VIIIA 18
Period 1	g H 1																	g He 2
2	m Li 3	m Be 4											sc B 5	sc C 6	g N 7	g O 8	g F 9	g Ne 10
3	m Na 11	m Mg 12											m Al 13	sc Si 14	sc P 15	sc S 16	sc Cl 17	g Ar 18
4	m K 19	m Ca 20	m Sc 21	m Ti 22	m V 23	m Cr 24	m Mn 25	m Fe 26	m Co 27	m Ni 28	m Cu 29	m Zn 30	m Ga 31	sm Ge 32	sm As 33	sc Se 34	l Br 35	g Kr 36
5	m Rb 37	m Sr 38	m Y 39	m Zr 40	m Nb 41	m Mo 42	m Tc 43	m Ru 44	m Rh 45	m Pd 46	m Ag 47	m Cd 48	m In 49	m Sn 50	sm Sb 51	sc Te 52	sc I 53	g Xe 54
6	m Cs 55	m Ba 56	m La 57 †	m Hf 72	m Ta 73	m W 74	m Re 75	m Os 76	m Ir 77	m Pt 78	m Au 79	l,m Hg 80	m Tl 81	m Pb 82	sm Bi 83	m Po 84	m At 85	g Rn 86
7	m Fr 87	m Ra 88	m Ac 89 ‡															

† Lanthanides

m Ce 58	m Pr 59	m Nd 60	m Pm 61	m Sm 62	m Eu 63	m Gd 64	m Tb 65	m Dy 66	m Ho 67	m Er 68	m Tm 69	m Yb 70	m Lu 71

‡ Actinides

m Th 90	m Pa 91	m U 92	m Np 93	m Pu 94	m Am 95	m Cm 96	m Bk 97	m Cf 98	m Es 99	m Fm 100	m Md 101	m No 102	m Lr 103

Table 1.7 *Division of the periodic table into (a) four blocks and (b) their electron configurations.*

(a)

Group	IA 1	IIA 2	IIIB 3	IVB 4	VB 5	VIB 6	VIIB 7		VIIIB 8 9 10		IB 11	IIB 12	IIIA 13	IVA 14	VA 15	VIA 16	VIIA 17	VIIIA 18
Period 1	$_1$H																	$_2$He
2	$_3$Li	$_4$Be											$_5$B	$_6$C	$_7$N	$_8$O	$_9$F	$_{10}$Ne
3	$_{11}$Na	$_{12}$Mg											$_{13}$Al	$_{14}$Si	$_{15}$P	$_{16}$S	$_{17}$Cl	$_{18}$Ar
4	$_{19}$K	$_{20}$Ca	$_{21}$Sc	$_{22}$Ti	$_{23}$V	$_{24}$Cr	$_{25}$Mn	$_{26}$Fe	$_{27}$Co	$_{28}$Ni	$_{29}$Cu	$_{30}$Zn	$_{31}$Ga	$_{32}$Ge	$_{33}$As	$_{34}$Se	$_{35}$Br	$_{36}$Kr
5	$_{37}$Rb	$_{38}$Sr	$_{39}$Y	$_{40}$Zr	$_{41}$Nb	$_{42}$Mo	$_{43}$Tc	$_{44}$Ru	$_{45}$Rh	$_{46}$Pd	$_{47}$Ag	$_{48}$Cd	$_{49}$In	$_{50}$Sn	$_{51}$Sb	$_{52}$Te	$_{53}$I	$_{54}$Xe
6	$_{55}$Cs	$_{56}$Ba	$_{57}$La†	$_{72}$Hf	$_{73}$Ta	$_{74}$W	$_{75}$Re	$_{76}$Os	$_{77}$Ir	$_{78}$Pt	$_{79}$Au	$_{80}$Hg	$_{81}$Tl	$_{82}$Pb	$_{83}$Bi	$_{84}$Po	$_{85}$At	$_{86}$Rn
7	$_{87}$Fr	$_{88}$Ra	$_{89}$Ac‡															

Transition elements — Main transition elements

s-block   *d*-block   *p*-block

f-block

†Lanthanides	$_{58}$Ce	$_{59}$Pr	$_{60}$Nd	$_{61}$Pm	$_{62}$Sm	$_{63}$Eu	$_{64}$Gd	$_{65}$Tb	$_{66}$Dy	$_{67}$Ho	$_{68}$Er	$_{69}$Tm	$_{70}$Yb	$_{71}$Lu
‡Actinides	$_{90}$Th	$_{91}$Pa	$_{92}$U	$_{93}$Np	$_{94}$Pu	$_{95}$Am	$_{96}$Cm	$_{97}$Bk	$_{98}$Cf	$_{99}$Es	$_{100}$Fm	$_{101}$Md	$_{102}$No	$_{103}$Lr

f-block

(b)

Block	Group	Electron configuration†
s	1(IA) and 2(IIA)	Outer electron configurations: inert gas structures plus outer ns^1(IA) or ns^2(IIA) elements
p	13(IIIA) to 18(VIIIA)	Outer electron configuration: ns^2np^x where x = 1 to 6.
d	3(IIIB) to 12(IIB)	Outer electron configurations: $(n-1)d^x ns^y$, where x = 1 to 10, y = 1 or 2‡
f	Lanthanides and actinides	Two s-electrons in outer shell, f-electrons in inner shell: $(n-2)f^x ns^2$, where x = 1 to 14‡‡

† See Table 1.5. ‡ With the exception of palladium $4d^{10}$.
‡‡ With the exception of thorium $6d^2 7s^2$.

can see from Table 1.5, the electron configurations of the element iron ([Ar]$3d^6$ $4s^2$), cobalt ([Ar]$3d^7$ $4s^2$), and nickel ([Ar]$3d^8$ $4s^2$) differ, because they have different numbers of electrons in their $3d$ sublevels; however, the outermost level for these transition metals is $4s$ sublevels having the same number of two electrons except for chromium and copper. This can be applied to the transition elements in periods 5 through 7. Thus the transition metals exhibit somewhat similar chemical and physical behaviour.

In the periodic table, metallic elements tend to be those on the left and towards the bottom of the table; non-metallic elements are towards the top and the right. From the viewpoint of research scientists and engineers engaged in liquid metallic processing, metallic elements account for approximately 80 per cent of all the elements in the periodic table. Let us here add an important matter in liquid metallic processing science and engineering. Semiconductors, such as germanium, silicon, and tellurium, in the solid state show a remarkable increase in their thermal and electrical conductivities on melting, and they become metallic liquids. Similarly, semimetals (antimony and bismuth) also show an increase in their electronic transport property values mentioned above on melting. This phenomenon of increasing conductivities can be attributed to increase in the number of conduction electrons on melting.

We give now a brief explanation for the behaviour of electrons in the outer energy levels of the metallic elements.

Metals are characterized by high electrical and high thermal conductivities. We can understand many thermophysical properties of metals, in particular the electronic transport properties, in terms of the free electron model. According to the free electron model, when metallic atoms are assembled to make a metal's condensed phase, the valence electrons, i.e. electrons in the outermost energy level, of the atoms become conduction electrons and are free to move through the volume of the condensed phase. The characteristic feature of metallic bonding is the lowering of the kinetic energy of the valence electrons in the metal's condensed phase as compared to the free atom. (The interaction of the conduction electrons with the ion cores of the original atoms is neglected in the free electron approximation.)

The alkali atoms have only one valence electron (called s-electron) on each atom; in the alkali metals the s-electrons become conduction electrons. The Fermi surfaces of alkali metals are spherical, or nearly spherical, and the conduction electrons act as if they are free. The interatomic distances are relatively large in the alkali metals; at large interatomic distances the kinetic energy of the conduction electrons is lower. This leads to the weak bonding of the alkali metals. The alkali metals in the liquid state are called 'simple liquid metals'.

In transition metals there is additional binding from the inner electron shell. The additional binding forces (covalent-type bonds) are generated by the interactions among the inner d-electron shells. The copper group metals (Group 11, or Group IB) are also monovalent, but differ from the alkali metals by having a d-electron shell involved in the binding. Their cohesive energies are 3 to 4 times higher, compared with those in the alkali metals. Transition metals are characterized by high binding energy and high melting point. Transition metals in the liquid state are known as 'complex liquid metals'.

1.6.3 Atomic and Ionic Radii and Ionization Energy

Knowledge of the electron configuration of an element gives a basic understanding and further prediction of its chemical and physical properties. However, the atomic and/or ionic radii and the ionization energy of an element are also related to its chemical and physical properties.[24]

The atomic radii (or the ionic radii) of the elements are determined experimentally. The radii of isolated neutral atoms are calculated theoretically on the basis of quantum mechanics (see Figure 10.13). Distances between atoms (ions) in crystals (ionic crystals[25]) can be measured accurately by X-ray diffraction. Values of the atomic and ionic radii are obtained on the basis of the drastic but simple assumptions that the atoms (ions) are spherical with definite sizes (i.e. hard sphere), and come into contact with each other. Nevertheless, the concepts of both the atomic and ionic radii are useful for understanding the chemical and physical properties; in other words, values of the atomic and/or ionic radii account semi-quantitatively for some thermophysical properties of elements.

Figures 1.10 and 1.11 show values for the atomic and ionic radii of the metallic elements plotted against their atomic numbers. As is evident from these figures, both atomic and ionic radii of the metallic elements vary periodically with their atomic numbers: the metals in Group I (alkali metals) occupy the peaks and the transition metals in Groups 5 to 11 the valleys, of the curves in the periodic relationships. The values for atomic and ionic radii of elements depend on the method of determination. Therefore, the data on atomic and ionic radii plotted in Figures 1.10 and 1.11 must be considered approximate values only [2].

In any group in the periodic table, seeing that each element has one more energy level than the element above it, as we move down through a particular group, the atomic radii of the elements increase.

We next look at the change in values for the atomic radii of the metallic elements within each period. As we move from left to right across a period, the number of protons increases along with the number of electrons. As a result, the atomic radii would decrease with increasing atomic number (i.e. as the number of protons increases), because the increased number of protons pull hard on the electrons of the inner energy levels. This is the case in Periods 2 and 3, with Groups 1 to 6 in Periods 4 to 7, and with the lanthanoids. This phenomenon is called the lanthanoid (or lanthanide) contraction. Incidentally, the lanthanoid elements are generally trivalent; europium and ytterbium in the metallic state are divalent, which leads to their anomalous properties in the lanthanoid series (see Figures 1.10 and 1.11). Cerium would be tetravalent. However, the atomic radii change very little between Groups 5 to 11, and increase, contrary to our expectations, from Group 12 and over in Periods 4 to 6. As for transition metals, the atomic radii change little, because electrons are filling the d and f subshells.

---

[24] The ionization energy is sometimes called the ionization potential or the ionization voltage.

[25] There are four principal types of crystal: molecular crystals, ionic crystals, metallic crystals, covalent crystals.

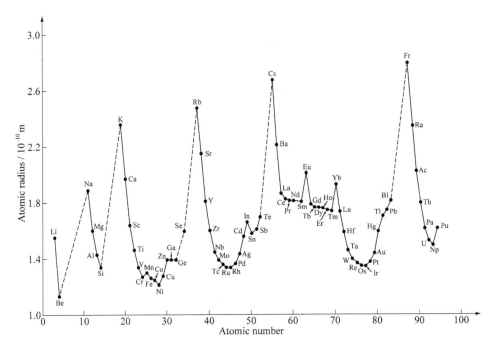

Figure 1.10 *Atomic radius plotted against atomic number for a large number of metallic elements. (Data from Benenson et al. [2]).*

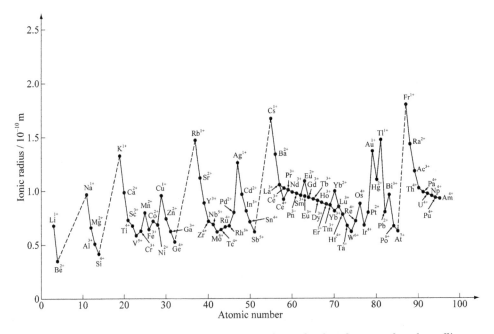

Figure 1.11 *Positive ionic radius plotted against atomic number for a large number of metallic elements. (Data from Benenson et al. [2]).*

Concerning the periodic variation in the positive ionic radii with atomic number, there is, broadly speaking, a trend similar to that in the atomic radii. We should note, however, that in the copper group metals, the positive ion cores are relatively larger.

The ionization energy of an element is the minimum energy required to remove an electron from a given atom or molecule to infinity. It is the least energy that causes the ionization of that element.[26] When a neutral atom A becomes ionized, we can write the following equation:

$$ A \longrightarrow A^{1+} + e^{1-} $$

$$
\begin{array}{ccc}
\text{an isolated} & \text{a positive} & \text{an electron} \\
\text{ground-state} & \text{ion} & \\
\text{neutral atom} & &
\end{array}
\tag{1.23}
$$

where the positive ion and the electron are far enough apart for their electrostatic interaction to be negligible. The energy necessary to remove the least strongly bound electron is called the first ionization energy. The first ionization energy corresponds to the formation of the ground state (i.e. the lowest stable energy state of an atom) of the singly charged ion.

It is also possible to remove electrons from inner energy levels, or inner orbitals, in which their binding energy is greater. The minimum energy to remove the second least strongly bound electron from a neutral atom is called the second ionization energy.[27] Similarly, third, and higher ionization energies can also be measured.

Figure 1.12 gives values for the first ionization energy of the metallic elements against their atomic numbers. As seen, there is a periodic variation in the values of ionization energy; the periodic Group 12, i.e. zinc group metals, occupy the peaks, while the periodic Groups 1 and 13 metals occupy the valleys of the curve.

The farther away from the nucleus an electron is, the less it is attracted by the positive charge of the nucleus (i.e. protons). Thus, in any group of the periodic table, the ionization energy tends to decrease as we move down the group. Within each period, ionization energy increases from left to right, with some exceptions. The ionization energies of the transition metals in each period do not very much. The transition metals in Period 6 (from tantalum to gold) have relatively high ionization energies.

If some conditions are met, the atomic and/or ionic radii, and ionization energy of an element, determine its chemical and physical properties. For example, the atomic and ionic radii of lithium are both very small (the smallest) in Group 1 metals, i.e. alkali metals (see Figures 1.10 and 1.11). Because of these physical quantities, lithium obeys the diagonal relationship,[28] and some of its properties are similar to those of magnesium.

---

[26] The process of forming ions is called ionization.

[27] The second ionization energy is often taken to be the minimum energy required to remove an electron from the singly charged ion.

[28] A relationship within the periodic table by which certain elements in the second period have a close chemical similarity to their diagonal neighbours in the next group of the third period. A notable example is a similarity between lithium and magnesium. The reason for this relationship is a combination of the trends to increase size down a group and to decrease size along a period.

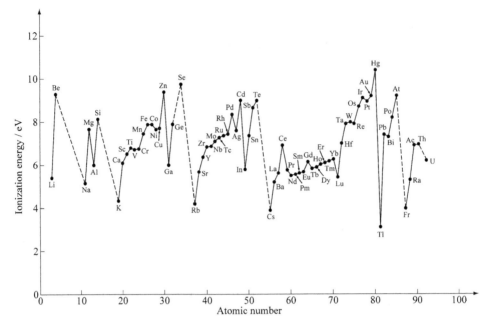

Figure 1.12 *The first ionization energy plotted against atomic number for a large number of metallic elements. (Data from Benenson et al. [2]).*

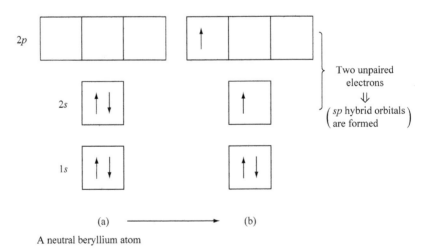

Figure 1.13 *Electron configurations of beryllium atom ($_4$Be). The arrow ↑ or ↓ indicates the spin orientation of the electrons.*

We give another example. Beryllium has not only extremely small atomic and ionic radii (which are the smallest, of all metallic elements), but also possesses a high ionization energy (see Figures 1.10 to 1.12). For this reason, the ionization of beryllium will not be caused easily, so that unpaired electrons are produced, as shown in Figure 1.13. Thus, beryllium is more apt to form a covalent bond,[29] rather than an ionic bond,[30] using these unpaired electrons, i.e. using the *sp* hybrid orbitals formed by the unpaired electrons (see Figure 1.13).

1.7   Other Important Matters in Studying Metallic Liquids

1.7.1   The Reliability of Experimental Data

At the end of this first chapter, the matter of experimental data must be addressed wholeheartedly. We repeat here that accurate and reliable data for the thermophysical properties of metallic liquids are required, not only for both the execution of computer experiments and the development of mathematical models, but also for the direct solution of industrial problems in high temperature processing operations. Nevertheless, even now, accurate and reliable data for thermophysical properties of liquid metallic elements are still not particularly abundant. For example, Table 3.1 shows that fairly large discrepancies exist between experimental values of the volume changes during the melting of some metals, such as aluminium, iron, nickel, magnesium, and zinc. In discussing the nature and behaviour of metallic liquids, density or number density is an indispensable basic quantity. Furthermore, these are important engineering metals. Table 1.8 also reveals discrepancies in the values of evaporation enthalpy for some 15 liquid metallic elements. The evaporation enthalpy $\Delta_1^g H_b$ is very useful thermophysical quantity; unfortunately, $\Delta_1^g H_b$ is not known for all liquid metallic elements. In addition, it should be pointed out that experimental data for some important thermophysical properties of liquid metallic elements, e.g. sound velocity and diffusivity, are relatively scarce.

1.7.2   Microscopic, Mesoscopic, and Macroscopic Approaches

Although we have placed emphasis on the importance of microscopic, or atomistic, approach to a study of metallic liquids from the standpoint of materials process science, the concept of a liquid as a continuum is still very useful. For example, Stokes' law, based on treating a liquid as a continuous medium, is often used to calculate the terminal velocity of particles (e.g. non-metallic inclusions such as Al_2O_3 or bubbles in a fluid medium). However, colloidal particles, in the range of 10^{-9} to 10^{-6} m in diameter,[31] suspended in

---

[29] A covalent bond is a chemical bond in which two or more atoms share electrons.
[30] An ionic bond is a chemical bond in which one atom transfers one or more electrons to another atom and the resulting ions are hold together by the attraction of opposite charges.
[31] 10^3 to 10^9 atoms are contained.

Table 1.8 *Discrepancies in values for the evaporation enthalpy of liquid metallic elements listed in the literatures.*

Element		$\Delta_l^g H_b$ / kJ mol^{-1}			
		A[†]	B[‡]	C[††]	D[‡‡]
Antimony	Sb	128	165	67.9	–
Bismuth	Bi	152	179	151.5	–
Caesium	Cs	65.9	66.5	55.7	67.78
Gallium	Ga	254	270	250.3	267
Gold	Au	325	343	324.5	310.5
Hafnium	Hf	661	571	–	–
Iridium	Ir	750	612	563.8	–
Lithium	Li	142	148	158.9	148
Manganese	Mn	230	220	255.7	225
Mercury	Hg	57.2	59.1	61.35	58.1
Molybdenum	Mo	538	590	594.3	590
Platinum	Pt	447	469	510.6	447
Potassium	K	77.4	79.1	88.1	77.4
Sodium	Na	–	99.2	106.8	89.1
Titanium	Ti	430	426	397 ± 20	–

Sources of data: [†] W. Benenson et al. [2]; [‡] Iida and Guthrie [12]; [††] Japan Institute of Metals [13]; [‡‡] Nagakura et al. [4].

a liquid medium, are dominated by Brownian movement. This is a visible demonstration of molecular bombardment by the continually, irregularly moving molecules (atoms, ions) of the liquid. The smaller the particles are, the more extensive the motion is. This is a mesoscopic phenomenon. Thus, Stokes' law is not applicable to the small particles, such as colloids, and mesoscopic approaches are needed.

....

REFERENCES

1. W.H. Giedt, *Thermophysics*, Van Nostrand Reinhold Company, New York, 1971, p.58.
2. W. Benenson, J.W. Harris, H. Stocker, and H. Lutz (eds.), *Handbook of Physics*, Springer-Verlag, New York, 2002.

3. W.F. Gale and T.C. Tolemeier (eds.), *Smithells Metals Reference Book*, 8th ed., Elsevier Butterworth-Heineman, Oxford, 2004, 14–1.
4. S. Nagakura, H. Inokuchi, H. Ezawa, H. Iwamura, F. Sato, and R. Kubo (eds.), *Iwanami Dictionary of Physical Sciences (Iwanami Rikagaku Jiten)*, 5th ed., Iwanami Shoten Publishers, Tokyo, 1998.
5. F.A. Lindemann, *Phys. Z.*, **11** (1910), 609.
6. J.W.M. Frenken and J.F. van der Veen, *Phys. Rev. Lett.*, **54** (1985), 134.
7. M. Polcik, L. Wilde, and J. Hasse, *Surf. Sci.*, **405** (1998), 112.
8. R. Kojima and M. Susa, *High Temp-High Press.*, **34** (2002), 639.
9. R. Kojima and M. Susa, *Sci. Technol. Adv. Mater.*, **5** (2004), 497; 677.
10. R. Trittibach, C. Grütter, and J.H. Bilgram, *Phys. Rev. B*, **50** (1994), 2526.
11. J.A. Barker and D. Henderson, *Sci. Am.*, **245** (1981), 130.
12. T. Iida and R.I.L. Guthrie, *The Physical Properties of Liquid Metals*, Clarendon Press, Oxford, 1988, p.8.
13. The Japan Institute of Metals, *Metals Data Book*, 4th ed., Maruzen Company, Tokyo, 2004, 11.

2

Structure and Pair Potential Energy

2.1 Introduction

In obtaining direct structural information on liquids, the results of X-ray and neutron diffraction experiments have been proved to be of value.[1] We find that the X-ray diffraction pattern of the (perfect) crystalline solid consists of symmetrically sharp peaks. This result is due to a regular periodic array of atoms in three dimensions, i.e. the long-range order existing in the crystalline solid. For the case of a gas, however, the X-ray diffraction patterns show a continuous scattering intensity with no maxima. This can be explained by a lack of any regular atomic arrangement in a gas at low number density. According to experimental diffraction patterns for liquids, which have been time and space averaged, liquids exhibit a few maxima and minima, as shown in Figure 2.1. This can be interpreted as showing that atoms in the liquid state are randomly distributed in a nearly close-packed arrangement. Thus, we presume that liquids have a certain amount of short-range order, which is a necessary consequence of high packing density, but have long-range disorder owing to thermal excitation and motion.

The structures of liquids closely resemble those of amorphous solids. The latter are generally obtained through rapid quenching from the liquid state. The amorphous solid provides a 'snapshot' of the atomic configuration of the liquid state. However, liquid structures averaged over time and over all atoms are slightly different from those of an amorphous solid, on account of significant differences in the degree of atomic motion between the two states.

---

[1] The techniques of neutron diffraction have not been used widely compared with those of X-rays; neutron diffraction experiments are carried out using a flux of thermal neutrons from a nuclear reactor. X-rays are scattered by the electrons in atoms, whereas neutrons are scattered by the nuclei of atoms. Therefore, X-rays are not suitable for investigating light elements (e.g. hydrogen, helium). However, experiments using a beam of neutrons give diffraction patterns from such light elements, and furthermore, valuable information on the thermal motions of atoms.

The Thermophysical Properties of Metallic Liquids: Volume 1 – Fundamentals. First Edition.
Takamichi Iida and Roderick I. L. Guthrie. © Takamichi Iida and Roderick I. L. Guthrie 2015.
Published in 2015 by Oxford University Press.

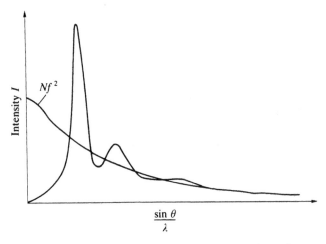

Figure 2.1 *X-ray diffraction pattern for a liquid. The Nf^2 curve corresponds to an (ideal) gas. I is the intensity of X-rays scattered through an angle 2θ, λ is the wavelength of the incident beam, N is the total number of atoms, and f is the atomic scattering factor.*

2.2 Distribution Functions and Pair Potentials

2.2.1 Pair Distribution Function and Radial Distribution Function

The description of liquids in terms of models has so far proved unsuccessful. However, in order to discuss the structure of liquids, it is first necessary to have a way of describing this structure mathematically. For this purpose, the pair distribution function $g(r)$ is of central importance to the modern theory of liquids, and the structures and properties of liquids in equilibrium are best described in terms of this function. As already mentioned, however, it should be kept in mind that $g(r)$ (i.e. a mathematical method of describing the spatial arrangement of the atoms in a liquid, or the structure of a liquid) is averaged over time and over all atoms, and does not give a 'snapshot' of an atomic distribution at any particular point in time.

Let us now consider a monatomic liquid in equilibrium, where atoms are spherical with interatomic forces acting between atomic centres (i.e. a simple liquid). Then consider any atom with its centre at the point $r = 0$. The number of atoms dN in the spherical shell between radial distances r and $r + dr$ from the origin atom is

$$dN = 4\pi r^2 dr \left(\frac{N}{V}\right) g(r) \tag{2.1}$$

where N is the total number of atoms in the volume V. This pair distribution function $g(r)$ depends on the magnitude of r but not its direction, since real (monatomic) liquids are generally isotropic.

If there is no correlation between atoms, $g(r)$ is everywhere equal to unity (i.e. $g(r) \neq f(r)$). Consequently, in the case of the ideal gas, the number of atoms in the spherical shell between r and $r + dr$ is

$$dN = 4\pi r^2 dr\, n_0 \qquad (2.2)$$

where $n_0\ (= N/V)$ is the mean number density.

We now consider the pair distribution function in real liquids. The pair distribution function curve as a function of distance becomes complicated on account of the existence of atomic interactions, i.e. the forces of attraction or repulsion between atoms in the condensed state. For larger r, as we might expect, the value for $g(r)$ tends towards unity, because the interactions between the central and outlying atoms weaken rapidly with distance. In other words, at large r, the probability of finding another atom in the spherical shell between r and $r + dr$ is independent of the presence of the reference atom. This corresponds to complete disorder. On the other hand, for small values of r, that is, for values of r less than the atomic diameter, the probability must tend to zero, because two atoms cannot overlap.

Consequently, the pair distribution function $g(r)$ may be defined as follows: $g(r)$ is proportional to the probability of finding another atom at a distance r from the reference atom located at $r = 0$.[2] In this respect, the function $4\pi r^2 n_0 g(r)$ is generally called the radial distribution function (r.d.f.).

A typical curve for the pair distribution function is given in Figure 2.2. As can be seen from this figure, there are a few maxima and minima in $g(r)$, and their amplitudes decrease rapidly towards unity with increasing r. This departure of $g(r)$ from unity shows

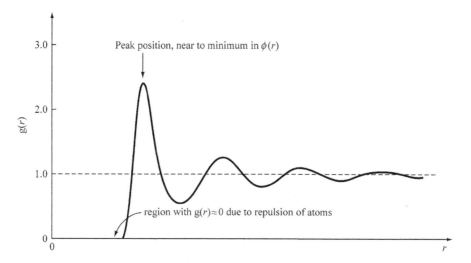

Figure 2.2 *Typical pair distribution function for a simple liquid.*

---

[2] The pair distribution function does not consider interaction between the pair of atoms and other atoms surrounding them.

the existence of short-range order around the reference atom. A pronounced first peak in the $g(r)$ curve is located roughly on the minimum of the pair potential (see Figures 2.5(e) and (f)).

We mention here an important point of the pair distribution function $g(r)$ curve. The pair distribution function for any monatomic liquid, no matter what the interatomic interaction is like, is very similar to that which would be expected if the attractive forces of atoms in a liquid were entirely neglected (i.e. hard-spheres). Indeed, the experimentally obtained $g(r)$ curves for monatomic liquids (e.g. liquid metals, liquid argon) have the same general form; in other words, the form of $g(r)$ is determined largely by the repulsive forces of atoms.

A typical form of the r.d.f. is shown in Figure 2.3. Since the pair distribution function approaches unity for larger r, the radial distribution function draws close to the parabolic curve of $4\pi r^2 n_0$. The hatched area, or the area under the principal peak, or the main peak, can be interpreted as the number of nearest-neighbour atoms and the so-called first coordination number. The first coordination number Z can be expressed mathematically by

$$Z = 2 \int_{r_0}^{r_m} 4\pi r^2 n_0 g(r) \mathrm{d}r \tag{2.3}$$

Since the properties of liquid can be described approximately in terms of the first coordination number, it is frequently used as an important parameter. At the present time, however, we have no absolute method for estimating the first coordination number. It is necessary to note that its values will vary depending on the methods used in its estimation.

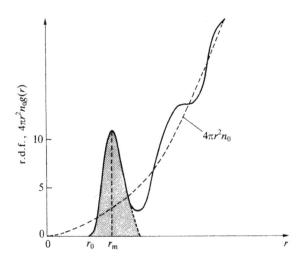

Figure 2.3 *Typical curve for the radial distribution function (r.d.f.). r_0 and r_m represent the beginning and first peak values of radial distance in the r.d.f. curve, respectively.*

2.2.2 Theoretical Calculation of $g(r)$

According to statistical mechanical theory, the pair distribution function is described through the formula:

$$g(r) = V^2 \frac{\int^{(N-2)} \cdots \int e^{-\Phi/kT} dv_3 \cdots dv_N}{\int^{(N)} \cdots \int e^{-\Phi/kT} dv_1 \cdots dv_N} \tag{2.4}$$

where V represents the total volume of the system, dv a volume element, and Φ the total potential energy of the system. The pair distribution function in the equation is expressed in terms of the total interatomic potential energy Φ. However, it is impossible, in practice, to calculate the pair distribution function by direct solution of this multi-integral equation. Consequently, several approximate equations, e.g. Born–Green (or Yvon–Born–Green), Percus–Yevick, and hypernetted chain equations, have been suggested. These equations give us integral formulae connecting the pair distribution function $g(r)$ and the pair potential (energy) $\phi(r)$.[3] As already mentioned, this approach is called the pair theory of liquids. In practice, these equations are sometimes used in reverse, so as to calculate $\phi(r)$ from $g(r)$ which can be measured experimentally.

The theoretical calculation of the pair distribution function has proved to be a difficult task because of the mathematical complexity and lack of available information regarding the pair potential ϕ_{ij}. As yet, in the theoretical calculation of $g(r)$ there are no satisfactory results for engineering applications.

2.2.3 Experimental Determination of $g(r)$

The experimental investigation of liquid structures can be carried out through the use of X-ray or neutron diffraction techniques.

The pair distribution function is obtained from measured intensities (see Figure 2.4), by using the formula:

$$g(r) = 1 + \frac{1}{2\pi^2 n_0 r} \int_0^\infty Q\left(\frac{I}{Nf^2} - 1\right) \sin(Qr) dQ \tag{2.5}$$

$$Q = \frac{4\pi \sin\theta}{\lambda}$$

where 2θ is the scattering angle (of X-rays), λ is the wavelength of the incident beam, f is the atomic scattering factor, i.e. the Fourier transform of the electron density in the atom, and I is the intensity of reflected beams from the liquid. This formula may also be interpreted as the definition of $g(r)$. The determination of $g(r)$ from observed intensities

---

[3] It is assumed that the total potential Φ can be decomposed into a sum of values of pair potential ϕ_{ij} (the potential developed between pairs of atoms i, j), such that $\Phi(\mathbf{r}_1 \cdots \mathbf{r}_N) = \sum \phi(r_{ij})$ for $i < j$, where three-body (or three-particle) correlations are neglected.

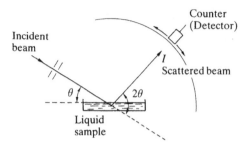

Figure 2.4 *Schematic representation of a scattering experiment.*

is subject to some problems. These stem mainly from experimental difficulties and from the methods used to treat experimental data. The relative intensity of (I/Nf^2) is defined as a (liquid) structure factor or an interference function (denoted by $S(Q)$ in this book, i.e. $I/Nf^2 \equiv S(Q)$), which is frequently used in discussions concerning the structures of non-crystalline materials.

2.2.4 The Mutual Potential Energy of Two Atoms (Molecules): the Pair Potential

We may now introduce the pair potential $\phi(r)$. This represents the potential energy between an atom and its surrounding neighbours. All equilibrium properties of liquids can be expressed directly using the pair distribution function $g(r)$ in conjunction with $\phi(r)$. For this reason, $g(r)$ and $\phi(r)$ can be regarded as being the most fundamental of liquid properties. Basically, pair potentials $\phi(r)$ should be derivable, in theory, from quantum mechanics or electron theory. However, strictly speaking at the present time, quantum mechanical calculations of pair potentials are practically impossible apart from those of simple atoms such as hydrogen and helium. Consequently, various model pair potentials are frequently used as an alternative in numerical calculations, as follows:

(i) The hard sphere potential (Figure 2.5(a)):

$$\phi(r) = +\infty \quad \text{for } r < \sigma$$

$$\phi(r) = 0 \qquad \text{for } r > \sigma \tag{2.6}$$

where σ is the diameter of the hard sphere. This is the simplest model. Attractive forces are entirely neglected in this model.

The following model potentials, i.e. (ii) the inverse power and (iii) Sutherland potentials, provide a somewhat more realistic presentation of the repulsive or attractive energy.

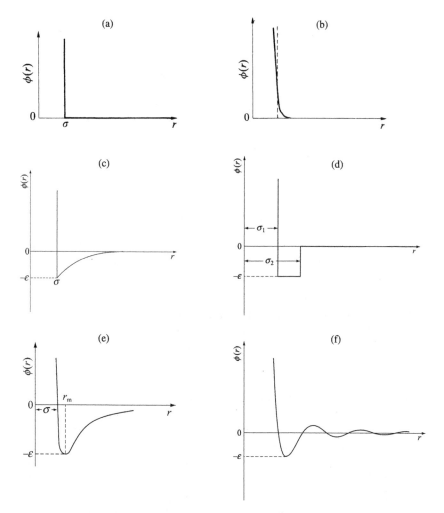

Figure 2.5 *Schematic representation of the pair potentials. ε is the depth of the attractive well.*

(ii) The inverse power potential (Figure 2.5(b)):

$$\phi(r) = \varepsilon\left(\frac{\sigma}{r}\right)^{n} \tag{2.7}$$

where ε, σ, and n are parameters. If n tends to infinity, the inverse power potential becomes that of the hard sphere. The repulsive exponent n will give a general indication of the extent of an atom's hardness or softness (the hardness or softness of the pair interatomic potential, to be exact).

(iii) The Sutherland potential (Figure 2.5(c)):

$$\phi(r) = +\infty \qquad \text{for } r < \sigma$$

$$\phi(r) = -\varepsilon\left(\frac{\sigma}{r}\right)^m \qquad \text{for } r > \sigma \qquad (2.8)$$

where $-\varepsilon$ is a minimum value of the potential, and $m \geq 3$. This model is to consider hard spheres which exert attractive forces on each other.

(iv) The square-well potential (Figure 2.5(d)):

$$\phi(r) = +\infty \quad \text{for } r < \sigma_1$$

$$\phi(r) = -\varepsilon \quad \text{for } \sigma_1 < r < \sigma_2 \qquad (2.9)$$

$$\phi(r) = 0 \quad \text{for } r > \sigma_2$$

where σ_1, σ_2, and ε are parameters; the square-well potential has three adjustable parameters. This is the simplest model potential including both attractive and repulsive contributions, and is a good compromise between mathematical simplicity and realism.

(v) The Lennard-Jones potential (Figure 2.5(e))
This is widely used for molecular 'simple liquids',

$$\phi(r) = 4\varepsilon\left[\left(\frac{\sigma}{r}\right)^n - \left(\frac{\sigma}{r}\right)^m\right] \qquad (2.10)$$

where ε is the well depth and σ is the distance at which $\phi(r) = 0$. The Lennard-Jones (L-J) potential can be considered as a semi-empirical expression. The parameters n and m are generally taken as 12 and 6 (the L-J (6,12) potential),[4] although other combinations are frequently used. Calculations of many macroscopic properties of liquids have been made on the basis of L-J potential (e.g. inert gas liquids, nonpolar liquids), partly because of its mathematical simplicity.

(vi) The effective ion–ion potential (Figure 2.5(f))
This potential is used for liquid metals. In the case of liquid metals, the pair potential corresponds to the effective ion–ion potential given by:

$$\phi(r) = \frac{A}{r^3}\cos(2k_F r) \qquad (2.11)$$

---

[4] The L-J (6,12) potential is

$$\phi(r) = 4\varepsilon\left[\left(\frac{\sigma}{r}\right)^{12} - \left(\frac{\sigma}{r}\right)^6\right].$$

When $r = r_m$, $d\phi(r)/dr = 0$. From the above equation, we obtain

$$r_m = 1.12\sigma, \quad (r_m^6 = 2\sigma^6)$$

$$\sigma = 0.89\, r_m$$

where A is a parameter and k_F is the radius of the Fermi sphere or the Fermi wave vector. In a liquid metal, the two charged species are positive ions and conduction electrons. The ions are immersed in a sea of an effective potential generated by the distribution of conduction electrons. Consequently, the pair potential (i.e. the effective ion–ion potential) for metals (unlike the Coulomb potential)[5] has oscillations resulting from the presence of conduction electrons. However, the amplitude of the oscillatory potential curve is very small.[6]

We mention again that the distribution function curve $g(r)$ of the liquid metals is very similar to that of condensed matter of neutral atoms (e.g. argon) with closed electron shells; the relationship between the oscillatory tail of $\phi(r)$ for liquid metals and their thermophysical properties is not yet clear.

The parameters appearing in Eqs. (2.6) to (2.11) can be determined by experiments involving X-ray and/or neutron scattering.

2.2.5 Some Equations in Terms of $g(r)$ and $\phi(r)$

In order to facilitate an understanding of the concept of the pair distribution function $g(r)$, let us derive the internal energy, which is a very important thermophysical property.

(1) Total internal energy

For a monatomic liquid at temperature T, the (total) internal energy U_1 is given by

$$U_1 = \frac{3}{2}NkT + <\Phi> \qquad (2.12)$$

where $<\Phi>$ is the mean total potential energy.

The term $3NkT/2$ in Eq. (2.12) represents the liquid's total kinetic energy. Let us derive $<\Phi>$.

(a) As mentioned previously, the average number of atoms within a radial distance between r and $r + dr$ of a given reference atom is $n_0 g(r)(4\pi r^2 dr)$.

(b) The average potential energy of interaction with these neighbours is $n_0 g(r)\phi(r)(4\pi r^2 dr)$.

(c) Integrating over r to obtain the total potential energy and dividing by 2 to avoid counting each pair interaction twice over, result in

$$<\Phi> = \frac{n_0 N}{2}\int_0^\infty g(r)\phi(r)4\pi r^2 dr.$$

---

[5] The Coulomb potential energy is inversely proportional to the distance between charged particles.
[6] Ion–ion repulsion in a liquid metal is regarded as the principal factor determining the ionic arrangement (or the liquid metal's structure).

Consequently, we obtain

$$U_1 = \frac{3}{2}NkT + \frac{n_0 N}{2} \int_0^\infty g(r)\phi(r)4\pi r^2 dr \tag{2.13a}$$

or

$$U_1 = \frac{3}{2}NkT + \frac{2\pi N^2}{V} \int_0^\infty g(r)\phi(r)r^2 dr \tag{2.13b}$$

Although Eq. (2.13) has been derived through descriptive arguments, it is also possible to arrive at it through the partition function in statistical mechanics, as given bellow (i.e. Eq. (2.17)).

(2) Heat capacity at constant volume
Combining Eq. (2.13) with the thermodynamic formula for heat capacity at constant volume C_V immediately gives

$$C_V = \left(\frac{\partial U}{\partial T}\right)_V = \frac{3}{2}Nk + \frac{2\pi N^2}{V} \int_0^\infty \left[\frac{\partial g(r)}{\partial T}\right]_V \phi(r)r^2 dr \tag{2.14}$$

Since the assumption is made that $\phi(r)$ is independent of temperature, only information on the variation of $g(r)$ with temperature at constant volume is required.

According to statistical mechanical theory, if the partition function Z given by the following equation is known, the thermodynamic properties of a system of N molecules (atoms) comprising a liquid can be calculated (in general, the procedures for their derivations are complicated).

$$Z = \frac{1}{N!}\left(\frac{2\pi mkT}{h^2}\right)^{3N/2} \int \cdots \int \exp\left(-\frac{\Phi}{kT}\right) dv_1 \cdots dv_N \tag{2.15}$$

where h is the Planck constant.
As examples, a few equations are given for thermodynamic properties.
The Helmholtz free energy F

$$F = -kT \ln Z \tag{2.16}$$

The (total) internal energy U

$$U = kT^2 \left(\frac{\partial \ln Z}{\partial T}\right)_V \tag{2.17}$$

The entropy S

$$S = k \ln Z + kT \left(\frac{\partial \ln Z}{\partial T} \right)_V$$
(2.18)

The enthalpy H

$$H = kT^2 \left(\frac{\partial \ln Z}{\partial T} \right)_V + kTV \left(\frac{\partial \ln Z}{\partial V} \right)_T$$
(2.19)

The equation of state

$$P = kT \left(\frac{\partial \ln Z}{\partial V} \right)_T$$
(2.20a)

$$\frac{PV}{NkT} = 1 - \frac{2\pi N}{3kTV} \int_0^\infty \frac{\partial \phi(r)}{\partial r} g(r) r^3 \, dr$$
(2.20b)

$$\frac{PV}{NkT} = 1 + \frac{2\pi N}{3V} \sigma^3 g(\sigma), \text{ for a hard-sphere fluid}$$
(2.21)

Incidentally, the partition function Z is one of the most important functions in statistical mechanics.

2.3   The Structure of Liquid Metallic Elements

A considerable amount of structural information about metallic liquids has been accumulated in the last half century. The values of pair distribution functions obtained from X-ray scattering experiments are shown in Figures 2.6 to 2.9 for sodium, lead, copper, and iron near their melting points. All of the distribution curves have the same

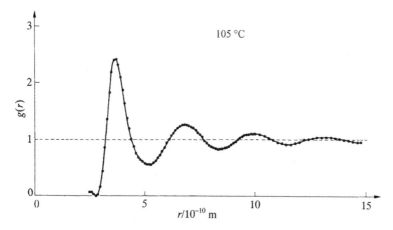

Figure 2.6 *Pair distribution function g(r) of liquid sodium near the melting point.*

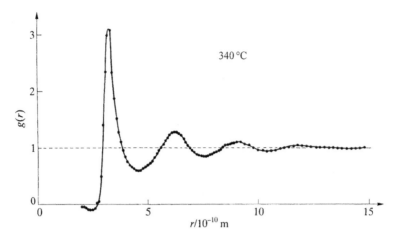

Figure 2.7 *Pair distribution function g(r) of liquid lead near the melting point.*

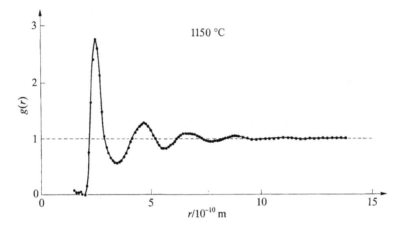

Figure 2.8 *Pair distribution function g(r) of liquid copper near the melting point.*

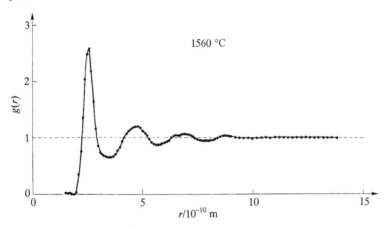

Figure 2.9 *Pair distribution function g(r) of liquid iron near the melting point.*

general shape (as noted earlier, although the form of the pair potential of a metal differs considerably from that of a monatomic, non-metallic liquid, pair distribution functions obtained from experiments show that the shape of $g(r)$ for liquid metals closely resembles non-metallic liquids such as argon); namely, the values of $g(r)$ show deviations from unity at distances approximately one (a pronounced peak), two, three, and four atomic diameters, and become equal to about unity beyond four atomic distances. These figures also show that the value of $g(r)$ at distances of approximately one-and-a-half atomic diameters (i.e. $r \approx 1.5\sigma$) are very small. The deviations in the curves from unity imply that short-range order holds to three or four atomic diameters in liquid metals.

Structure factors for these metals are shown in Figures 2.10 to 2.13. Many workers have investigated the structure of liquid metals from the standpoint of the hard-sphere

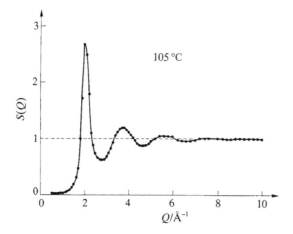

Figure 2.10 *Structure factor S(Q) of liquid sodium near the melting point.*

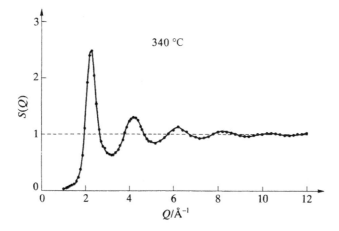

Figure 2.11 *Structure factor S(Q) of liquid lead near the melting point.*

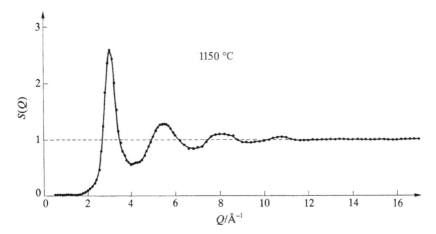

Figure 2.12 *Structure factor S(Q) of liquid copper near the melting point.*

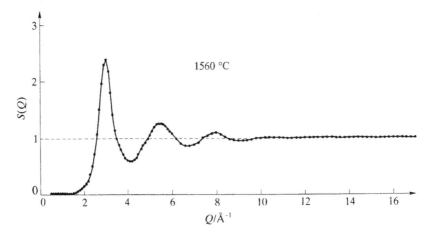

Figure 2.13 *Structure factor S(Q) of liquid iron near the melting point.*

model. They found that the hard-sphere potential (though a crude approximation), through suitable choice of hard-sphere diameter and packing fraction, could reproduce measured structure factors. The results imply that the characteristic features of the structure factor of liquid metal are largely determined by ion–ion repulsions, and that the effect of the long-range oscillations in the pair potential on the structure is very small for liquid metals. In practice, even though liquid metals have very complex interatomic interactions, these simplifications can be made.

As further support, we may note that the random packing of hard sphere also leads to reasonably good agreement with experimental data for the pair distribution function.

The best value for the packing fraction that is compatible with experimental data for the $S(Q)$ curves is 0.45 to 0.46 for the majority of liquid metals near their melting points.

The pair distribution functions obtained from scattering data for mercury and tin near their melting point temperatures are shown in Figures 2.14 and 2.15, respectively. These curves of $g(r)$ are somewhat different from those of the simple metals such as sodium; namely, in the case of mercury there seems to be a slight asymmetry in the first peak. To this type of $g(r)$ curve belong indium, thallium, gadolinium, and terbium. On the other hand, in the case of $g(r)$ curve for tin (250 °C) compared with that for lead (340 °C), there is a small hump or shoulder to the right of the first peak, as the large-scale graph in Figure 2.16 shows. To this type belong gallium, silicon, germanium,

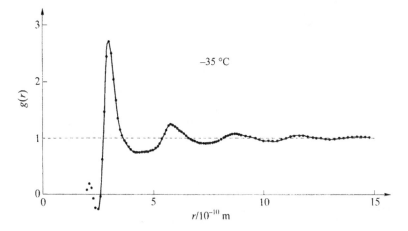

Figure 2.14 *Pair distribution function g(r) of liquid mercury near the melting point.*

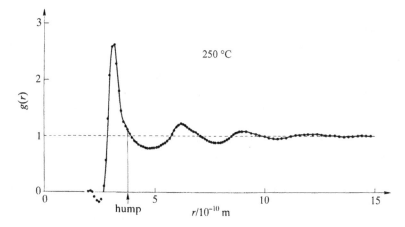

Figure 2.15 *Pair distribution function g(r) of liquid tin near the melting point.*

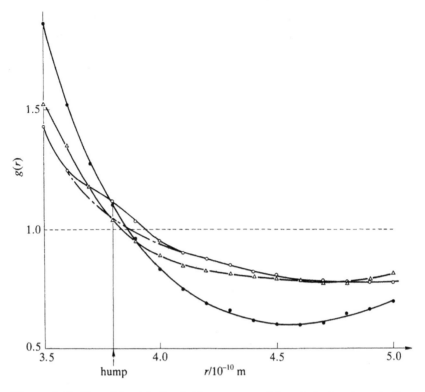

Figure 2.16 *Plots of the right-hand side component of the principal peak, or the first peak, in pair distribution functions g(r) for liquid tin and lead. ○ tin (250 °C), △ tin (1100 °C), ● lead (340 °C).*

antimony, and bismuth. These metallic elements also exhibit a small hump to the right of the main peak in $S(Q)$.

As mentioned previously, the structure of liquid metals can be described by the random packing of hard spheres. In a stricter sense, however, atoms in the liquid metals just mentioned can retain slightly-directional bonding, characteristic of their solid crystalline structures, despite thermal agitation in the liquid state.

As an example of the temperature dependence of the pair distribution function, experimental results for liquid aluminium and tin are shown in Figures 2.17 and 2.18, respectively.

In discussing the properties of metallic liquids, structural information is of considerable importance. Structural data on $g(r)$ are listed in Table 2.1 (see Figure 2.19). These structural data on $g(r)$ will be used for calculating a metallic liquid's viscosity in Chapter 7. In view of the importance of the first peak (or the main peak) in the $g(r)$ curve, values of $g(r_m)$ for some metallic liquids at various temperatures, i.e. the temperature dependence of $g(r_m)$, are given in Figure 2.20(a–c).

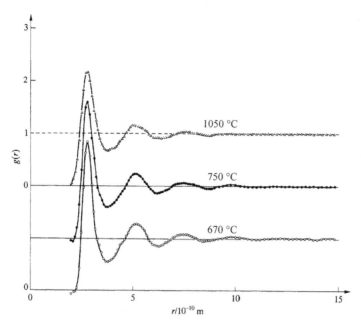

Figure 2.17 *Temperature dependence of the pair distribution function for liquid aluminium.*

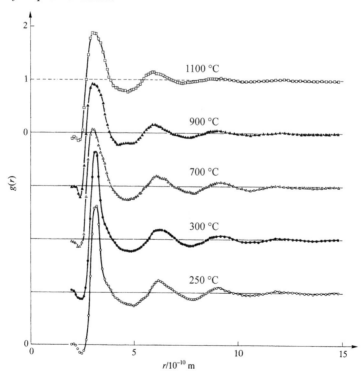

Figure 2.18 *Temperature dependence of the pair distribution function for liquid tin.*

Table 2.1 *Structural information on the main peaks, or the first peaks, of g(r) of liquid metals and semimetals near their melting points.*

Metallic element		Temperature °C	r_0 10^{-10} m	r_m 10^{-10} m	$g(r_m)$
Aluminium	Al	670	2.28	2.78	2.83
Antimony	Sb	660	2.58	3.26	2.31
Bismuth	Bi	300	2.78	3.34	2.56
Cadmium	Cd	350	2.54	3.00	2.82
Cobalt	Co	1550	1.88	2.48	2.37
Copper	Cu	1150	2.06	2.50	2.76
Gallium	Ga	50	2.38	2.78	2.62
Gold	Au	1150	2.34	2.80	2.77
Indium	In	160	2.70	3.14	2.66
Iron	Fe	1560	1.98	2.56	2.54
Lead	Pb	340	2.76	3.26	3.07
Magnesium	Mg	680	2.52	3.10	2.46
Mercury	Hg	−35	2.62	3.00	2.71
Nickel	Ni	1500	1.88	2.46	2.36
Potassium	K	70	3.60	4.56	2.35
Silver	Ag	1000	2.34	2.82	2.58
Sodium	Na	105	2.92	3.68	2.42
Thallium	Tl	315	2.74	3.22	2.75
Tin	Sn	250	2.68	3.14	2.62
Zinc	Zn	450	2.16	2.66	2.42

Data are from Waseda [1].

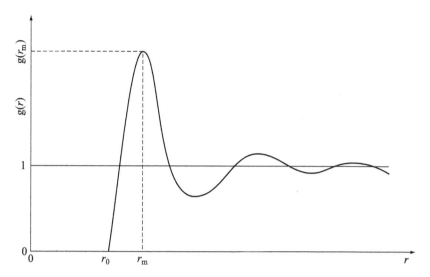

Figure 2.19 *Plot of g(r) curve vs. distance. r_0 and r_m denote the beginning and first peak (or main peak) values of radial distance in the g(r) curve, respectively. $g(r_m)$ denotes the value of g(r) at r_m.*

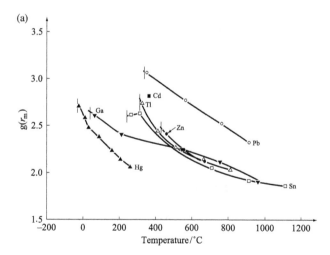

Figure 2.20 *(a) Temperature dependence of $g(r_m)$ (see Figure 2.19) for several liquid metals. |, melting point. (b) Temperature dependence of $g(r_m)$ for several liquid metals and semimetals. |, melting point. (c) Temperature dependence of $g(r_m)$ for several liquid transition metals. |, melting point.*

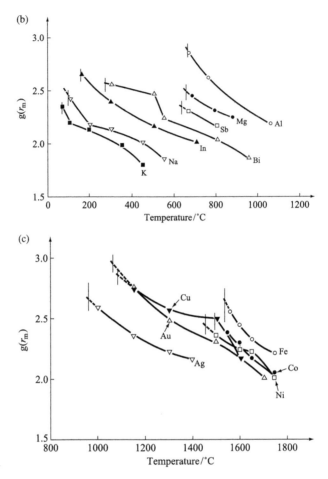

Figure 2.20 *(continued)*

2.4   The Structure of Liquid Alloys

Theoretical treatments for the structure of a liquid alloy (even a binary alloy) can only be achieved with considerable difficulty. In a binary alloy of components 1 and 2, three pair distribution functions, i.e. the partial distribution functions $g_{11}(r)$, $g_{12}(r)$, and $g_{22}(r)$, are required for a complete description of its structure.

In terms of general notation, the partial pair distribution function $g_{\alpha\beta}(r)$ corresponds to the probability of finding an atom β at a distance r from an origin atom α. In other words, if there is an atom α at $r = 0$, the number of atoms β in the spherical shell between distances r and $r + dr$ from the reference atom α is $4\pi r^2 \, dr n_0 x_\beta g_{\alpha\beta}(r)$, where x_β is the molar fraction of atom β. The partial distribution function of a binary system is defined by

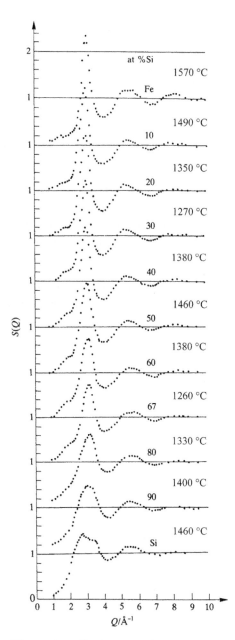

Figure 2.21 *Total structure factors S(Q) for liquid iron–silicon alloys. The experiments, using an X-ray diffraction technique, were carried out at temperatures about 30 to 50 °C above their liquidus temperatures.*

$$g_{\alpha\beta}(r) = 1 + \frac{1}{2\pi^2 n_0 r} \int_0^\infty Q[S_{\alpha\beta}(Q) - 1] \sin(Qr) dQ \qquad (2.22)$$

where $S_{\alpha\beta}(Q)$ represents the partial structure factor.

The structure factor of a binary alloy system obtained from scattering experiments, i.e. the total structure factor $S(Q)$, can be expressed in terms of its three partial structure factors:

$$S(Q) = W_{11}S_{11}(Q) + W_{22}S_{22}(Q) + 2W_{12}S_{12}(Q) \qquad (2.23)$$

where $W_{\alpha\beta} = x_\alpha x_\beta f_\alpha f_\beta / < x_\alpha f_\alpha + x_\beta f_\beta >^2$ in which x is the molar fraction and f is the atomic scattering factor.

We give an example of some structural information about binary liquid alloys. Figures 2.21 to 2.24 show the experimental data for liquid iron–silicon alloys. According to the experimental results, in the concentration range up to 40 at. % Si, the overall features of $S(Q)$ and $g(r)$ together with Q_1, $S(Q_1)$, W_f, r_1, and Z are practically equivalent to those for pure liquid iron. For the alloys in the concentration range beyond 50 at. % Si, the first peak of the $S(Q)$ curve becomes broader and more asymmetric, while the first

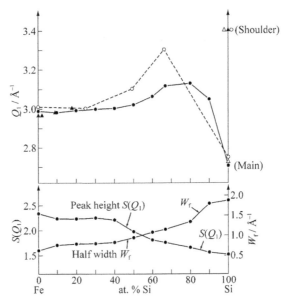

Figure 2.22 *Composition dependence of position Q_1, the height $S(Q_1)$, and the half width W_f of the first peak of $S(Q)$ for liquid iron–silicon alloys.* ▲△ *Waseda [1] (▲ 1600 °C, △ liquidus +30 °C);* ○ *Vatolin et al. (liquidus +30 °C; quoted in [2]);* ● *Kita et al. [2] (liquidus +30 to 50 °C).*

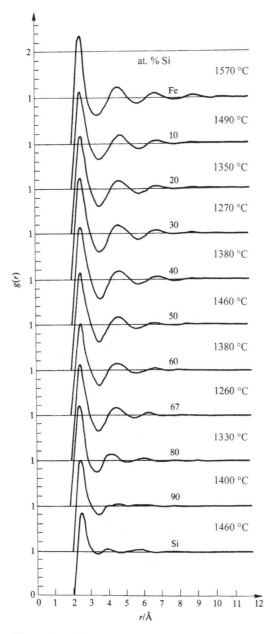

Figure 2.23 *Total pair distribution functions g(r) for liquid iron–silicon alloys calculated from the experimental data shown in Figure 2.21.*

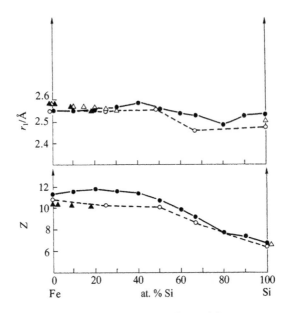

Figure 2.24 *Composition dependence of the nearest-neighbour distance r_1, i.e. the peak position in r.d.f. curves, and the first coordination number Z for liquid iron–silicon alloys. ▲△ Waseda [1] (▲ 1600 °C, △ liquidus +30 °C); ○ Vatolin et al. (liquidus +30 °C; quoted in [2]); ● Kita et al. [2] (liquidus +30 to 50 °C).*

peak of the $g(r)$ curve becomes narrower, and subsequent peaks more strongly damped. Further, the first coordination number Z decreases.

These experimental results suggest that the structure of liquid iron–silicon alloys remains nearly close-packed up to 40 at. % Si and that it then gradually changes to a more open, lower-coordinated structure at higher silicon levels.

In the last 20 years or so, the structural studies of liquid semiconductors, especially silicon, have been carried out intensively in the context of semiconductor materials processing operations. It is well known that commercial applications of semiconductors began with single-crystal growth of silicon from the melt via the Czochralski process, and that amorphous semiconductor silicon is made from slightly undercooled melts. Thus, a detailed knowledge of the structure and thermophysical properties of liquid semiconductors, including undercooled regions, has been required for the manufacture of high-quality semiconductor materials, such as silicon crystals without defects.

Figures 2.6 to 2.18 and Figure 2.20 are drawn using the experimental data of Waseda [1]. Figures 2.21 to 2.24 are taken from Kita et al. [2] (see also the special feature by Hibiya and Egry [3]).

REFERENCES

1. Y. Waseda, *The Structure of Non-Crystalline Materials, Liquids and Amorphous Solids*, McGraw-Hill, New York, 1980, Appendices 8 and 9.
2. Y. Kita, M. Zeze, and Z. Morita, *Trans. ISIJ*, **22** (1982), 571.
3. T. Hibiya and I. Egry (eds.), *Meas. Sci. Technol.*, **16** (2005), 317.

3

Density

3.1 Introduction

In discussing the nature and behaviour of metallic liquids, density is an indispensable fundamental quantity or property. From a practical point of view, density data for metallic liquids provide essential information on topics ranging from mass balance calculations in refining operations or the kinetics of slag/metal reactions to thermal natural convection phenomena in furnaces and ladles. Similarly, the physical separation of metallic liquids and overlaying slags, and the terminal velocity of non-metallic inclusions through metallic liquids, are essentially dominated by differences in densities between the two phases. As a final example of the importance of density, a detailed knowledge of volume changes in metallic materials at their melting point temperatures is of critical importance in the understanding of solidification processes or melt growth of crystals.

From a more fundamental standpoint, the density or number density of a liquid is needed in the description of the radial distribution function. Since determinations of this and almost all of the basic thermophysical properties of metallic liquids (e.g. viscosity, surface tension) require density data, the property of density is of primary importance.

It is not surprising, therefore, that the densities of metallic liquids have long been of interest from both the technological and scientific points of view. Density measurements have been carried out over the past 150 years on a number of metallic liquids, i.e. liquid metals, semimetals, semiconductors, and alloys (mainly on binary alloy systems).

However, at present, reliable density data exist only for the common low melting point metallic liquids. More accurate density data on various metallic liquids are still needed. Unfortunately, it is impossible to obtain accurate and reliable experimental data on the density of all metallic liquids, since some of them are too chemically reactive, too refractory, or too scarce. As a result, values of density and its temperature dependence, or volume expansivity, must be predicted for metallic liquids for which experimental data are lacking. To obtain reliable predictions of density, theoretical and phenomenological studies of density are required. Nevertheless, such approaches to density have been very limited to date, because it is extremely difficult to accurately predict density values. Indeed, densities are generally provided merely as experimental input data for the computation of other thermophysical properties.

The Thermophysical Properties of Metallic Liquids: Volume 1 – Fundamentals. First Edition.
Takamichi Iida and Roderick I. L. Guthrie. © Takamichi Iida and Roderick I. L. Guthrie 2015.
Published in 2015 by Oxford University Press.

3.2 Two Categories of Thermophysical State Properties

In general, we distinguish two categories of the properties, or the state properties,[1] of a system:[2] extensive and intensive properties.

3.2.1 Thermodynamic State Properties[3]

3.2.1.1 Extensive Properties

Properties proportional to the quantity of material in a system (or properties which are dependent upon the size of the system) are defined as the extensive properties; mass, volume, and energy are extensive thermophysical properties. If the quantity of material is multiplied, then all extensive quantities are multiplied.

In the case of heterogeneous total systems, the extensive properties of the total system are composed additively from the corresponding properties of the individual phases.

3.2.1.2 Intensive Properties

Properties independent of the quantity of material (or properties independent of the size of the system) and not additive for the various phases of the system are defined as the intensive properties; density, temperature, pressure, and molar quantities are intensive properties. Intensive properties may be defined locally, i.e. they may vary in space.

Quotients of two extensive quantities are intensive quantities. Specific quantities (e.g. specific heat, specific heat capacity, specific volume) are intensive properties. The product of an extensive property and an intensive property is an extensive property.

3.2.2 Density and Molar Volume

3.2.2.1 Density

Density (or mass density) can be defined as the mass per unit volume of a substance, or the ratio of its mass to its volume (i.e. the quotient of two extensive quantities: mass and volume). In SI units, it is measured in $kg\ m^{-3}$.

The relative density (formerly called specific gravity) is the density of a substance divided by the density of water.

Density, in general, expresses the closeness of any linear, superficial, or space distribution. For example, number density is the number of atoms (molecules) per unit volume; electron density is the number of electrons per unit volume.

---

[1] A state property is defined as a physical quantity that specifies a macroscopic property as uniquely as possible (e.g. temperature, pressure, volume, vapour pressure, surface tension, viscosity, etc.).
[2] A system is any portion of the universe selected for consideration, or an arbitrary assembly of matter with properties (volume, energy, particle number, etc.).
[3] Thermodynamic (state) properties may be defined and measured only in equilibrium.

3.2.2.2 *Molar Volume*

In the theoretical treatment of metallic liquids, molar volume as well as density is frequently used for describing their properties.

A molar quantity G_{mol} is defined by the quotient of an extensive quantity G and the number of moles n: $G_{mol} = G/n$.

Thus the molar volume V_{mol} is given by

$$V_{mol} = \frac{V}{n} \tag{3.1}$$

where V is the volume. The molar volume is the volume per mole, or the volume occupied by one mole (of a substance). Similarly, the molar heat capacity is the heat capacity per mole.

The molar volume is also given by

$$V_{mol} = \frac{M}{\rho} \tag{3.2}$$

namely the molar volume is the quotient of the molar mass M and the density ρ.

3.3 Volume Change on Melting

The volume displays a step-like behaviour on melting, i.e. $\Delta_s^l V_m \neq 0$ ($\Delta_s^l V_m > 0$ or $\Delta_s^l V_m < 0$), as illustrated in Figure 3.1. The entropy also displays a step-like behaviour on melting, i.e. $\Delta_s S_m > 0$ (see Richards' rule). The volume and entropy jumps are characteristic of a first-order phase transition.[4]

The majority of metals exhibit a volume increase of 3.8 per cent, on average, during the solid–liquid phase transition. Volume contraction, however, during melting occurs in gallium, silicon, germanium, bismuth, cerium, and plutonium. The volume increase is unusually small for lanthanum, praseodymium, neodymium, and neptunium. The data for volume and entropy changes during melting are shown in Table 3.1.

On the basis of the volume change on melting, metallic elements can strictly divided into two groups: Group 1, exhibiting volume increases during melting, and Group 2 characterized by volume decreases during melting. Most metals are in Group 1. According to Wittenberg and DeWitt [1] (see Table 3.1), close-packed metals, i.e. face-centred cubic (fcc) and hexagonal close-packed (hcp), exhibit an average volume increase of approximately 4.6 per cent and an average entropy change of approximately 9.6 J mol^{-1} K^{-1}. For the body-centred cubic (bcc) metals, these values are somewhat smaller, the average volume increase being 2.7 per cent, and $\Delta_s^l S_m$ being 7.1 J mol^{-1} K^{-1}.

---

[4] The heat capacity and the compressibility of a substance approach infinity at the first-order phase transition.

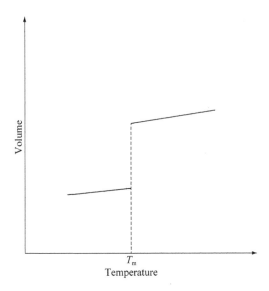

Figure 3.1 *A first-order phase transition characterized by a jump in volume, i.e.* $\Delta_s^l V \neq 0$.

Although all the lanthanoid metals, except erbium, thulium, and lutetium (which are hcp), and four actinoid metals have bcc allotropes in equilibrium with the liquid, these elements were not included in the above averages.

The Group 2 metals which contract on fusion appear to be split into subgroups. The semimetals (gallium, silicon, germanium, and bismuth)[5] form the first, in which the $\Delta_s^l S_m$ values are much larger than those of the Group 1 elements. The other subgroup is composed of the elements cerium and plutonium, for which the $\Delta_s^l S_m$ values are smaller than for the Group 1 elements. Although the early lanthanoid elements, i.e. lanthanum, praseodymium, and neodymium, do not contract during melting, their values for $\Delta_s^l V_m$ and $\Delta_s^l S_m$ are less than those in Group 1. The remaining elements in the lanthanoid series appear to have nearly normal values. Similarly, in the actinoid elements, the values for uranium are nearly normal, those for neptunium and plutonium are less than normal, and those for americium are, again, nearly normal.

The anomalous behaviour of the elements in Group 2 can be explained as follows (cf. Wittenberg and DeWitt [1]). For the semimetals, the rigid and directional bonds of the solid are apparently broken on melting, and the atoms become more spherical and pack closer together. On the other hand, in cerium, praseodymium, uranium, and plutonium, their vacancies increase slightly, with corresponding reductions in their metallic radii.

---

[5] Grouping due to Wittenberg and DeWitt [1]. The present authors' grouping is given in Table 1.6.

Table 3.1 *Volume and entropy changes associated with the melting of various metallic elements (metals, semimetals, semiconductors).*

Element	Temperature °C	Entropy $\Delta_s^l S_m$ $\text{J mol}^{-1}\,\text{K}^{-1}$	Rel. volume change on melting $\Delta_s^l V_m / V_m$ %			
			W	B[a]	M[b]	U[c]
Body-centred cubic structures (*A* 2)[d]						
Li	180.5	6.61	2.74	1.5	–	1.65
Na	96.5	7.03	2.6	2.5	–	2.5
K	63.5	6.95	2.54	2.4	–	2.55
Rb	38.9	7.24	2.3	–	–	2.5
Cs	28.6	6.95	2.6	–	–	2.6
Tl	302	7.7	2.2	–	–	–
Fe	1536	7.61	3.6	–	4.55	–
Face-centred cubic structures (*A* 1)[d]						
Cu	1083	9.71	3.96	–	4.21	4.51
Ag	960.7	9.16	3.51	5.0	–	3.30
Au	1063	9.25	5.5	5.19	–	5.1
Al	658	11.6	6.9	6.6	7.48	6.0
Pb	327	7.99	3.81	3.6	–	–
Ni	1454	10.1	6.3	–	4.94	–
Pd	1552	(9.6)	5.91	–	–	–
Pt	1769	(9.6)	6.63	–	–	–
In[†] (*A* 6)[d]	156.6	7.61	2.6	2.5	–	2.7
Hexagonal close-packed structures (*A* 3)[d]			–			
Mg	651	9.71	2.95	4.2	3.27	3.05
Zn	419	10.7	4.08	6.9	1.92	4.2
Cd	321	10.4	3.4	4.7	–	4.7
Complex structures						
Se	217	16.2	16.8	–	–	–
Te	451	24.2	4.9	–	–	–
Sn	232	13.8	2.4	2.6	–	–
Ga[†]	29.8	18.5	–2.9	–3.0	–	–3.2
Si	1410	29.8	–9.5	–	–10.4	–

continued

Table 3.1 *(continued)*

Element	Temperature °C	Entropy $\Delta_s^l S_m$ J mol^{-1} K^{-1}	Rel. volume change on melting $\Delta_s^l V_m/V_m$ %			
			W	B[a]	M[b]	U[c]
Ge	934	30.5	−5.1	−	−	−
Bi	271	20.8	−3.87	−	−	−
Sb	630.5	22.0	(−0.995, +1.1)[‡]	−0.94	−	−
Hg	−38.9	9.79	3.64	3.6	−	3.7
Lanthanoid elements						
La	920	5.61	0.6	−	−	−
Ce	800	4.85	−1.0	−	−	−
Pr	935	5.73	0.02	−	−	−
Nd	1024	5.48	0.9	−	−	−
Sm	1072	6.40	3.6	−	−	−
Eu	826	9.33	4.8	−	−	−
Gd	1312	6.40	2.1	−	−	−
Tb	1356	6.53	3.1	−	−	−
Dy	1407	8.54	4.9	−	−	−
Ho	1461	9.41	7.5	−	−	−
Er	1497	11.2	9.0	−	−	−
Tm	1545	9.67	6.9	−	−	−
Yb	824	6.86	4.8	−	−	−
Lu	1652	7.15	3.6	−	−	−
Actinoid elements				−	−	−
U	1133	8.62	2.2	−	−	−
Np	640	5.69	1.5	−	−	−
Pu	640	3.18	−2.4	−	−	−
Am	1176	9.92	2.3	−	−	−

Data, except for those bearing the superscripts a, b, or c, are taken from Wittenberg and DeWitt [1].
[a] Benenson et al. [2].
[b] Mills [3].
[c] Ubbelode [4].
[d] See, for example, Barrett [5].
[†] According to Iida and co-workers' experimental data [6] In 2.3%, Ga −3.1%.
[‡] Sb expands slightly on melting and has a small positive values for $\Delta_s^l V_m$ [7].

3.4   Theoretical, Semi-Empirical, and Empirical Analyses of Liquid Density

3.4.1   Theoretical Analyses of Density—Computer Experiments

As mentioned earlier, there has been little theoretical interest in the property of 'density' itself. Density has only been discussed as a part of studies of the structure and thermophysical properties of liquids. However, density or packing density,[6] or packing fraction, is a parameter of fundamental importance in explaining the behaviour and properties of liquids. The behaviour (or motion) of the particles depends upon their initial velocities and upon their tightness of packing. Computer experiments enable the dynamic trajectory, or history, of each particle to be recorded; calculations for small assemblies of hard spheres[7] and of particles interacting with model pair potentials (e.g. square-well potentials, Lennard-Jones potentials) have been made (e.g. Figure 1.8). The behaviour and properties of small assemblies in the order of 100 particles prove not very different from those of much larger ones.

Wood and Jacobson [8] and Alder and Wainwright [9] were the first to begin computer experiments using modern computers on packing density.[8] Alder and Wainwright went on to examine the behaviour of hard-sphere systems as a function of the packing fraction by means of a mathematical model of atomic collisions. They studied a two-dimensional system containing 870 hard-disc particles. The simultaneous motions of the particles were obtained through mathematical solutions with the aid of a high-speed digital computer. In their work, the equation of state for the hard-disc particles was set up as a function of the reduced area A^*. This reduced area was defined as being the ratio of the total area of the particle system A to its area at close packing A_0, i.e. $A^* \equiv A/A_0$. The authors concluded that a fluid and a solid phase (in a van der Waals-like loop) coexisted in the region of A^* from 1.26 to 1.33. Furthermore, they were able to accurately predict the transport coefficients of the hard-sphere fluid as a function of the atom's packing fractions up to 0.49. According to their computations, transport coefficients changed very rapidly at the highest (packing) densities.[9] With increasing η , the fluid viscosity increased rapidly, while self-diffusivity decreased rapidly. These results suggest that the hard-sphere model exhibits the phenomenon of a liquid (fluid)–solid transition, which is equivalent to that of a real substance. This characteristic of hard-sphere systems (the so-called Alder transition) implies that the repulsive force between atoms represents the principal factor governing liquid–solid transformations and atomic arrangements. Note that one-component hard-sphere systems have two different equilibrium phases: a fluid phase and a solid phase (a crystalline phase). Thus there is no distinction between a liquid and a

---

[6] The packing fraction is sometimes called the packing density.

[7] If the particles are hard spheres, their energies are entirely kinetic energies. Thus the total internal energy U of the assembly is simply the sum of the kinetic energies of the individual particles: $U = (3/2)NkT$, where N is the number of hard-sphere particles.

[8] Some reliable results were obtained.

[9] In discussing the behaviour and properties of hard-sphere particles, the packing density, or the packing fraction η, is usually used instead of mass density.

gas in the hard-sphere systems. To separate the liquid state from the gas, attractive forces would be needed in the model (cf. the results of Barker and Henderson in Figure 1.8).

The occurrence of a phase transition in the hard-sphere system depends only on packing fraction. However, this phase transition in real liquids depends not only on the packing fraction but also on absolute temperature, because the repulsive interactions between atoms are not perfectly elastic.

Subsequent theoretical considerations by Hoover and Ree [10] have indicated that while the value of packing fraction for hard spheres is 0.49 at the point of freezing, this value is increased to 0.54 at the point of melting (a hysteresis phenomenon). The fluid was found to be stable for all packing fractions equal to, or less than, 0.49.

Over the past half century, considerable interest in the structures and properties of amorphous solids, or metallic glasses, has developed. Woodcock [11] and Hudson and Andersen [12] have demonstrated that the hard-sphere fluid can undergo a glass-like transition when it is compressed to a high density. The estimated value of the packing fraction at the glass transition is 0.53. This value is substantially lower than 0.64, the packing fraction of a dense, randomly packed array of hard spheres.

As mentioned, repulsive forces between atoms (ions) in liquids are, in fact, softer than those in the hard-sphere model. Notwithstanding this, several properties of liquid metals can be explained successfully by treating ions (of metallic liquids) as hard spheres. Examples include excess entropies, and self-diffusivities, together with the structure factors and pair distribution functions.

During the 1960s, progress was made in a pseudopotential theory (i.e. an effective potential introduced so as to treat the behaviour of conduction electrons, or valence electrons) of liquid metals. The pseudopotential methods are supported by a large amount of empirical experience as well as by theoretical arguments (e.g. the electrical resistivity of liquid metals, particularly simple liquid metals).

At the present time, however, it is utterly impossible to predict accurate density values for metallic liquids from the standpoint of materials process science and engineering.

3.4.2 Empirical or Semi-Empirical Models for the Density of Liquid Metallic Elements

Of the several empirical methods for predicting the critical densities of metals, the most familiar is the law of Cailletet and Mathias, or the law of rectilinear diameters. This law states that the average density of a liquid and its vapour, or half the sum of the densities of a liquid and its saturated vapour, decrease linearly as the temperature rises up to the critical point (i.e. a material constant). This is equivalent to saying that

$$\frac{\rho_l + \rho_g}{2} = a - bT \tag{3.3}$$

holds over the entire temperature range of a liquid, where ρ_l is the liquid's density, ρ_g is the vapour's density, T is the absolute temperature, and a and b are constants. At the critical point, the densities of liquid and saturated vapour are equal, i.e. $\rho_l = \rho_g = \rho_c$, where ρ_c represents the fluid's density at its critical temperature T_c. Equation (3.3) can be rewritten in dimensionless form using critical density ρ_c:

$$D^* = \frac{\rho_l + \rho_s}{2\rho_c} = 1 + \mu(1 - T^*) \tag{3.4}$$

where D^* is the reduced rectilinear diameter, T^* $(= T/T_c)$ is the reduced temperature, and μ $(\equiv bT_c/\rho_c)$ is a constant whose value is considered smaller than unity.

The validity of Eqs. (3.3) and (3.4) has been confirmed experimentally for a large variety of thermally stable organic and inorganic liquids such as benzene, water, and argon. In the case of liquid metals, few data on critical temperatures and critical densities have been obtained because property measurements at high temperatures and high pressures are extremely difficult, requiring special materials and techniques.

Grosse [13] proposed a method for estimating critical temperatures of liquid metals, based on the assumption that the molar entropies of vaporization of all liquid metals, i.e. $\Delta_l^g S = \Delta_l^g H/T$, can be expressed as a universal function of the reduced temperature T^*, namely that the curve of $\Delta_l^g S$ *vs.* T^* is the same for all metals as that for mercury.[10] Later, Dillon et al. [14] obtained experimental data for the critical constants of the liquid alkali metals. Their results demonstrate that a universal relationship can only be regarded as a rough approximation, as seen from Figure 3.2. However, using experimentally

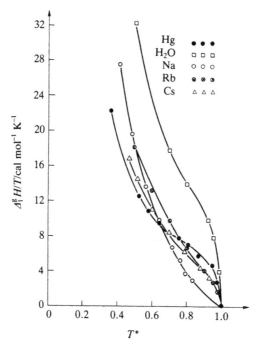

Figure 3.2 *Variation of entropy of evaporation, or vaporization, with reduced temperature for alkali metals, mercury, and water (after Dillon et al. [14]).*

[10] At that time, only the critical constants for mercury had been determined experimentally.

determined entropies of vaporization for other metals, we may read off rough values for their critical temperatures from the curves given in Figure 3.2.

Once the critical temperature of a metal is known, its corresponding critical density can be estimated using the law of rectilinear diameters, i.e. by extrapolating the experimental data on densities up to the critical temperature. An example of the density and rectilinear diameter plot is shown in Figure 3.3. A generalized correlation for plotting reduced density $\rho^*(\equiv \rho/\rho_c)$ vs. reduced temperature T^*, is given in Figure 3.4.

Experimental observations for critical data on caesium, mercury, and rubidium are as follows [15]: caesium, $T_c/°C = 1651$, $P_c/10^6 \text{Pa} = 9.25$, $\rho_c/10^3 \text{ kg m}^{-3} = 0.38$; mercury, $T_c = 1478$, $P_c = 167.3$, $\rho_c = 5.8$; rubidium, $T_c = 1744$, $P_c = 12.45$, $\rho_c = 0.29$.

Since the density of a liquid ρ_l is much greater than the density of a vapour ρ_g, i.e. $\rho_l \gg \rho_g$, over a wide temperature range, we can, using Eq. (3.4), write the following expression:

$$\rho^* \approx 2 + \{(\rho_m^* - 2)/(1 - T_m^*)\}(1 - T^*) \tag{3.5}$$

where $\rho_m^*(\equiv \rho_m/\rho_c)$ represents the reduced density at the metal's melting point T_m. McGonigal [16] has presented an alternative method for estimating critical densities using the dimensionless variable Δ. On the basis of Eq. (3.3) or (3.4), this dimensionless quantity was defined as

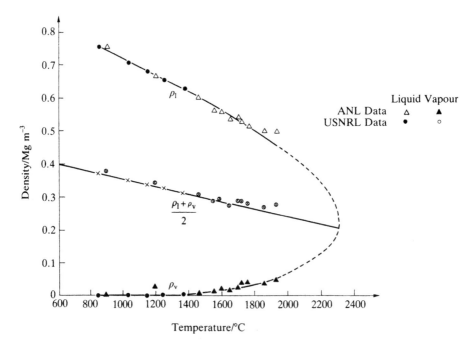

Figure 3.3 *Density and rectilinear diameter plot for sodium. Rectilinear diameter line is given by $(\rho_l + \rho_v)/2$, where the experimental points are given as follows: $\otimes \otimes \otimes$, Argonne National Laboratory data; × × ×, US Naval Research Laboratory data. Estimated critical constants: $\rho_c = 0.21 \times 10^3 \text{kg m}^{-3}$, $T_c = 2300°C$, $P_c = 3.55 \times 10^7$ Pa (after Dillon et al. [14]).*

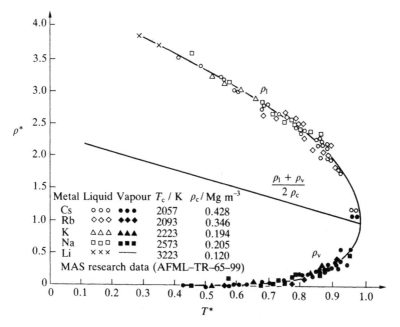

Figure 3.4 *Generalized correlation of reduced density with reduced temperature (after Dillon et al. [14]).*

$$\Delta \equiv \left(\frac{dD^*}{dT^*}\right)\Big/\left(\frac{D^*}{T^*}\right) \tag{3.6}$$

Correlations between $\Delta^* (\equiv \Delta / \Delta_c$, where Δ_c is the value of Δ at the critical point) and T^* are shown in Figure 3.5. McGonigal reported that the agreement among the liquid metals shown in Figure 3.5 can be considered to be good, excepting the Δ^* values of the periodic Group IIB (or the periodic Group 12) metals. Δ^* can be expressed mathematically as

$$\Delta^* = \frac{T^*}{1 + \mu(1 - T^*)} \tag{3.7}$$

However, since μ is not a universal constant for all liquid metals, Δ^* cannot be regarded as a universal quantity.

Incidentally, another relationship, known as Thiesen's equation, has been suggested:

$$\rho_g - \rho_l \propto (T_c - T)^\beta, \quad (\beta \approx 1/3) \tag{3.8}$$

Although critical data (critical temperature, critical pressure, and critical density) are very important thermophysical properties (or quantities) of metallic liquids, experimentally derived critical data are extremely scarce. Hess and Schneidenbach [17] have proposed a method for the estimation of critical data for various fluid metals.

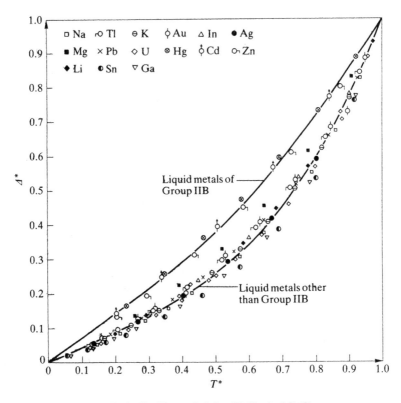

Figure 3.5 Δ^* *vs. T^* for liquid metals (after McGonigal [16]).*

3.4.3   Relationship between Molar Volume of Liquid Metallic Elements and Other Physical Quantities

3.4.3.1   Relationship with Atomic Number

The relationship of periodic variation in the molar volumes of the element with its atomic number, i.e. the number of electrons, or protons, in the neutral atom, has been known for a long time. The trend in the molar volumes of liquid metallic elements at their melting point temperatures with their atomic number is plotted in Figure 3.6. It is clear from this figure that the liquid metallic elements' molar volumes obey a periodic law: the Group 1 metals (i.e. alkali metals) occupy the peaks, whereas the Group VIIIB transition metals occupy the valleys of the curve.

3.4.3.2   Relationship between Molar Volumes and Evaporation Enthalpies

The enthalpy of evaporation for liquid metallic elements is a very useful thermophysical quantity. There is a positive correlation between the reciprocal of molar volumes and the enthalpies of evaporation of liquid metallic elements; the correlation can be divided into several groups (see Figure 4.7).

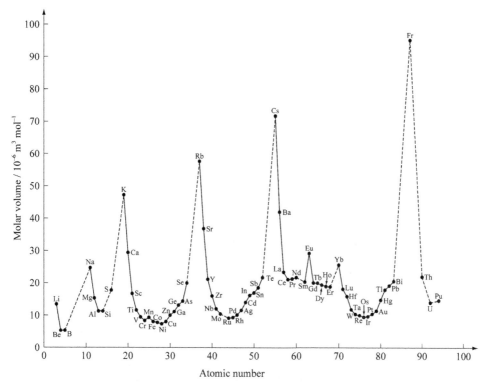

Figure 3.6 *The periodicity of molar volumes of liquid metallic elements (liquid metals, semimetals, and semiconductors) at their melting point temperatures. Data are given in Table 17.4.*

3.5 Models for the Temperature Dependence of the Density of Liquid Metallic Elements, and their Assessment

3.5.1 Models for the Temperature Dependence of Density

3.5.1.1 *Relationship between Volume Expansivity (or the Temperature Dependence of Density) and Melting Point Temperature*

The volume expansivity α is defined as: $\alpha = V^{-1}(\partial V / \partial T)_P = -\rho^{-1}(\partial \rho / \partial T)_P$, where V is the volume and ρ is the density.

It has long been known that both solid and liquid metallic elements tend to follow a simple relationship between their volume expansivities, or their temperature dependence of densities, and their melting temperatures. For the volume expansivities of liquid metallic elements at their melting point temperatures, α_m, this relationship is expressed as

$$\alpha_m = \frac{0.09}{T_m} \tag{3.9}$$

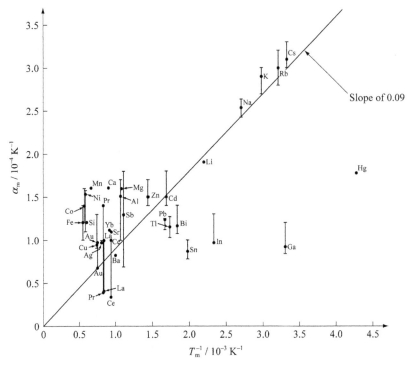

Figure 3.7 *The isobaric volume expansivities of liquid metallic elements vs. the reciprocal of their absolute melting temperatures. In the figure, the bars represent a general feeling for the reliability of the data, i.e. the range bars indicate the extremes of experimental values for Λ or $\alpha_m (\equiv -\Lambda/\rho_m)$ which Steinberg [18] considered reasonable. Incidentally, the range bars in Figures 3.8, 3.10, and 3.11 also represent a general feeling for the reliability of the data.*

Figure 3.7 shows this relationship; as seen, roughly speaking, very large discrepancies exist between calculated and measured densities for liquid metallic elements. About 20 metallic elements lie within the ± 30 per cent error band; this relationship is not particularly satisfactory as it has too many exceptions.

3.5.1.2   Steinberg Model

According to Steinberg [18], a simple relationship between the volume expansivity at 293 K, α_0, and the melting point temperature:

$$\alpha_0 = \frac{0.06}{T_m} \tag{3.10}$$

is valid for solid metals and semimetals to ± 43 per cent. If Eq. (3.10) is restricted to fcc, bcc, and hcp structures only, then the total error is reduced to ± 26 per cent. (The volume expansivity of an element in the liquid state is greater than that of the element in the solid state.) On the basis of this equation, Steinberg proposed an empirical expression for the temperature dependence $\Lambda [\equiv (\partial P / \partial T)_P]$ of liquid density for metallic elements, showing that

$$\frac{\Lambda T_b}{D_{00}} = -0.23, \quad \left(or - \Lambda = 0.23 \frac{D_{00}}{T_b} \right) \tag{3.11}$$

where D_{00} is a scale factor defined by $D_{00} = \rho_m - \Lambda T_m$, namely D_{00} is the density determined by extrapolating from ρ_m (melting point density) to absolute zero. Figure 3.8 shows a plot of Λ against D_{00} / T_b for a large number of liquid metallic elements, and shows a linear correlation between the two variables. Incidentally, the slope of 0.23, indicated in Figure 3.8, i.e. the numerical factor of 0.23 appearing in Eq. (3.11), was determined so as to give the best fit to data (excluding semimetals, Group IIB, and Group VIII metals). The dashed lines represent the ± 20 per cent error band. The range bars

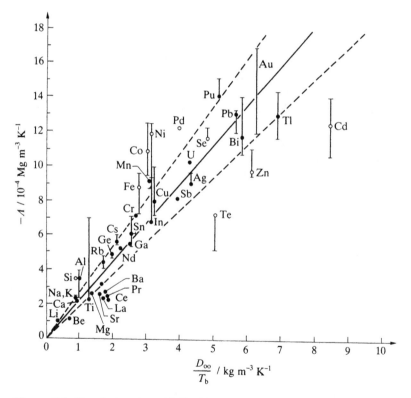

Figure 3.8 *Correlation of Λ with D_{00} / T_b for 41 elements. Open circles represent semimetals and Groups VIII and IIB metals. Mercury and platinum points fall outside the range of the figure. Their ordinates are 24.2 and 28.8 and their abscissas 22.6 and 6.05, respectively. The broken lines represent ± 20 per cent error band (after Steinberg [18]).*

give a general feeling for the reliability of the data. In addition, a horizontal bar on the tin data point gives the range in the abscissa if T_b is varied.

Equation (3.11) has been used to predict Λ for high melting point metals for which no experimental data are available.

By rewriting Eq. (3.11),[11] we have an equation for the volume expansivity of liquid metallic elements, as follows:

$$\alpha_m = \frac{0.23}{T_b - 0.23\,T_m} \tag{3.12}$$

3.5.1.3 *Models in Terms of Common Parameters* $\xi_T^{1/2}$ *and* $\xi_E^{1/2}$

The volume expansivity of crystalline solids can be calculated using the Grüneisen relation [19–21], which is expressed in the form

$$\alpha = \frac{\gamma_G \kappa_T C_V}{V} \tag{3.13}$$

giving

$$\gamma_G = -\frac{d(\log \nu)}{d(\log V)} = -\frac{d(\log \Theta)}{d(\log V)}$$

where γ_G is the Grüneisen constant, κ_T is the isothermal compressibility $[= -V^{-1}(\partial V/\partial p)_T]$, C_V is the constant-volume heat capacity, V is the volume, ν is the frequency of atoms (lattice vibration), and Θ is the Debye characteristic temperature.

As already mentioned, a solid and a liquid are similar in the manner in which the atoms in both states crowd together. This means that the atomic arrangement in the liquid is a fairly regular one, very like that in the solid. The solid melts into a liquid with the result that there is local, or short-range, order in the liquid (but no long-range order throughout the liquid); hence each atom in the liquid, which is surrounded by approximately ten nearest neighbours, vibrates as that in the solid. Thus the Grüneisen relation is also considered to be applicable to liquid metallic elements.

On the basis of this idea, the present authors have proposed two models for the volume expansivity of liquid metallic elements by combining the Grüneisen relation with two common parameters, denoted by $\xi_T^{1/2}$ and $\xi_E^{1/2}$. These are useful in discussing anharmonic motions of atoms (see Chapter 5): substituting Eq. (5.35) into Eq. (3.13), we have[12]

$$\alpha_m = \left.\frac{A_T}{\xi_T^{1/2} T_m}\right|_{T=T_m} \tag{3.14}$$

$$A_T \equiv \left(3.69 - \frac{1.25}{\xi_T^{1/2}}\right) \times 10^{-2}\frac{C_P}{R} = \left(4.44 - \frac{1.51}{\xi_T^{1/2}}\right) \times 10^{-3} C_P$$

[11] $\alpha_m \equiv -\Lambda/\rho_m$
[12] Combining Eqs. (3.13), (5.32), and (5.35), and well-known thermodynamic relations, i.e. Eqs. (3.2), (5.1), and (5.2), we have Eq. (3.14). Similarly, combining Eqs. (3.13), (5.26), and (5.31), and well-known thermodynamic relations, we obtain Eq. (3.15).

where C_P is the constant-pressure heat capacity and R is the molar gas constant.
Similarly, substitution of Eq. (5.31) in Eq. (3.13) gives

$$\alpha_m = \left.\frac{A_E}{\xi_E^{1/2}\,\Delta_l^g H_b}\right|_{T=T_m} \tag{3.15}$$

$$A_E \equiv \left(1.18 - \frac{0.474}{\xi_E^{1/2}}\right)C_P$$

3.5.1.4 Relationship between Temperature Dependence and Atomic Number

In Figure 3.9, the temperature dependence of the density of liquid metallic elements
is shown as a function of their atomic number. It can be seen that there is a periodic
variation in Λ values. Broadly speaking, close-packed (i.e. fcc and hcp) metals in the
solid state at their melting point temperatures lie on the peaks, while bcc metals lie in

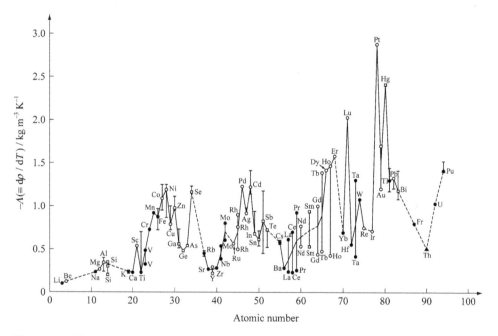

Figure 3.9 *Temperature dependence of the liquid metallic density, $\Lambda[\equiv (\partial\rho/\partial T)_P]$, plotted against
atomic number Z. The symbols refer to the crystal structures in the solid state at the melting
temperature.* ○, *close-packed structure;* ●, ▲, *body-centred cubic structure* (▲, *estimated by Steinberg
[18]);* □, *other complex structures. The bars indicate the extremes of experimental values which are
estimated to be reasonable. Points linked by a vertical line represent two or three different
experimental values for a single metallic element.*

the valleys, of the curve. In more detail, four peaks,[13] or main local maxima, can be seen; reported data for (the third corresponding to) the lanthanoid elements remain erratic. Of the other three peaks, it can be seen that the peaks split into two, namely the Group 11 metals, or copper group metals, sink and consequently the Groups 10 and 12 metals lie on the peaks of the curve; the Group 14 elements (except for lead having a close-packed structure) seem to lie in the valleys of the curve. Evidently, a correlation exists between the values of Λ in the liquid state and the crystal structures (of metallic elements) in the solid state at their melting point temperatures. If we can provide a theoretical explanation of this correlation, it could make for great progress in the field of materials process science. Table 3.2 lists the crystal structures of solids from the standard temperature (273.15 K = 0 °C) to their melting point temperatures at one standard atmosphere.

Table 3.2 *Crystal Structures of solids.*[†]

Element		Cryst. struc.
Actinium	Ac	fcc
Aluminium	Al	fcc
Americium	Am	h
Antimony	Sb	r
Arsenic	As	r
Barium	Ba	bcc
Beryllium	Be	h
Bismuth	Bi	r
Boron	B	r
Cadmium	Cd	h
Caesium	Cs	bcc
Calcium	Ca	fcc[1]
Carbon	C	h,d
Cerium	Ce	fcc[2]
Chromium	Cr	bcc
Cobalt	Co	h
Copper	Cu	fcc

---

[13] Conceivably, five peaks: aluminium could occupy the peak.

Table 3.2 *(continued)*

Element		Cryst. struc.
Dysprosium	Dy	h
Erbium	Er	h
Europium	Eu	bcc
Francium	Fr	bcc
Gadolinium	Gd	h
Gallium	Ga	o
Germanium	Ge	d
Gold	Au	fcc
Hafnium	Hf	h^3
Holmium	Ho	h
Indium	In	t
Iridium	Ir	fcc
Iron	Fe	bcc
Lanthanum	La	h^4
Lead	Pb	fcc
Lithium	Li	bcc
Lutetium	Lu	h
Magnesium	Mg	h
Manganese	Mn	bcc
Mercury	Hg	r
Molybdenum	Mo	bcc
Neodymium	Nd	h
Neptunium	Np	o^5
Nickel	Ni	fcc
Niobium	Nb	bcc
Osmium	Os	h
Palladium	Pd	fcc

continued

Table 3.2 *(continued)*

Element		Cryst. struc.
Phosphorus (yellow)	P	c
Platinum	Pt	fcc
Plutonium	Pu	m[6]
Polonium	Po	m
Potassium	K	bcc
Praseodymium	Pr	h[7]
Promethium	Pm	h
Protactinium	Pa	t
Radium	Ra	bcc
Rhenium	Re	h
Rhodium	Rh	fcc
Rubidium	Rb	bcc
Ruthenium	Ru	h
Samarium	Sm	r
Scandium	Sc	h
Selenium	Se	h
Silicon	Si	d
Silver	Ag	fcc
Sodium	Na	bcc
Strontium	Sr	fcc[8]
Sulphur	S	o,m,r
Tantalum	Ta	bcc
Technetium	Tc	h
Tellurium	Te	h
Terbium	Tb	h
Thallium	Tl	h[9]

Table 3.2 *(continued)*

Element		Cryst. struc.
Thorium	Th	fcc[10]
Thulium	Tm	h
Tin	Sn	t,d
Titanium	Ti	h[11]
Tungsten	W	bcc
Uranium	U	o[12]
Vanadium	V	bcc
Ytterbium	Yb	fcc[13]
Yttrium	Y	H
Zinc	Zn	H
Zirconium	Zr	h[14]

[†] Data are taken from Benenson et al. [2]. Crystal structure: c cubic, bcc body-centred cubic, fcc face-centred cubic, h hexagonal, t tetragonal, o orthorhombic, r rhombohedral, m monoclinic, d diamond.

The metals, bearing the superscripts 1 to 14, have body-centred cubic structures above the following temperatures, which are stable up to their melting point temperatures (Nagakura et al. [22]: 1) 450 °C, 2) 726 °C, 3) 1760 °C, 4) 865 °C, 5) 500 °C, 6) 476±5 °C, 7) 798 °C, 8) 621 °C, 9) 230 °C, 10) 1380 °C, 11) 882 °C, 12) 778 °C, 13) 732 °C, 14) 862 °C.

3.5.2 Assessment of Models for Volume Expansivity

Table 3.3 gives a comparison of comparatively accurate experimental values for the volume expansivity of 16 liquid metallic elements selected, with those calculated from Eqs. (3.9), (3.12), (3.14), and (3.15); the corresponding δ_i, Δ and S values needed for statistical assessment of the models (see Section 1.5) are listed in Table 3.4. Incidentally, it is estimated that their experimental uncertainties lie in the range of ± 3 to ± 18 per cent around the mean. The data, except for $(C_P)_m$, used for calculating volume expansivities using these four equations are given in Tables 1.1, 5.5, 17.1, and 17.3. With the exception of the periodic Group 12, or the Group IIB (zinc group), metals, the Steinberg model represented by Eq. (3.12) performs extremely well for the liquid metallic elements listed in Table 3.4, in spite of its simplicity in terms of using well-known physical quantities of only their melting and boiling point temperatures.

Table 3.3 *Comparison of comparatively accurate experimental values for the volume expansivity of liquid metallic elements selected, with those calculated from four equations.*

Element		$(\alpha_m)_{exp}$	Range	$(\alpha_m)_{cal}$, 10^{-4} K^{-1}				$(C_P)_m^{\dagger}$
		10^{-4} K^{-1}	10^{-4} K^{-1}	Eq. (3.9)	Eq. (3.12)	Eq. (3.14)	Eq. (3.15)	Jmol^{-1} K^{-1}
Bismuth	Bi	1.17	1.07–1.40	1.65	1.34	1.48	1.22	30.5
Cadmium	Cd	1.52	1.42–1.76	1.51	2.55	1.35	1.59	29.7
Caesium	Cs	3.1	3.0–3.3	2.98	2.63	3.94	3.98	37
Copper	Cu	0.95	0.90–1.26	0.66	0.91	0.75	0.74	33.0[a]
Gallium	Ga	0.92	0.84–1.2	2.97	0.96	1.96	0.75	27.7
Indium	In	0.97	0.97–1.3	2.09	1.02	1.63	0.87	29.7
Lead	Pb	1.24	1.12–1.25	1.50	1.22	1.34	1.06	31.5
Lithium	Li	1.9	–	1.98	1.52	2.18	1.54	30.3
Mercury	Hg	1.78	–	3.84	3.99	2.58	2.35	28.5
Potassium	K	2.9	2.7–3.0	2.67	2.41	3.08	2.97	32.1
Rubidium	Rb	3.0	2.8–3.2	2.88	2.59	3.49	3.24	34.0
Silver	Ag	0.98	0.98–1.0	0.73	1.07	0.80	0.83	33.5[b]
Sodium	Na	2.54	2.43–2.64	2.43	2.15	2.79	2.60	31.9
Thallium	Tl	1.15	1.03–1.27	1.56	1.43	1.42	1.17	30.5
Tin	Sn	0.87	0.77–1.0	1.78	0.83	1.40	0.69	29.7
Zinc	Zn	1.5	1.4–1.64	1.30	2.25	1.29	1.56	31.3[c]

[†] Data, except for those bearing the superscripts a, b, or c, are taken from Gale and Tolemeier [23].
[a] Data form Baykara et al. [24].
[b] Data from Kubaschewski et al. [25].
[c] Data from Mills [3].

Figures 3.10 and 3.11 provide comparisons of experimental and calculated values (from Eqs. (3.14) and (3.15)) for the volume expansivity of the 16 liquid metallic elements selected. Although many approximations are involved in the derivations of Eqs. (3.13) to (3.15), as seen Figure 3.10 and Table 3.4, with the exception of gallium, indium, mercury, and tin, the model represented by Eq. (3.14) gives good results with Δ (12) and S(12) values of 16.0 per cent and 0.170, respectively. As is clear from Figure 3.11 and Table 3.4, the model described in terms of the common

Table 3.4 *Values of parameters δ_i, Δ, and S obtained from four equations.*

Element		δ_i, %			
		Eq. (3.9)	Eq. (3.12)	Eq. (3.14)	Eq. (3.15)
Bismuth	Bi	−29.1	−12.7	−20.9	−4.1
Cadmium	Cd	0.7	−40.4	12.6	−4.4
Caesium	Cs	4.0	18	−21	−22
Copper	Cu	44	4.4	27	28
Gallium	Ga	−69	−4.2	−53	23
Indium	In	−54	−4.9	−40	11
Lead	Pb	−17.3	1.6	−7.5	17.0
Lithium	Li	−4.0	25	−13	23
Mercury	Hg	−53.9	−55.4	−31.0	−24.3
Potassium	K	8.6	20	−5.8	−2.4
Rubidium	Rb	4.2	16	−14	−7.4
Silver	Ag	34	−8.4	23	18
Sodium	Na	4.5	18.1	−9.0	−2.3
Thallium	T1	−26.3	−19.6	−19.0	−1.7
Tin	Sn	−51	4.8	−38	26
Zinc	Zn	15	−33	16	−3.8
	$\Delta(16)$ %	26.2	17.9	21.9	13.7
	$S(16)$	0.340	0.229	0.253	0.167

parameter $\xi_E^{1/2}$ performs well with the liquid metallic elements selected, giving Δ (16) and $S(16)$ values of 13.7 per cent and 0.167, respectively; for most of these liquid metallic elements, calculated values fall, or almost fall, within the range of uncertainties in the experimental measurements. Nevertheless, the agreement between calculated and experimental values for several metals (i.e. the value of $|\delta_i|$ is approximately > 20 per cent) is unsatisfactory. Some corrections are needed for these metallic elements, and the correlation shown in Figure 3.9 may give some hints regarding this issue.

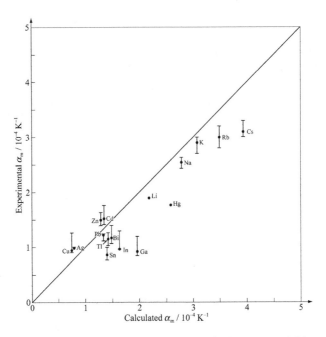

Figure 3.10 *Comparison of experimental volume expansivities of 16 liquid metallic elements selected with those calculated from Eq. (3.14).*

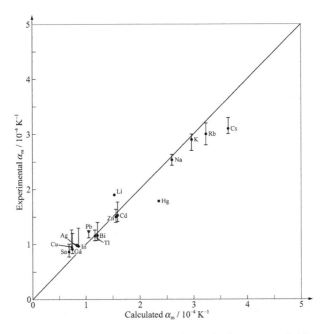

Figure 3.11 *Comparison of experimental volume expansivities of 16 liquid metallic elements selected with those calculated from Eq. (3.15).*

3.6 Methods of Density Measurement

Various methods exist for measuring the density of metallic liquids. The essential points of these techniques are described below under the following subheadings:

(a) Archimedean method (direct, indirect)
(b) Pycnometric method
(c) Dilatometric method
(d) Maximum bubble pressure method
(e) Manometric method
(f) Sessile drop method
(g) Oscillating (or levitated) drop method
(h) Gamma radiation attenuation method

To obtain accurate and reliable density data (in general, thermophysical property data) for metallic liquids, it is necessary to understand the special features of each method of density measurement (thermophysical property measurement) and to employ the most appropriate one for the system of interest.

3.6.1 Archimedean Method

This technique is based on the well-known principle of Archimedes, and involves its direct, and indirect, application.

3.6.1.1 *Direct Archimedean Method*

A solid sinker or bob of known weight (in vacuum or, more generally, in air) w_1, is suspended by a wire attached to the arm of a balance. When the sinker is immersed in the liquid metallic specimen, as shown in Figure 3.12(a), a new weight w_2, or an apparent loss of weight $\Delta w (= w_1 - w_2)$, is observed. The difference in the two weights, i.e. the apparent loss of the weight of the immersed sinker, originates mainly in the buoyant force exerted by the liquid metallic sample. The density of the liquid metallic specimen ρ is given by

$$\rho = \frac{\Delta w + s}{g(V + v)} \tag{3.16}$$

$$s = 2\pi r \gamma \cos \theta$$

where g is the acceleration of gravity, V is the volume of the sinker, v is the volume of the immersed suspension wire, s is the surface tension correction, i.e. the force acting against the suspension wire of radius r in a liquid for which the surface tension is γ and the contact angle between the wire and the liquid is θ.

From the standpoint of metrology, and in order to obtain accurate data, the volumes of the sinker and the immersed suspension wire or rod must always be corrected for thermal expansion to operating temperature, since their volumes are, in general, determined

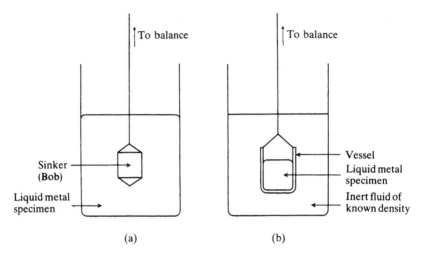

Figure 3.12 *Outline of the Archimedean method.*

experimentally at room temperature. The weight loss of the sinker must also be corrected for surface tension. The effect of s and v in Eq. (3.16) can be minimized by using a fine suspension wire, but this is not always practicable in the case of density determinations of metallic liquids at high temperatures. According to Veazey and Roe's report [26], the magnitude of s is of the order of 0.5 per cent of the total buoyancy Δw for typical metals and suspensions.

Some workers, e.g. Mackenzie [27] and Martinez and Walls [28], have eliminated the surface tension correction experimentally by using two sinkers of different size. These are suspended by wires of the same diameter. Density measurements of metallic liquids have frequently been made by this latter procedure (the 'double-sinker method') because of a lack of accurate data on surface tensions and contact angles. In addition, the effect of v can be eliminated if the two sinkers are immersed to the same depth within the liquid metallic specimen.

From the standpoint of materials science and engineering, the important problem with this method is one of materials. Graphite, Pyrex, refractory metals (e.g. molybdenum, tungsten), or metal-coated alumina (Al_2O_3) or zirconia (ZrO_2) can be used for the sinker. While metals are the most suitable for the suspension wire, problems of chemical reaction between the metal wire and the liquid metallic sample may occur.

In the case of the direct Archimedean method, a large volume of the liquid metallic specimen is required. Nevertheless, this technique has frequently been employed for density measurements of metallic liquids because it is relatively simple from the experimental point of view and allows for continuous measurements over wide temperature ranges.

3.6.1.2 *Indirect Archimedean Method*

In this technique, a liquid metallic specimen is contained in a vessel, which is weighed while immersed in an inert fluid of established density, as shown in Figure 3.12(b).

An advantage of this technique is that it provides continuous measurements of density from solid to liquid, or from liquid to solid; volume changes during melting (or solidifying) can be observed by this method. However, the method involves one great difficulty; accurate density values of the liquid which is inert to the contacting materials (i.e. to both the liquid metallic specimen and to the vessel or crucible) are required at the given temperature.

Gebhardt et al. [29] determined density values for aluminium and aluminium–copper alloys from 100 to 900 °C using this technique. Iida et al. [6] measured density values for gallium-tin and indium-tin alloy systems from solid to liquid at temperatures of 0 to 177 °C and 77 to 197 °C, respectively, using this method. In these experiments, silicon oil (dimethyl polysiloxane) was used as the inert liquid of known density. According to their experimental results, the density changes on melting for pure gallium and indium are an increase of 3.1 per cent and a decrease of 2.3 per cent, respectively (see Table 3.1), and the melting point densities for pure liquid gallium and indium are 6.13×10^3 kg m^{-3} and 7.03×10^3 kg m^{-3}, respectively. They estimated the experimental uncertainty to be about 0.2 to 0.3 per cent.

3.6.2 Pycnometric Method

A liquid metallic specimen is filled in a vessel or pot of known volume, as shown in Figure 3.13. Upon freezing, the solid metallic specimen contained within the pot is weighed at room temperature. One of the characteristics of the pycnometric method is that accurate, absolute determinations are possible. The principle and construction of the pycnometer are both quite simple. However, accurate density measurements of metallic liquids at high temperatures (above 1100 °C) are considerably more difficult because of limitations in vessel materials. The following characteristics are required for the materials of

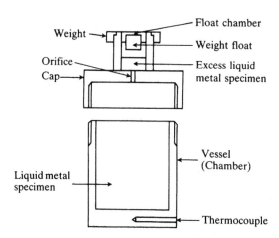

Figure 3.13 *Cross section of a pycnometer.*

a pycnometer: (a) excellent machinability, (b) no reaction with the liquid metallic specimen, (c) refractoriness, (d) a low and well-known coefficient of thermal expansion. In practice, no materials exist which have precisely all such properties. Nevertheless, refractory materials such as graphite, quartz, metals (e.g. tantalum), and metal oxides (e.g. alumina, zirconia) have been used for the pycnometer.

Pycnometric techniques are unsuitable for the continuous measurement of density since a separate filling and weighing is necessary at each temperature.

Experimental details of the pycnometric method are adequately described in a review article by Crawley [7].

3.6.3   Dilatometric Method

In this technique for density measurement, an accurately weighed specimen is contained in a dilatometer (see Figure 3.18). This is a calibrated vessel fitted with a long, narrow neck. Changes in volume of the specimen with temperature are continuously monitored through changes in liquid meniscus height. By calibrating the dilatometer beforehand with a liquid of known density, e.g. mercury, the specimen's density and thermal expansivity are obtained. The merits of this method are that a large amount of specimen is not needed and that it provides continuous density measurements. As a result, dilatometric methods have proven popular.

While accurate density measurements require precise measurements of liquid volume, these are sometimes difficult for metallic liquids, owing to their meniscus shapes. Obviously, the volumetric uncertainty due to the meniscus effect can be decreased by employing a dilatometer with a fine-bore neck. In such experiments, careful degassing of the liquid metallic specimen and keeping its surface (or meniscus) clean, and perfectly free from any contamination, are essential, particularly for dilatometers with narrow capillary necks. Moreover, the choice of materials for the dilatometer is extremely important. The same material characteristics as those for a pycnometer are needed for the dilatometer.

For transparent dilatometers such as Pyrex or quartz, it is possible to observe the position of the liquid meniscus directly using a cathetometer. On the other hand, in the case of an opaque dilatometer of metal oxide, the meniscus level can be measured indirectly using an electrical contact method [30].

When an inert liquid, such as a mineral oil, is used as the meniscus-indicating medium in the capillary tube of the dilatometer, measurements of density from solid to liquid (or from liquid to solid), of the specimen's volume change on fusion [31], and of volume change on mixing [32], using a U-shaped cell, can be obtained.

Density data obtained by dilatometric methods are accurate, and the direct observation method provides particularly accurate and reliable density values.

3.6.4   Maximum Bubble Pressure Method

When an inert gas is passed through a capillary tube immersed to a certain depth in a liquid, bubbles of gas will detach from the tip of the capillary (see Figure 3.14).

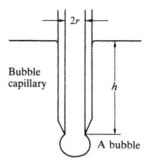

Figure 3.14 *A bubble tube with a conically ground tip.*

The density of the liquid specimens can be determined by measuring the over-pressure required to just detach a bubble of the inert gas from the tip of the capillary. The maximum bubble pressure P_m at depth h is equal to the sum of the pressure needed to maintain the column of liquid ρgh and the pressure needed to create a bubble's new surface $2\gamma / r$ (see Eq. (6.72)) so that,

$$P_m = \rho gh + \frac{2\gamma}{r} \qquad (3.17)$$

where γ is the surface tension, and r is the radius of the capillary tube. This requires that spherical-shape bubbles of radius r detach from the tips of the capillary. In other words, corrections must be made for non-spherical bubbles. The second term in the above equation (i.e. the surface tension effect) can be eliminated by making determinations of the maximum bubble pressure at different depth immersion. Hence, if the maximum bubble pressures are P_{m1} and P_{m2}, at the depths of immersion h_1 and h_2, respectively, the density of the liquid specimen is given by

$$\rho = \frac{P_{m1} - P_{m2}}{g(h_1 - h_2)} \qquad (3.18)$$

It is obvious from this equation that the maximum bubble pressure P_m and depth of immersion h or displacement $\Delta h \, (= h_1 - h_2)$ require accurate measurements for reliable density data. Furthermore, a correction must be made for any expansion of the capillary tube.

The technique allows density measurements to be carried out at high temperatures and over wide temperature ranges [33]. Many workers have employed the method for density measurements of metallic liquids. However, it is not as accurate as the pycnometric method, since precise measurements of maximum bubble pressure are, in practice, difficult. Similarly, Irons and Guthrie [34] have shown that non-wetting effects in gas/liquid metal systems can lead to spreading away from an inside nozzle perimeter and across the surfaces of the capillary tube. These can reduce P_m values. In addition, a rather large volume of liquid sample is needed.

Further information about this method is given in Subsection 6.9.2.

3.6.5   Manometric Method

This method requires an application of the U-tube manometer. The manometric method consists of measuring the difference in the heights of two columns of the liquid specimen contained within a U-tube. The height differential is maintained through a gas pressure difference between the two arms of the U-tube. This pressure difference is measured using another manometer, or by electric contact. The manometric technique has characteristics that are common to the maximum bubble pressure method in measuring the pressures, and common to the dilatometric method in the measurements of liquid levels. The sources of error in the manometric method are similar to those in the dilatometric and the maximum bubble pressure method.

Densities of liquid iron and iron–carbon alloys were measured using the manometric method based on X-ray transmission technique [35].

3.6.6   Sessile Drop Method

The sessile drop method and the oscillating (or levitated) drop method are known generically as the 'liquid drop method'. The liquid drop methods involve photographing profiles of a liquid metallic drop. Following solidification, the metallic specimen is then weighed and the volume of the liquid metallic drop calculated through geometrical analysis of the photographs.

The liquid drop methods can be used for high temperature liquid metallic systems. A liquid metallic drop resting on an inert, non-wetting refractory plate or substrate of a smooth horizontal surface acquires a shape such as that shown in Figure 3.15. The volume of this liquid drop can then be calculated from the numerical table of Bashforth and Adams [36] by determining every size, i.e. $X, \cdots, \theta$, indicated in Figure 3.15 (see also Subsection 6.9.1). To obtain accurate values for drop volumes, great efforts have been expended in producing truly symmetrical liquid drops. However, precise volume

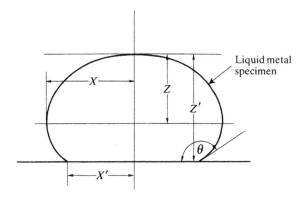

Figure 3.15 *A liquid metal drop on a plate (the sessile drop method).*

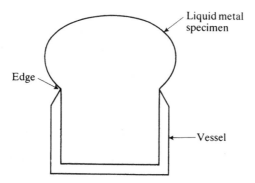

Figure 3.16 *Outline of the large drop method.*

measurements seem to be difficult; their errors are estimated to be about ± 0.25 to ± 2 per cent.

To prevent a sessile drop of asymmetrical shape forming, the so-called 'large drop method' has been used, which is outlined in Figure 3.16. A sessile drop surmounting a cylindrically ground and edged vessel is apt to be perfectly symmetrical. As can be seen from Figure 3.16, this method may be considered to be a combination of the pycnometric and sessile drop techniques.

3.6.7 Oscillating (or Levitated) Drop Method

The principle of this technique is to float, or levitate, a metallic drop. Levitation can be achieved using electromagnetic, electrostatic, aerodynamic, and acoustic techniques.

El-Mehairy and Ward [37] developed this method, i.e. the containerless method, in order to eliminate the problem of any chemical reaction between liquid metallic drops and ceramic substrates or vessels. Later, some workers examined this technique for density measurements of liquid iron, and its alloys. The accuracy attainable with this method is generally inferior to the pycnometric, the dilatometric, and the Archimedean methods. Nevertheless, levitated drop techniques do have the advantage of being compatible with highly reactive and high melting point, metallic liquids.

More recently, containerless oscillating drop techniques have made thermophysical property measurements of high melting point metals possible [38,39]. Ishikawa et al. [40] and Paradis et al. [41] carried out noncontact density measurements of high melting point metals in the liquid and undercooled states, using a ground-based vacuum electrostatic levitator. The following is an outline of the experimental procedure performed by Ishikawa et al. and Paradis et al. Upon melting, the liquid metal sample took on a spherical shape due to surface tension, and its images and its temperatures were simultaneously recorded. After the experiment, the sample radius was extracted from each digitized video image and matched to a temperature profile. Since the liquid sample was axisymmetric and because its mass was known, the density was found as a function of temperature. From these density data, the volume expansivities were also calculated. Incidentally, the radiance temperature was measured with two single-color pyrometers, and was calibrated to true temperature using the known melting point temperature of the metal sample. In their experiments, they have estimated that the uncertainty of the density measurements is less than 2 per cent from the resolution of the video grabbing capability (640 × 480 pixels) and from the uncertainty in mass measurement (±0.0001g).

3.6.8   Gamma Radiation Attenuation Methods

The principle of this method is based on attenuation of a gamma ray (γ-ray) beam passing through matter. The experimental arrangement for the gamma radiation attenuation technique is shown schematically in Figure 3.17. The γ-ray beam from the source of radiation (e.g. ^{60}Co, ^{137}Cs) [42] in a lead container is collimated before entering a liquid metallic specimen. After the incident beam is attenuated according to the mass of the liquid metallic sample, the intensity of the emergent beam is recorded by a radiation counter. For a γ-ray beam of intensity I_0 penetrating a specimen of length x, the emergent radiation intensity I is given by

$$I = I_0 \exp(-\alpha_a \rho x) \tag{3.19}$$

where α_a is the absorption coefficient per unit mass. In determining the liquid metallic density, values for α_a can first be obtained from absorption measurements on the metallic specimen in its solid form, where its density will be well known or easily measured.

Basin and Solov'ev [43] obtained density data for liquid lead using this technique. According to their experimental results, the differences in the density values for liquid

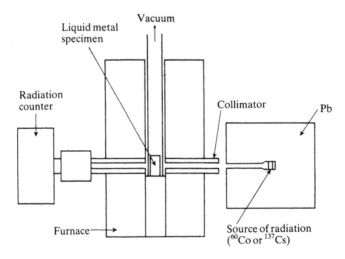

Figure 3.17 *Diagram of the experimental set-up for the gamma radiation attenuation method.*

lead determined by gamma radiation attenuation and the pycnometric methods are 0.16, 0.10, and 0.05 per cent at the melting point (327.5 °C), 400, and 500 °C, respectively [7]. This technique, therefore, seems to provide accurate density data. It can be seen from Eq. (3.19) that the accuracy of this technique depends on the sensitivity of the radiation counter.

However, we should note that it is still technically difficult to obtain reliable density data at high temperatures using this method. On the other hand, an advantage of the technique is that surface tension effects and chemical contamination of the liquid metal's surface are not involved, since the γ-ray beam penetrates the bulk of the specimen.

3.7 Experimental Data for the Density and Molar Volume of Metallic Liquids

3.7.1 Experimental Data for the Density of Liquid Metallic Elements

Periodic variation in the molar volumes of liquid metallic elements at their melting point temperatures with their atomic numbers is already shown in Figure 3.6, where experimental data (i.e. experimental density data, because of the molar volume $V = M/\rho$) for 70 liquid metallic elements are plotted; their experimental liquid density data are listed in Chapter 17. (The volume expansivities of 16 liquid metallic elements are given in Table 3.3.)

It is found experimentally that the temperature dependence Λ of density for liquid metallic elements is linear. Thus, density data can be adequately represented by the equations

$$\rho = \rho_m + \Lambda(T - T_m), \quad \Lambda \ [\equiv (\partial\rho/\partial T)_P] \tag{3.20}$$

The liquid molar volume for metallic elements can be approximately expressed as

$$V = V_m\{1 + \alpha_m(T - T_m)\}, \quad \alpha \left[\equiv -\rho^{-1}(\partial\rho/\partial T)_P\right] \tag{3.21}$$

However, for aluminium, gallium, and antimony, there are sufficient data to indicate that Λ is not constant over the whole liquid range. In general, the observed linear temperature variation of liquid metallic densities may not be precisely true. If these measurements had been made over a wide temperature range, i.e. from the melting point up to a certain ratio of the reduced temperature, some curvatures would have been probable towards the critical point temperatures [44] (see Figures 3.3 and 3.4).

3.7.2 Experimental Data for the Molar Volume of Liquid Alloys

For a binary system, the molar volume V_A is defined by

$$V_A = \frac{x_1 M_1 + x_2 M_2}{\rho_A} \tag{3.22}$$

where x_1, x_2, M_1, and M_2 are the molar fractions and molar masses of components 1 and 2, respectively, and ρ_A is the alloy density. In studies of real alloy systems, the excess volume V^E is frequently used. For a binary alloy system,

$$V^E = V_A - V_{\text{ideal}}$$

$$= \frac{x_1 M_1 + x_2 M_2}{\rho_A} - \left(\frac{x_1 M_1}{\rho_1} + \frac{x_2 M_2}{\rho_2} \right) \tag{3.23}$$

In addition, as large differences in the molar volumes of two metals (or metallic elements) forming a liquid alloy can have an important effect on its properties, related size factor parameters are also used in theoretical descriptions.

The molar volumes for most binary liquid alloys are actually linear functions of composition. While many liquid alloys show small deviations from this linear relationship, maximum values of V^E are generally ± 3 per cent, or more usually, within ± 2 per cent.

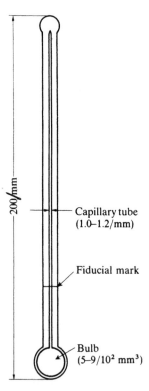

200/mm

Capillary tube
(1.0–1.2/mm)

Fiducial mark

Bulb
(5–9/10² mm³)

Figure 3.18 *Cross section of a dilatometer with a narrow capillary tube (after Morita et al. [46]).*

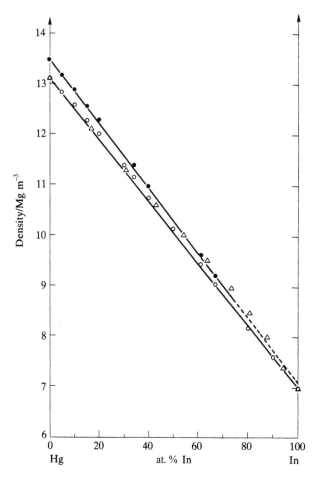

Figure 3.19 *Isothermal densities of liquid mercury–indium alloys at 50 °C and 200 °C (after Morita et al. [46]). ●, 50 °C, ○, 200 °C (Morita et al. [46]); △, 200 °C (Predel and Eman [47]).*

As a result, the molar volumes of binary liquid alloys may be approximately evaluated by assuming additivity of component molar volumes. For a multicomponent system, molar volumes may be estimated by

$$V \approx \sum x_i V_i \tag{3.24}$$

where the subscript i refers to the component. Some exceptions are compound-forming alloys such as iron–silicon, mercury–sodium, mercury–potassium, sodium–lead, sodium–bismuth, aluminium–copper, magnesium–lead, and sodium–indium [7, 45].

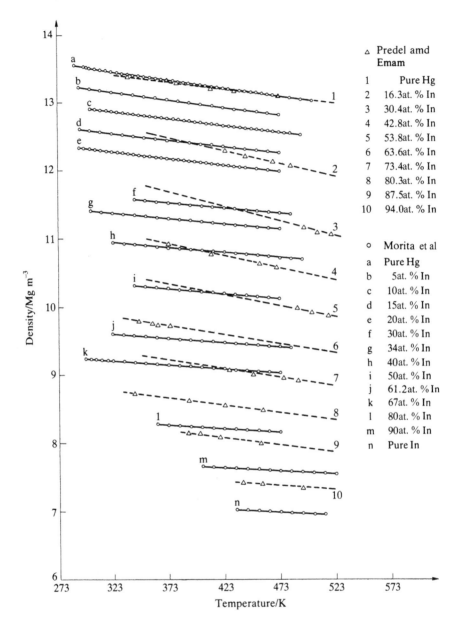

Figure 3.20 *Densities of liquid mercury–indium alloys (after Morita et al. [46]).*

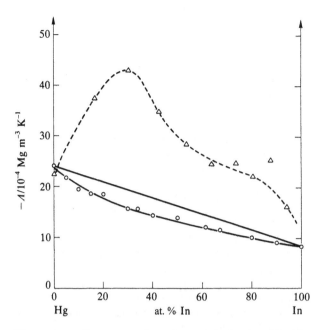

Figure 3.21 *Temperature dependence of the density of liquid mercury–indium alloys (after Morita et al. [46]).* ○ *Morita et al. [46]); △, Predel and Eman [47].*

These systems show large negative V^E values which are in the order of –6 to –27 per cent. Incidentally, eutectic density minima and rapid changes in density with composition have been reported for a few systems. However, more careful investigations are necessary before the validity of these density anomalies is established. These include the discontinuities in temperature variations of metallic liquids.

Morita et al. [46] obtained accurate density data for liquid mercury–indium alloys, using a dilatometer with a capillary neck, as shown in Figure 3.18. In Figures 3.19 to 3.21, their results are compared with Predel and Eman's data [47]. As can be seen from Figure 3.19, the isothermal densities by Morita et al. and Predel and Eman both show positive deviations from the additivity of densities, but the maximum values of the deviations are, at 200 °C, 1.2 per cent and 4 per cent, respectively. Furthermore, the values for Λ also indicate rather large discrepancies, as shown in Figures 3.20 and 3.21. As seen, these experimental results are inconsistent. In addition, experimental data for the composition dependence of the molar volumes of a liquid indium - tin alloy system are also inconsistent: the result of Iida et al. [6] coincides qualitatively with that of Berthou and Tougas [48], showing a positive deviation from the additivity of densities, but, in contrast, that of Predel and Eman [49], is showing a negative deviation, although the density data for low melting point systems are considered to be accurate and reliable.

....

REFERENCES

1. L.J. Wittenberg and R. DeWitt, *J. Chem. Phys.*, **56** (1972), 4526.
2. W. Benenson, J.W. Harris, H. Stocker, and H. Lutz (eds.), *Handbook of Physics*, Springer-Verlag, New York, 2002.
3. K.C. Mills, *Recommended Values of Thermophysical Properties for Selected Commercial Alloys*, Woodhead Publishing, Cambridge, 2002.
4. A.R. Ubbelode, *Melting and Crystal Structure*, Clarendon Press, Oxford, 1965, p.170.
5. C.S. Barrett, *Structure of Metals*, McGraw-Hill, New-York, 1966, Table A–6.
6. T. Iida, Y. Kita, Y. Kikuya, T. Kirihara, and Z. Morita, *J. Non-Cryst. Solids*, **117/118** (1990), 567.
7. A.F. Crawley, *Int. Met. Rev.*, **19** (1974), 32.
8. W.W. Wood and J.D. Jacobson, *J. Chem. Phys.*, **27** (1957), 1207.
9. B.J. Alder and T.E. Wainwright, *J. Chem. Phys.*, **27** (1957), 1208; *Phys. Rev.*, **127** (1962), 359; T.E. Wainwright and B.J. Alder, *Nuovo Cimento*, **9** (Suppl. 1) (1958), 116.
10. W.G. Hoover and F.H. Ree, *J. Chem. Phys.*, **47** (1967), 4873; **49** (1968), 3609.
11. L.V. Woodcock, *J. Chem. Soc. Faraday Trans.*, II, **72** (1976), 1667.
12. S. Hudson and H.C. Andersen, *J. Chem. Phys.*, **69** (1978), 2323.
13. A.V. Grosse, *J. Inorg. Nucl. Chem.*, **22** (1961), 23.
14. I.G. Dillon, P.A. Nelson, and B.S. Swanson, *J. Chem. Phys.*, **44** (1966), 4229.
15. F. Hensel, *Physica Scr.*, **T25** (1989), 283.
16. P.J. McGonigal, *J. Phys. Chem.*, **66** (1962), 1686.
17. H. Hess and H. Schneidenbach, *Z. Metallkd.*, **87** (1996), 979.
18. D.J. Steinberg, *Met. Trans.*, **5** (1974), 1341.
19. N.F. Mott and H. Jones, *The Theory of the Properties of Metals and Alloys*, Clarendon Press, Oxford, 1936, p.17.
20. L.S. Darken and R.W. Gurry, *Physical Chemistry of Metals*, McGraw-Hill, New York, 1953, p.158.
21. C. Kittel, *Introduction to Solid State Physics*, 2nd ed., John Wiley & Sons, Inc., 1956, p.153.
22. S. Nagakura, H. Inokuchi, H. Ezawa, S. Iwamura, F. Sato, and R. Kubo (eds.), *Iwanami Dictionary of Physical Sciences (Iwanami Rikagaku Jiten)*, 5th ed., Iwanami Shoten Publishers, Tokyo, 1998.
23. W.F. Gale and T.C. Tolemeier (eds.), *Smithells Metals Reference Book*, 8th ed., Elsevier Butterworth-Heineman, Oxford, 2004, 14–12.
24. T. Baykara, R.H. Hauge, N. Norem, P. Lee, and J.L. Margrave, *II Ciocco Workshop Proc.*, 1993.
25. O. Kubaschewski, C.B. Alcock, and P.J. Spencer, *Materials Thermochemistry*, 6th ed., Pergamon Press, Oxford, 1993, p.258 (Table 1).
26. S.D. Veazey and W.C. Roe, *J. Mater. Sci.*, **7** (1972), 445.
27. J.D. Mackenzie, *J. Phys. Chem.*, **63** (1959), 1875.
28. J. Martinez and H.A. Walls, *Met. Trans.*, **4** (1973), 1419.
29. E. Gebhardt, M. Becker, and S. Dorner, *Z. Metallkd.*, **44** (1953), 573.
30. A.R. Keskar and S.J. Hruska, *Met. Trans.*, **1** (1970), 2357.
31. R.H. Perkins, L.A. Geoffrion, and J.C. Biery, *Trans. Met., Soc. AIME*, **233** (1965), 1703.
32. O.J. Kleppa, *J. Phys. Chem.*, **64** (1960), 1542.
33. L.D. Lucas, *Techniques of Metals Research* (ed. R.F. Bunshah), Interscience Publishers, New York, 1970, Vol. 4, Part 2, p.219.
34. G.A. Irons and R.I.L. Guthrie, *Can. Met. Q.*, **19** (1981), 381.

35. K. Ogino, A. Nishiwaki, and Y. Hosotani, *J. Japan Inst. Metals*, **48** (1984), 996;1004.
36. F. Bashforth and J.C. Adams, *An Attempt to Test the Theories of Capillary Action*, Cambridge University Press, 1883.
37. A.E. El-Mehairy and R.G. Ward, *Trans. Met. Soc. AIME*, **227** (1963), 1226.
38. D.M. Herlach, R.F. Cochrane, I. Egry, H.J. Fecht, and A.L. Greer, *Int. Mater. Rev.*, **38** (1993), 273.
39. T. Hibiya and I. Egry, *Meas. Sci. Technol.*, **16** (2005), 317.
40. T. Ishikawa, P.-F. Paradis, T. Itami, and S. Yoda, *Meas. Sci. Technol.*, **16** (2005), 443.
41. P.-F. Paradis, T. Ishikawa, R. Fujii, and S. Yoda, *Appl. Phys. Lett.*, **86** (2005), 04190–1.
42. G. Döge, *Z. Naturforsch.*, **21a** (1966), 266.
43. A.S. Basin and A.N. Solov'ev, *Zhur. Priklad. Mekh. Tekhn. Fiziki*, (1976), No. 6, 83.
44. S. Jüngst, B. Knuth, and F. Hensel, *Phys. Rev. Lett.*, **55** (1985), 2160.
45. Y. Marcus, *Introduction to Liquid State Chemistry*, John Wiley & Sons, London, 1977, Chap. 8.
46. Z. Morita, T. Iida, and Y. Matsumoto, *The 140th Committee the Japan Society for the Promotion of Science (JSPS)*, Rep. No.177, Dec. 1985.
47. B. Predel and A. Emam, *Mater. Sci. Eng.*, **4** (1969), 287.
48. P.E. Berthou and K. Tougas, *Met. Trans.*, **1** (1970), 2978.
49. B. Predel and A. Emam, *J. Less.-Comm. Met.*, **18** (1969), 385.

4

Thermodynamic Properties

From the viewpoint of thermophysics, this chapter deals with some of the more important thermodynamic (or thermophysical) properties of liquid metallic elements in the area of materials process science. These comprise evaporation enthalpy, or the latent heat of vaporization, vapour pressure, and heat capacity. Theoretical and empirical equations describing these thermodynamic properties are therefore presented together with experimental data. As experimental methods associated with these data are well documented in other books, [1–4] they are not reported here.

4.1 Evaporation Enthalpy of Liquid Elements

4.1.1 Introduction

General speaking, the enthalpy H, which is a thermodynamic function (or property) of a system, is equal to the sum of its energy, or its internal energy U, and the product of its pressure P and volume V (i.e. $H = U + PV$); the enthalpy (J in SI units) is of great importance for the description of processes proceeding at constant pressure.

The evaporation enthalpy can be formulated in terms of the pair distribution function $g(r)$ and the pair potential (the intermolecular potential energy, or the mutual potential energy of two molecules) $\phi(r)$, based on the pair theory of liquids.

The evaporation enthalpy is indispensable in discussing the thermodynamic or thermophysical properties of liquids. Thus, several important relationships between the evaporation enthalpy and other thermophysical properties or quantities are given in this section.

4.1.2 Equation in terms of $g(r)$ and $\phi(r)$

The evaporation enthalpy, $\Delta_l^g H$, or the latent heat of vaporization, is given by

$$\Delta_l^g H = U_g - U_l + P(V_g - V_l) \tag{4.1}$$

The Thermophysical Properties of Metallic Liquids: Volume 1 – Fundamentals. First Edition.
Takamichi Iida and Roderick I. L. Guthrie. © Takamichi Iida and Roderick I. L. Guthrie 2015.
Published in 2015 by Oxford University Press.

in which the subscripts g and l refer to a gas and a liquid, respectively. In the above equation, the product of pressure and the change of volume, i.e. $P(V_g - V_l)$, is the work of expansion against an external pressure during vaporization. If the gas behaves as an ideal, monatomic molecule (for metallic elements, this may be estimated to a good approximation), U_g and $P(V_g - V_l)$, respectively, can be expressed as

$$U_g = \frac{3}{2} NkT \tag{4.2}$$

and

$$P(V_g - V_l) \approx PV_g = NkT \tag{4.3}$$

Thus, from Eqs. (2.13b) and (4.1) to (4.3), we obtain

$$\Delta_l^g H = \frac{3}{2} NkT - \left[\frac{3}{2} NkT + \frac{2\pi N^2}{V} \int_0^\infty g(r)\phi(r)r^2 \mathrm{d}r \right] + NkT$$

$$= NkT - \frac{2\pi N^2}{V} \int_0^\infty g(r)\phi(r)r^2 \mathrm{d}r \tag{4.4}$$

At the boiling point, the molar evaporation enthalpy $\Delta_l^g H_b$ is

$$\Delta_l^g H_b = RT_b - <\Phi>_b \tag{4.5a}$$

$$<\Phi>_b = \frac{2\pi N_A^2}{V} \int_0^\infty g(r)\phi(r)r^2 \mathrm{d}r \bigg|_{T=T_b}$$

where V is the molar volume, or

$$-\frac{<\Phi>_b}{\Delta_l^g H_b} = 1 - \frac{RT_b}{\Delta_l^g H_b} \tag{4.5b}$$

In Figure 4.1, values of $(-<\Phi>_b / \Delta_l^g H_b)$ for 62 metallic elements are plotted as a function of their atomic number. The figure shows a periodic variation in $(-<\Phi>_b / \Delta_l^g H_b)$ values with increase in their atomic number. The highest values are apparent for the *d*-block main transition metals (see Table 1.7), excluding manganese, their ratios of the mean total potential energy to the evaporation enthalpy lie between 0.92 and 0.94 (on average, 0.930 for 22 main transition metals (cf. Table 1.7), and 0.916 for 62 metallic elements). Incidentally, the figure shows that the periodic Group IVA metallic elements (excluding lead) occupy the minor peaks, and the Group IA metals (i.e. alkali metals) occupy the major and the Group IIB metals the minor valleys of the curve. The periodic variation in $RT_b / \Delta_l^g H_b$ values for 62 metallic elements with their atomic numbers is also given in Figure 4.2.

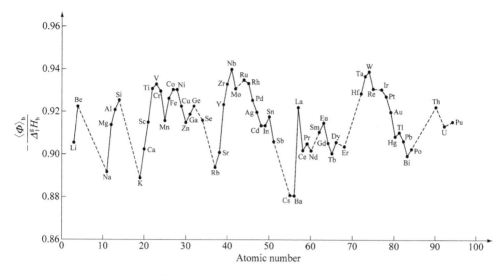

Figure 4.1 *Values of $<\Phi>_b / \Delta_l^g H_b$ plotted against atomic number.*

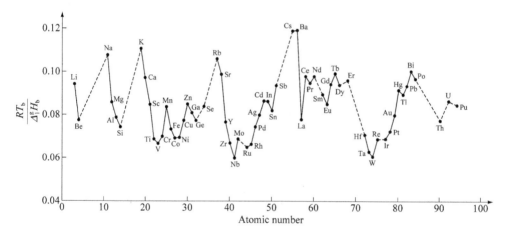

Figure 4.2 *Values of $RT_b / \Delta_l^g H_b$ plotted against atomic number.*

4.1.3 Relationships between Evaporation Enthalpy and Thermophysical Properties or Quantities

4.1.3.1 Evaporation Enthalpy $\Delta_l^g H_b$ vs. Cohesive Energy E_c^0 at 0 K at 101.325 kPa (1 atm)

In discussing the physical properties of solids, the cohesive energy is an essential quantity. The cohesive energy (denoted by E_c^0) is defined as the energy required to form separated neutral atoms in their ground electronic state from a solid at 0 K at 101.325 kPa (i.e. standard pressure).

Figure 4.3 shows a plot of $\Delta_l^g H_b$ against E_c^0 for a large number of metallic elements (57 metals, and two semiconductors, i.e. germanium and silicon). The data for E_c^0 are given in Table 4.1. As can be seen, there is a good linear relationship between the two variables. Thus, the relationship is represented by a simple equation, as follows:

$$\Delta_l^g H_b = 0.889 E_c^0 \tag{4.6}$$

or

$$\frac{\Delta_l^g H_b}{E_c^0} \approx 0.9 \tag{4.7}$$

The numerical factor of 0.889 was determined so as to give the minimum S value (see Appendix 2) for the elements illustrated in Figure 4.3.

Table 4.2 compares experimental values of $\Delta_l^g H_b$ for 59 metallic elements with those calculated from Eq. (4.6), together with δ_i, Δ, and S values needed for statistical assessment of the relationship; Eq. (4.6) performs excellently with various metallic

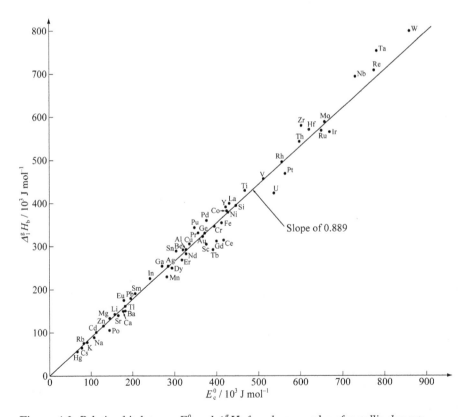

Figure 4.3 *Relationship between E_c^0 and $\Delta_l^g H_b$ for a large number of metallic elements.*

elements, which gives $\Delta(59)$ and $S(59)$ values of 4.8 per cent and 0.063, respectively. As is obvious from Figure 4.3 and Table 4.2, the maximum positive deviation from the relationship described by Eq. (4.6) is 11.7 per cent for plutonium; on the other hand the maximum negative deviation is −17.2 per cent for polonium. The uncertainties in the calculations of the evaporation enthalpy using Eq. (4.6) will probably fall within the range of uncertainties associated with experimental measurements.

Table 4.1 *Molar cohesive energies of crystalline elements[†] (i.e. energy required to form separated neutral atoms in their ground electronic state from the solid at 0 K at 1 atm (101.325 kPa)).*

Element		E_C^0 kJ mol^{-1}
Actinium	Ac	410
Aluminium	Al	327
Americium	Am	264
Antimony	Sb	265
Arsenic	As	285.3
Astatine	At	–
Barium	Ba	183
Berkelium	Bk	–
Beryllium	Be	320
Bismuth	Bi	210
Boron	B	561
Cadmium	Cd	112
Caesium	Cs	77.6
Calcium	Ca	178
Californium	Cf	–
Carbon	C	711
Cerium	Ce	417
Chromium	Cr	395
Cobalt	Co	424
Copper	Cu	336
Curium	Cm	385

continued

Table 4.1 *(continued)*

Element		E_C^0 kJ mol^{-1}
Dysprosium	Dy	294
Einsteinium	Es	–
Erbium	Er	317
Europium	Eu	179
Fermium	Fm	–
Francium	Fr	–
Gadolinium	Gd	400
Gallium	Ga	271
Germanium	Ge	372
Gold	Au	368
Hafnium	Hf	621
Holmium	Ho	302
Indium	In	243
Iridium	Ir	670
Iron	Fe	413
Lanthanum	La	431
Lawrencium	Lr	–
Lead	Pb	196
Lithium	Li	158
Lutetium	Lu	428
Magnesium	Mg	145
Manganese	Mn	282
Mendelevium	Md	–
Mercury	Hg	65
Molybdenum	Mo	658
Neodymium	Nd	328
Neptunium	Np	456

Table 4.1 *(continued)*

Element		E_C^0 kJ mol^{-1}
Nickel	Ni	428
Niobium	Nb	730
Nobelium	No	–
Osmium	Os	788
Palladium	Pd	376
Phosphorus	P	331
Platinum	Pt	564
Plutonium	Pu	347
Polonium	Po	144
Potassium	K	90.1
Praseodymium	Pr	357
Promethium	Pm	–
Protactinium	Pa	–
Radium	Ra	160
Rhenium	Re	775
Rhodium	Rh	554
Rubidium	Rb	82.2
Ruthenium	Ru	650
Samarium	Sm	206
Scandium	Sc	376
Selenium	Se	237
Silicon	Si	446
Silver	Ag	284
Sodium	Na	107
Strontium	Sr	166
Sulphur	S	275
Tantalum	Ta	782

continued

Table 4.1 *(continued)*

Element		E_C^0 kJ mol^{-1}
Technetium	Tc	661
Tellurium	Te	211
Terbium	Tb	391
Thallium	Tl	182
Thorium	Th	598
Thulium	Tm	233
Tin	Sn	303
Titanium	Ti	468
Tungsten	W	859
Uranium	U	536
Vanadium	V	512
Ytterbium	Yb	154
Yttrium	Y	422
Zinc	Zn	130
Zirconium	Zr	603

† Data are taken from Kittel [5].

In Figure 4.4, the periodic variation of $\Delta_l^g H_b$, together with E_c^0, is shown against atomic number. It can be seen that they display an analogous periodicity.

Equation (4.6) allows us to predict the value of $\Delta_l^g H_b$ for liquid metals (e.g. osmium, technetium, and ytterbium) from knowledge of the cohesive energy at 0 K at 101.325 kPa.

Of the thermophysical properties of liquid metallic elements, the enthalpy of evaporation, or the cohesive energy, is one of their most important quantities. The evaporation enthalpy is a direct measure of the cohesive energy of a liquid metallic element. It is evident, therefore, that various properties or quantities of liquid metallic elements are related to their evaporation enthalpy.

4.1.3.2 *Evaporation Enthalpy $\Delta_l^g H_b$ vs. Melting Point Temperature T_m*

A metallic element's melting temperature is also roughly related to the cohesive energy, or the evaporation enthalpy, of the element's condensed phase. As shown in Figure 4.5, the molar enthalpies of evaporation of liquid metallic elements at their boiling points are roughly proportional to their melting temperatures:

$$\Delta_l^g H_b \approx 2.40 \times 10^2\, T_m \tag{4.8}$$

Table 4.2 *Comparison of experimental values for the molar evaporation enthalpy of liquid metallic elements at their boiling point temperatures with those calculated from Eq. (4.6), together with δ_i, Δ, and S values.*

Element		$\Delta_l^g H_b / \text{kJ mol}^{-1}$		δ_i %
		Experimental	Calculated	
Aluminium	Al	294	291	1.0
Barium	Ba	151	163	−7.4
Beryllium	Be	294	284	3.5
Cadmium	Cd	100	99.6	0.4
Caesium	Cs	65.9	69.0	−4.5
Calcium	Ca	150	158	−5.1
Cerium	Ce	314	371	−15.4
Chromium	Cr	348	351	−0.9
Cobalt	Co	383	377	1.6
Copper	Cu	304	299	1.7
Dysprosium	Dy	251	261	−3.8
Erbium	Er	271	282	−3.9
Europium	Eu	176	159	10.7
Gadolinium	Gd	312	356	−12.4
Gallium	Ga	254	241	5.4
Germanium	Ge	333	331	0.6
Gold	Au	325	327	−0.6
Hafnium	Hf	571	552	3.4
Indium	In	226	216	4.6
Iridium	Ir	564	596	−5.4
Iron	Fe	354	367	−3.5
Lanthanum	La	400	383	4.4
Lead	Pb	180	174	3.4
Lithium	Li	142	140	1.4
Magnesium	Mg	132	129	2.3

continued

Table 4.2 *(continued)*

Element		$\Delta_1^g H_b / \text{kJ mol}^{-1}$		δ_i %
		Experimental	Calculated	
Manganese	Mn	230	251	−8.4
Mercury	Hg	57.2	57.8	−1.0
Molybdenum	Mo	590	585	0.9
Neodymium	Nd	284	292	−2.7
Nickel	Ni	380	380	0
Niobium	Nb	696	649	7.2
Palladium	Pd	361	334	8.1
Platinum	Pt	469	501	−6.4
Plutonium	Pu	344	308	11.7
Polonium	Po	106	128	−17.2
Potassium	K	77.4	80.1	−3.4
Praseodymium	Pr	333	317	5.0
Rhenium	Re	707	689	2.6
Rhodium	Rh	496	493	0.6
Rubidium	Rb	75.2	73.1	2.9
Ruthenium	Ru	568	578	−1.7
Samarium	Sm	192	183	4.9
Scandium	Sc	305	334	−8.7
Silicon	Si	395	396	−0.3
Silver	Ag	253	252	0.4
Sodium	Na	89.1	95.1	−6.3
Strontium	Sr	139	148	−6.1
Tantalum	Ta	753	695	8.3
Terbium	Tb	293	348	−15.8
Thallium	Tl	162	162	0
Thorium	Th	544	532	2.3

Table 4.2 *(continued)*

Element		$\Delta_l^g H_b$ / kJ mol^{-1}		δ_i %
		Experimental	Calculated	
Tin	Sn	291	269	8.2
Titanium	Ti	430	416	3.4
Tungsten	W	800	764	4.7
Uranium	U	423	477	−11.3
Vanadium	V	458	455	0.7
Yttrium	Y	393	375	4.8
Zinc	Zn	115	116	−0.9
Zirconium	Zr	582	536	8.6
			$\Delta(59)\%$	4.8
			$S(59)$	0.063

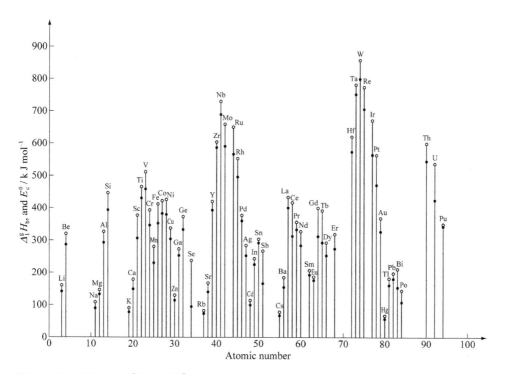

Figure 4.4 *Values of $\Delta_l^g H_b$ and E_c^0 plotted against atomic number.* •, $\Delta_l^g H_b$; ○, E_c^0.

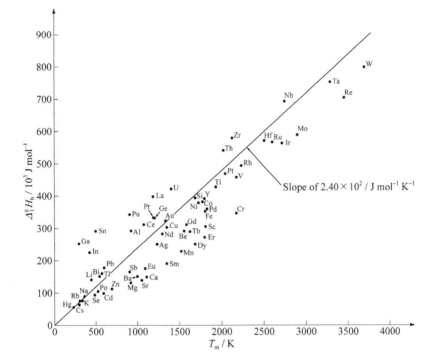

Figure 4.5 *Relationship between T_m and $\Delta_l^g H_b$ for a large number of metallic elements.*

Incidentally, the numerical factor, or the slope, 2.40×10^2 in J $mol^{-1}K^{-1}$ refers to the mean value determined from that of the respective elements plotted in Figure 4.5. Equation (4.8) gives $\Delta(62)$ and $S(62)$ values of 26.6 per cent and 0.456, respectively.

Equation (4.8) can be rewritten as

$$\frac{\Delta_l^g H_b}{R T_m} \approx C \qquad (4.9)$$

where R is the molar gas constant and C is a constant which is roughly the same for all metallic elements ($C \approx 29$). However, we can provide a rigorous expression for the relationship between $\Delta_l^g H_b$ and T_m.

Combining Eqs. (5.26) and (5.32) (see Chapter 5), we have

$$\Delta_l^g H_b = \frac{9.197^2 R}{2} \cdot \frac{\xi_T}{\xi_E} T_m = 3.516 \times 10^2 \xi T_m, \qquad \left(\xi \equiv \frac{\xi_T}{\xi_E}\right) \qquad (4.10)$$

where ξ is a dimensionless parameter which is obtained from sound velocity data. Figure 4.6 shows that the values of ξ also vary periodically with atomic number: the

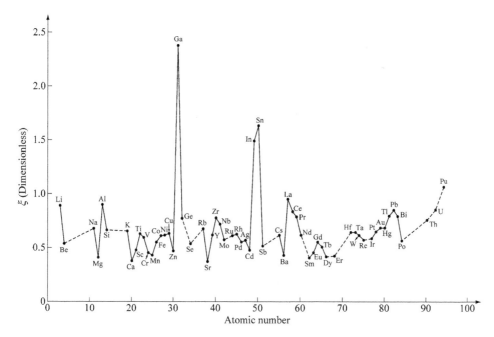

Figure 4.6 *Dimensionless parameter $\xi \, (\equiv \xi_T / \xi_E)$ plotted against atomic number for 62 liquid metallic elements.*

periodic Groups IA, IB, and IVB metals occupy the minor peaks, and Group IIA metals the major and Group IIB metals the minor valleys of the curve. In addition, gallium, indium, and tin exhibit very large values of ξ.

On the basis of Eq. (4.10) (utilizing the periodic variation in ξ values), we can predict values for the evaporation enthalpy $\Delta_l^g H_b$ of liquid metallic elements using their melting point temperature data.

4.1.3.3 Evaporation Enthalpy $\Delta_l^g H_b$ vs. Melting Enthalpy $\Delta_s^l H_m$

Table 4.3 shows the ratios of the molar enthalpies of melting to the molar enthalpies of evaporation for 63 metallic elements. From the table, we see that the enthalpies of melting for the majority of metals are between 2 and 5 per cent of the enthalpies of evaporation at their boiling point temperatures, i.e. $100\Delta_s^l H_m / \Delta_l^g H_b \approx 2$ to $5(\%)$; for semimetals, germanium, and silicon, the values are somewhat larger (see Table 4.3(b)). Thus, the change from solid to liquid involves a relatively small energy increase, as compared with the change from liquid to gas. Finally, it is worth noting that the metallic elements in the solid and liquid states have roughly the same cohesive energies.

Table 4.3 *Ratio of the molar enthalpy of melting, or the molar latent heat of fusion, $\Delta_s^l H_m$ to the molar enthalpy of evaporation, or the molar latent heat of evaporation, $\Delta_l^g H_b$ for metallic elements.*

(a) Metals

Metal		$100\Delta_s^l H_m / \Delta_l^g H_b$ %
Aluminium	Al	3.64
Barium	Ba	5.10
Beryllium	Be	4.25
Cadmium	Cd	6.30
Caesium	Cs	3.31
Calcium	Ca	5.77
Cerium	Ce	1.75
Chromium	Cr	4.20
Cobalt	Co	4.23
Copper	Cu	4.28
Dysprosium	Dy	4.42
Erbium	Er	7.34
Europium	Eu	5.23
Gadolinium	Gd	3.21
Gallium	Ga	2.22
Gold	Au	3.97
Hafnium	Hf	4.57
Indium	In	1.45
Iridium	Ir	3.99
Iron	Fe	4.38
Lanthanum	La	2.83
Lead	Pb	2.65
Lithium	Li	2.95
Magnesium	Mg	6.77

Table 4.3 *(continued)*

Metal		$100\Delta_s^l H_m / \Delta_l^g H_b$ %
Manganese	Mn	6.35
Mercury	Hg	4.14
Molybdenum	Mo	4.71
Neodymium	Nd	2.51
Nickel	Ni	4.68
Niobium	Nb	4.45
Palladium	Pd	4.63
Platinum	Pt	4.63
Plutonium	Pu	0.81
Polonium	Po	11.8
Potassium	K	3.01
Praseodymium	Pr	2.07
Rhenium	Re	4.68
Rhodium	Rh	4.52
Rubidium	Rb	2.93
Ruthenium	Ru	3.43
Samarium	Sm	4.49
Scandium	Sc	4.62
Silver	Ag	4.47
Sodium	Na	2.92
Strontium	Sr	5.93
Tantalum	Ta	4.78
Terbium	Tb	3.69
Thallium	Tl	2.60
Thorium	Th	2.54
Tin	Sn	2.43

continued

Table 4.3 *(continued)*

Metal		$100 \Delta_s^l H_m / \Delta_l^g H_b$ %
Titanium	Ti	3.60
Tungsten	W	4.41
Uranium	U	2.06
Vanadium	V	5.02
Yttrium	Y	2.90
Zinc	Zn	6.31
Zirconium	Zr	3.44

(b) Semimetals and semiconductors

Semimetal/semiconductor		$100 \Delta_s^l H_m / \Delta_l^g H_b$ %
Antimony	Sb	12.3
Bismuth	Bi	7.17
Boron	B	4.10
Germanium	Ge	10.4
Selenium	Se	5.72
Silicon	Si	12.7

4.1.3.4 *Evaporation Enthalpy $\Delta_l^g H_b$ vs. Reciprocal of Molar Volume V_m^{-1}*

Figure 4.7 gives a relationship between the molar enthalpies of evaporation and the reciprocal of molar volumes for a large number of metallic elements. Broadly speaking, it shows a positive correlation between the two variables. As indicated in Figure 4.7, the correlation could also be divided into several groups, which suggests that other parameters are probably involved. Nevertheless, the evaporation enthalpy, or the cohesive energy, which is a basic physical quantity in discussing the structures and properties of condensed matter, can be very roughly expressed in terms of molar volume alone. Incidentally, the reciprocal of molar volume can be written as

$$V_m^{-1} \propto (n_0)_m = N_A V_m^{-1} \tag{4.11}$$

where $(n_0)_m$ is the average number density at the melting point temperature and N_A is the Avogadro constant (6.022×10^{23} mol^{-1}).

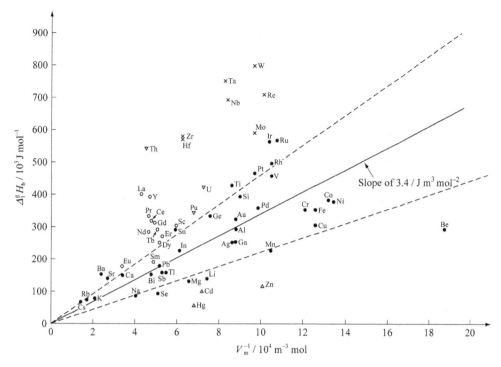

Figure 4.7 *Boiling point molar evaporation enthalpies vs. reciprocal of melting point molar volumes for a large number of liquid metallic elements.* ×, *early transition elements of Periods 5 and 6;* ○, *rare earth elements;* ▽, *actinide elements;* △, *Group IIB (zinc group) elements;* ●, *other elements. The broken lines denote a* ± 35 *per cent error band.*

4.2 Vapour Pressure of Liquid Metallic Elements

4.2.1 Introduction

The vapour pressure of a liquid metallic element is also related to the enthalpy of evaporation, or the cohesive energy. Form an engineering standpoint, knowledge of a liquid metallic element's vapour pressure is of obvious importance. For instance, it is sometimes of technological importance that a pure metal's loss by evaporation must be minimized during processing operations. Alternatively, rates of evaporation of liquid metallic elements or impurities are utilized in the pyrometallurgical production of mercury, cadmium, zinc, and magnesium, and in other refining processes. Since the equilibrium vapour pressure of a condensed phase varies exponentially with temperature, liquid metallic elements tend to evaporate more rapidly with rising temperature.

From the standpoint of thermodynamics, the application of the Clausius–Clapeyron relation to equilibrium between a condensed liquid phase and a gaseous phase provides an expression for changes in vapour pressure with temperature. However, the thermodynamic equation contains an integration constant which has to be determined experimentally.

The Sackur–Tetrode equation, derived from statistical mechanics, can also give the equilibrium vapour pressure of a condensed phase and contains no undetermined constant.

4.2.2   Theoretical Equations for Vapour Pressure

4.2.2.1   *Thermodynamic Equation*

As already mentioned, for two phases α and β in equilibrium with each other, the relation between temperature T and equilibrium vapour pressure P is given by the Clausius–Clapeyron equation, namely

$$\frac{dP}{dT} = \frac{H^\beta - H^\alpha}{T(V^\beta - V^\alpha)} = \frac{\Delta_\alpha^\beta H}{T \Delta_\alpha^\beta V} \tag{1.5'}$$

Now consider equilibrium between a liquid and a gaseous phase. Using the superscripts l for the liquid and g for the gas, we see from Eq. (1.5') that the equilibrium pressure varies with equilibrium temperature according to

$$\frac{dP}{dT} = \frac{H^g - H^l}{T(V^g - V^l)} = \frac{\Delta_l^g H}{T \Delta_l^g V} \tag{4.12}$$

This rigorous thermodynamic equation can be simplified using reasonable approximations: we can assume the gas is ideal (i.e. neglect the second virial coefficient, $\cdots$), and neglect the volume of liquid compared with that of the gas (i.e. $V^g \gg V^l$). For one mole of gas, we then have

$$V^g - V^l \approx V^g = \frac{RT}{P} \tag{4.13}$$

Substitution of Eq. (4.13) into Eq. (4.12) leads to

$$\frac{1}{P}\frac{dP}{dT} = \frac{d(\ln P)}{dT} = \frac{\Delta_l^g H}{RT^2} \tag{4.14}$$

or

$$\frac{d(\ln P)}{d(1/T)} = -\frac{\Delta_l^g H}{R} \tag{4.15}$$

From Kirchhoff's law,[1] we obtain

$$\Delta_l^g H = H_0 + \int_0^T \Delta_l^g C_P dT \tag{4.16}$$

---

[1] $\left(\frac{\partial \Delta H}{\partial T}\right)_P = \left(\frac{\partial H_2}{\partial T}\right)_P - \left(\frac{\partial H_1}{\partial T}\right)_P$, thus $\left(\frac{\partial \Delta H}{\partial T}\right)_P = C_{P_2} - C_{P_1} = \Delta C_P.$

where H_0 is the enthalpy of evaporation at 0 K, C_P is the heat capacity at constant pressure, and $\Delta_l^g C_P = C_P^g - C_P^l$. On combining this equation with the Clausius–Clapeyron equation represented by Eq. (4.14) and integrating, we then have

$$\ln P = -\frac{H_0}{RT} + \int_0^T \frac{\mathrm{d}T}{RT^2} \int_0^T \Delta_l^g C_P \mathrm{d}T + i \qquad (4.17)^2$$

where i is the constant of integration (the so-called vapour pressure constant), which must be determined experimentally. Assuming once more that the gas is ideal: $C_P^g = C_V^g + R = 5R/2$, and

$$\ln P = -\frac{H_0}{RT} + \frac{5}{2}\ln T - \frac{1}{R}\int_0^T C_P^l \mathrm{d}(\ln T) + \frac{1}{RT}\int_0^T C_P^l \mathrm{d}T + i \qquad (4.18)$$

To evaluate vapour pressures using Eq. (4.18), it is clearly necessary to know the integration constant i and the temperature dependence of heat capacity $\Delta_l^g C_P$. Since $\Delta_l^g H$ or $\Delta_l^g C_P$ varies slowly with temperature, we can approximate the change $(\Delta_l^g H)$ with the following simple algebraic expression for the temperature variation of $\Delta_l^g H$.

$$\Delta_l^g H = A + BT + CT^2 \qquad (4.19)$$

where A, B, and C are constants. On substituting this empirical equation into Eq. (4.14) and integrating,

$$\ln P = -\frac{A}{RT} + \frac{B}{R}\ln T + \frac{C}{R}T + D \qquad (4.20)$$

where D is the constant of integration.

Taking the enthalpy of evaporation as constant, Eq. (4.14) can be integrated immediately, to give

$$\ln P = -\frac{\Delta_l^g H}{RT} + \ln A \qquad (4.21)$$

or

$$P = A \exp\left(-\frac{\Delta_l^g H}{RT}\right) \qquad (4.22)$$

where $\ln A$ is a constant of integration. Equation (4.22) can provide approximate values for equilibrium vapour pressures over a wide range of temperature. The relationship shows that vapour pressures should increase exponentially with temperature and that their temperature sensitivities depend on their $\Delta_l^g H$ values.

4.2.2.2 *Statistical Mechanical Equation*

According to the theory of statistical mechanics, the entropy of a perfect gas S^g is given by

$$^2 \int_0^T \frac{\mathrm{d}T}{T^2}\int_0^T C_P \mathrm{d}T = \int_0^T \frac{C_P}{T}\mathrm{d}T - \frac{1}{T}\int_0^T C_P \mathrm{d}T = \int_0^T C_P \mathrm{d}(\ln T) - \frac{1}{T}\int_0^T C_P \mathrm{d}T$$

$$S^g = \frac{5}{2} Nk \ln T - Nk \ln P + Nk \left(\frac{5}{2} + i \right)$$ (4.23)

$$i = \ln \left\{ \frac{(2\pi m)^{3/2} k^{5/2}}{h^3} \right\}$$

where N is the number of atoms (or monatomic molecules), m is the mass of an atom, k is the Boltzmann constant, h is the Planck constant, and i is the chemical constant, or the vapour pressure constant. This relation was derived independently by Sackur and Tetrode, and is known as the Sackur–Tetrode equation. The only approximation made in deriving the Sackur–Tetrode equation is to treat the vapour as being an ideal gas with a constant heat capacity. This is generally a good approximation.

Let us now consider a liquid and its vapour in equilibrium at temperature T. If S^g and S^l are molar entropies of gas and liquid, we have

$$S^g - S^l = \frac{\Delta_l^g H}{T}$$ (4.24)

where $\Delta_l^g H$ is the molar evaporation enthalpy of liquid at temperature T. On combining this equation with Eq. (4.23) and solving for $\ln P$, we obtain

$$\ln P = -\frac{\Delta_l^g H}{RT} + \frac{5}{2} \ln T + \frac{5}{2} + \ln \left\{ \frac{(2\pi m)^{3/2} k^{5/2}}{h^3} \right\} - \frac{S^l}{R}$$ (4.25)

This equation is also often called the Sackur–Tetrode equation. The entropy of liquid S^l appearing in Eq. (4.25) can be calculated from experimental heat capacities using the relation:

$$S^l = \int_0^T \frac{C_P^l}{T} dT = \int_0^T C_P^l d(\ln T)$$ (4.26)

We should note that Eq. (4.25) contains no undetermined constants. Values calculated with this equation give excellent agreement with experimental data.

4.2.3   Empirical Equation for Vapour Pressure

It follows from Eq. (4.15) that if we plot $\ln P$ against $1/T$, the curve so obtained should at each point, have a slope equal to $\Delta_l^g H / R$. Figure 4.8 shows the plots of $\log P$,[3] or $0.4343 \ln P$, against $1/T$ for liquid mercury and sodium. Actually, in treating experimental systems, we obtain straighter lines than might have been expected in view of the wide range of temperatures seen in Figure 4.8. The explanation is that the variation in $\Delta_l^g H$ with respect to temperature is relatively small, and that those variations may be partly compensated by a certain degree of non-ideality in the gas.

---

[3] $\ln x = \dfrac{\log_{10} x}{\log_{10} e} = \dfrac{\log x}{0.4343}$, where $\ln x$ means $\log_e x$.

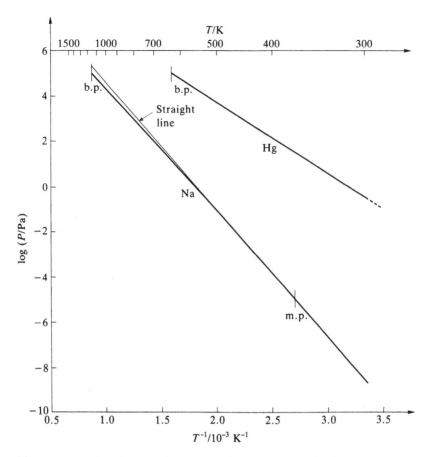

Figure 4.8 *log P, or 0.4343 lnP, vs. 1/T for liquid mercury and sodium.*

Saturated vapour pressures of metallic elements have frequently been represented by empirical formulae having the same form as Eq. (4.21), i.e. $\ln P = a - b/T$, where a and b are constants. The form of Eq. (4.21) provides a good approximation, and is useful for interpolation between tabulated values. Closer fits are obtained when a further term is added: $\ln P = a - bT + c\ln T$, where c is a constant.

Theoretical considerations for this type of formula have already been mentioned. To a good approximation, the temperature dependence of $\Delta_l^g H$ values for liquid metallic elements may be expressed by an expression of the form

$$\Delta_l^g H = \Delta_l^g H_m + \kappa (T - T_m) \qquad (4.27)$$

or

$$\left(\frac{\partial \Delta_l^g H}{\partial T}\right)_P = \kappa \qquad (4.28)$$

Values for the temperature dependence of $\Delta_l^g H$ (i.e. values for κ) appear to lie in the range of -5 to -15 J mol^{-1} K^{-1} for most liquid elements.

The temperature at which the vapour pressure of a saturated vapour is equal to one atmosphere (i.e. $P = 1$ atm $= 1.01325 \times 10^5$ Pa) is also known as the normal boiling point of a liquid T_b. We then have from Eq. (4.21) that

$$\frac{\Delta_l^g H_b}{RT_b} = \ln A \qquad (4.29)$$

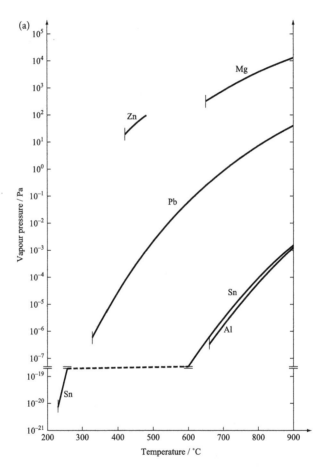

Figure 4.9 *(a) Equilibrium vapour pressures of liquid metals as a function of temperature. |, melting point. (b) Equilibrium vapour pressures of liquid metallic elements as a function of temperature. |, melting point. (c) Equilibrium vapour pressures of liquid metallic elements as a function of temperature. |, melting point.*

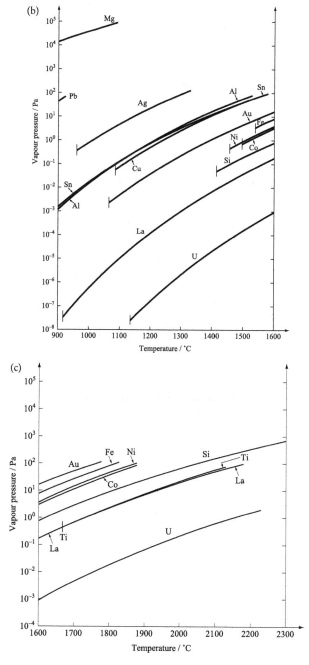

Figure 4.9 *(continued)*

This relationship is known as Trouton's rule (see Subsection 1.2.2.3). $\ln A$ is a constant that is approximately the same for all metallic elements.

To obtain evaporation enthalpies $\Delta_l^g H$, plots of $\ln P$ against $1/T$ can be constructed by measuring P at several known temperatures. The slope of such a plot is, from Eq. (4.15)

$$\text{Slope} = -\frac{\Delta_l^g H}{R} = \frac{d(\ln P)}{d(1/T)} = -\frac{T^2}{P}\frac{dP}{dT} \tag{4.30}$$

When Eq. (4.12) is combined with the above relation, we have

$$\text{Slope} = -\frac{T\,\Delta_l^g H}{P\,\Delta_l^g V} \tag{4.31}$$

4.2.4 Experimental Data for Vapour Pressure

Experimental values for the equilibrium vapour pressures of various metallic elements are shown graphically in Figures 4.9(a–c). Their governing equations are summarized in Tables 17.5(a) and (b).

Most normal metals are monatomic in the vapour form. However, the vapour species of semimetals and non-metals (e.g. phosphorus, sulphur) are often polymerized. For example, bismuth vapour contains both Bi and Bi_2 gas species. For most metals, even copper, silver, and gold, small amounts of dimer gas molecules have been observed.

4.3 Heat Capacity of Liquid Metallic Elements

4.3.1 Introduction

The heat capacity C of a substance is defined as the quantity δq of heat required to raise the temperature dT of the substance: $C = \delta q/dT$. In other words, the relation for the input of heat and temperature increase is determined by the heat capacity. The heat capacity is, therefore, a fundamental material property. For example, calculations of changes in enthalpy or entropy of a substance with temperature require knowledge of the temperature dependence of heat capacity. Similarly, the heat capacity of condensed matter is an important dominating factor for its volume expansivity.

Investigations of the heat capacities of solid elements have been made for many years. Dulong and Petit in 1819 proposed the empirical law that most solid metals have values of heat capacity lying around 26 J mol^{-1} K^{-1} (6.2 cal mol^{-1} K^{-1}) under the standard pressure of 101 325 Pa(= 1 atm) at room temperature.

The theoretical work of Einstein on the heat capacity of a solid element is well known. Einstein's theoretical approach is based on the innovative, simple assumption that each atom in a solid executes harmonic motion about a fixed point in space, or a lattice point. This is equivalent to the assumption that all atoms, arranged in the long-range order, vibrate independently. Despite this simplification, the values of heat capacity of a solid calculated by the Einstein model, using statistical thermodynamics and quantum mechanics, are in moderately good agreement with those measured.

By contrast, the behaviour, or the thermal motion, of the atoms in a liquid is much more complex, as a result of their cooperative interactions within a system of random order, or, at best, short-range order.

There are no theoretical models for the heat capacity of a metallic liquid for materials process science. (In the field of materials process science, both accuracy and universality are required of any model for predicting the thermophysical properties of liquid metallic elements.) Clearly, thermodynamics also gives no accurate predictions for the numerical values of heat capacity nor of their temperature dependence. Furthermore, no empirical relationship has been successful in providing reliable estimates of heat capacities for liquid metallic elements. Accurate experimental data for the heat capacities of metallic liquids are, therefore, essential.

4.3.2 Heat Capacity at Constant Volume and at Constant Pressure

For a given system, the value of heat capacity depends upon the constraints imposed on the system. The two cases of most common interest are for conditions of constant volume and constant pressure. These two heat capacities are defined by the following equations:

$$C_V = \frac{\delta q_V}{\mathrm{d}T} = \left(\frac{\partial U}{\partial T}\right)_V \tag{4.32}$$

$$C_P = \frac{\delta q_P}{\mathrm{d}T} = \left(\frac{\partial H}{\partial T}\right)_P = \left\{\frac{\partial(U + PV)}{\partial T}\right\}_P \tag{4.33}$$

where δq represents an infinitesimal quantity of heat added to a system, $\mathrm{d}T$ is the resulting infinitesimal temperature increase, U is the internal energy, H is the enthalpy, and subscripts V and P denote conditions of constant volume or constant pressure, respectively.

4.3.2.1 Constant-Volume Heat Capacity

We have already mentioned in Subsection 2.2.5.(2) that the heat capacity at constant volume can be expressed in terms of the pair distribution function $g(r)$ and the pair potential $\phi(r)$. On a molar basis, Eq. (2.14) becomes

$$C_V = \frac{3}{2}R + \frac{2\pi N_A^2}{V}\int_0^\infty \left[\frac{\partial g(r)}{\partial T}\right]_V \phi(r)r^2\mathrm{d}r \tag{4.34}$$

where C_V is the molar heat capacity at constant volume, N_A is the Avogadoro constant, and V is the molar volume.

Theoretical approaches to the subject are usually discussed on the basis of the condition of constant volume. Unfortunately, it is almost impossible to obtain experimental or theoretical values for $[\partial g(r)/\partial T]_V$ and $\phi(r)$.

We now discuss the heat capacity of monatomic solids in terms of atomic motion. At room temperature, classical theory is applicable to the subject of the heat capacity. If we consider that the atoms in the monatomic solids behave as harmonic oscillators about their equilibrium positions, the atoms have no translational energy, only vibrational. According to the equipartition law of energy, each vibrational degree of freedom involves both potential and kinetic energy, so that the average thermal energy is equal to kT ($= kT/2 + kT/2$) per degree of vibrational freedom. For each atom in the monatomic solids, there are three independent degrees of freedom, and each atom's three degrees of freedom can be expressed in terms of vibrational motion of the atom at a lattice point; therefore its average vibrational energy u is

$$u = 3kT \tag{4.35}$$

For one mole of the monatomic solids, this gives

$$U = N_A \times 3kT = 3RT \tag{4.36}$$

Hence, for molar heat capacity at constant volume, we have

$$C_V = \left(\frac{\partial U}{\partial T}\right)_V = 3R \cong 25\,\mathrm{Jmol^{-1}K^{-1}} \tag{4.37}$$

For an ideal gas in thermal equilibrium, each monatomic molecule has an average thermal energy of only three translations; the internal energy per molecule of the gas is $3kT/2$. The internal energy and the constant-volume heat capacity per mole are, respectively,

$$U = \frac{3}{2}RT \tag{4.38}$$

and

$$C_V = \frac{3}{2}R \tag{4.39}$$

The complexity of the thermal motion of atoms, or the interatomic potential energy, within a condensed matter system of random order (e.g. a metallic liquid) makes the numerical calculations of a liquid's heat capacity extremely difficult.

4.3.2.2 Constant-Pressure Heat Capacity

Most materials and chemical processes are carried out at constant pressure. Under constant pressure conditions and with no work other than reversible work of volume expansion or contraction,

$$\delta q_P = \mathrm{d}H = C_P \mathrm{d}T \qquad (4.40)$$

On integrating Eq. (4.40) between states (T_2, P) and (T_1, P), we have

$$\Delta H = H(T_2, P) - H(T_1, P) = \int_{T_1}^{T_2} C_P \mathrm{d}T \qquad (4.41)$$

According to the second law of thermodynamics, the corresponding increase in entropy is

$$\mathrm{d}S = \frac{\mathrm{d}H}{T} = \frac{C_P}{T} \mathrm{d}T \qquad (4.42)$$

Again, on integrating, we obtain

$$\Delta S = S(T_2, P) - S(T_1, P) = \int_{T_1}^{T_2} C_P \mathrm{d}(\ln T) \qquad (4.43)$$

ΔH and ΔS are called the enthalpy and entropy increments, respectively.

In general, heat capacity will vary with temperature. The value of ΔH is obtained by plotting C_P against T and evaluating the area under the curve; similarly the value of ΔS is calculated by plotting C_P against $\ln T$ and evaluating the area under the resulting curve.

In a phase transition, the heat capacity of a substance may become formally infinite, since heat is incorporated without leading to a change of temperature.

The values of the heat capacity of liquid alloys have often been estimated by proportional addition of the heat capacities of the constituent elements; this is known as the Neumann–Kopp law.

4.3.2.3 *Relation between C_P and C_V*

C_P and C_V are interrelated according to the following thermodynamic relation:

$$C_P - C_V = \frac{\alpha^2 V T}{\kappa_T} \qquad (4.44)$$

where α is the volume expansivity $(1/V)(\partial V / \partial T)_P$ and κ_T is the isothermal compressibility $-(1/V)(\partial V / \partial P)_T$. Equation (4.44) is useful, since, if we are in possession of any five of these measurable quantities, the sixth can be deduced.

From Eq. (4.44), we have for one mole of an ideal gas

$$C_P - C_V = R \qquad (4.45)$$

4.3.3   Empirical Representation of Heat Capacity

Accurate data for the heat capacities of liquid metallic elements are not abundant. The variation in the heat capacities of most liquid metallic elements with temperature appears to be small. Moreover, there are insufficient experimental data available to allow the temperature dependence of C_P to be determined. Generally speaking, the influence of temperature on the heat capacities of liquid metallic elements tends to be small. For most liquid metallic elements, therefore, the values of constant-pressure heat capacity are assumed to be constant over relatively wide ranges of temperature.

The temperature dependence of the heat capacity C_P of liquid metallic elements, like that of solid elements, can be represented by the following empirical equation.

$$C_P = a + bT + cT^{-2} + dT^2 \tag{4.46}$$

in which a, b, c, and d are constants. Figures 4.10(a) and (b) show the variation of C_P with T for several liquid metallic elements.

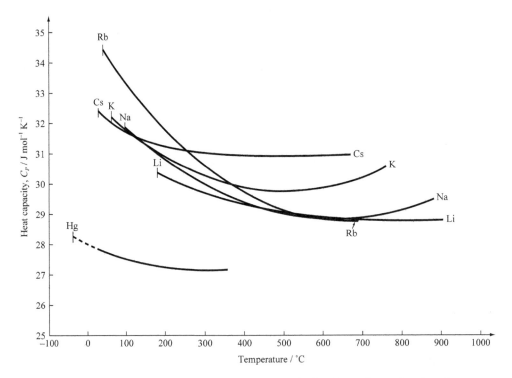

Figure 4.10 *(a) Molar heat capacities at constant pressure for liquid metals as a function of temperature. |, melting point. (b) Molar heat capacities at constant pressure for liquid metallic elements as a function of temperature. |, melting point.*

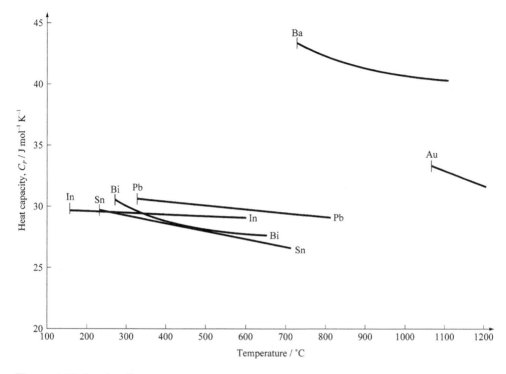

Figure 4.10 *(continued)*

According to experimentally derived data, liquid metallic elements have C_P values ranging between 25 and 50 J mol^{-1} K^{-1}. Values of constant-pressure heat capacity for a large number of liquid metallic elements are listed in Table 17.6(a) and (b).

....

REFERENCES

1. O. Kubaschewski and C.B. Alcock, *Metallurgical Thermochemistry*, 5th ed., Pergamon Press, Oxford, 1979.
2. O. Kubaschewski, C.B. Alcock, and P.J. Spencer, *Materials Thermochemistry*, 6th ed., Pergamon Press, Oxford, 1993.
3. K.C. Mills, in *Fundamentals of Metallurgy*, edited by S. Seetharaman, Woodhead Publishing, Cambridge, 2005.
4. H. Fukuyama and Y. Waseda (eds.), *High-Temperature Measurements of Materials*, Springer, Berlin, 2009.
5. C. Kittel, *Introduction to Solid State Physics*, 7th ed., John Wiley & Sons, Inc., 1996, p.57.

5

Velocity of Sound

5.1 Introduction

The velocity of sound in liquid metallic elements is one of the most basic thermo-dynamic properties. Sound velocity data can provide much useful information on the properties and behaviour of liquid metallic elements. The values of isentropic compress-ibility (sometimes referred to as adiabatic compressibility), isothermal compressibility, and constant-volume heat capacity of liquid metallic elements, can all be readily deter-mined using well-known thermodynamic formulae, provided that sound velocities can be measured (or sound velocity data are given), and that data for their constant-pressure heat capacities, densities, and volume expansivities are available. In addition, in the last several years or so, a new approach to the velocity of sound in liquid metallic elem-ents, based on the theory of liquids, has been developed in the area of materials process science. In this new approach, a new dimensionless parameter (denoted by ξ) was intro-duced as a correction factor to include in expressions for the velocity of sound in a liquid metallic element. This new dimensionless parameter is element specific, and gives an indication of an atom's hardness or softness as well as atomic arrangement, or me-tallic liquid structure. Furthermore, the new parameter is also useful in discussions of anharmonic effects of atomic motion in liquid metallic elements. These physical con-cepts of the nature and behaviour of liquid metallic atoms are important considerations that have been neglected in all previous models of metallic liquids used in the field of materials process science. Several thermophysical properties of liquid metallic elements such as thermal expansion, evaporation enthalpy, surface tension, and viscosity, can be easily expressed in terms of the new parameter, as revealed through data for the velocity of sound; this approach allows for much better calculations, or predictions, for these thermophysical properties at high temperatures.

Blairs [1] published a review of the velocity of sound in liquid metallic elements in 2007. This review comprises theoretical models or equations, semi-empirical relations, references on experimental methods, and experimental data for the velocity of sound in liquid metallic elements. As such, it is valuable for materials process scientists. In this chapter, therefore, we place special emphasis on the recently developed new approach to treating the velocity of sound in liquid metallic elements.

The Thermophysical Properties of Metallic Liquids: Volume 1 – Fundamentals. First Edition.
Takamichi Iida and Roderick I. L. Guthrie. © Takamichi Iida and Roderick I. L. Guthrie 2015.
Published in 2015 by Oxford University Press.

5.2 Thermodynamic Relationship between Sound Velocity and Compressibility

The relationship between the velocity of sound U and the isentropic compressibility $\kappa_S (\equiv -V^{-1}(\partial V/\partial P)_S)$ is given by

$$U = \left(\frac{1}{\rho \kappa_S}\right)^{1/2} = \left(\frac{B_S}{\rho}\right)^{1/2} \tag{5.1}$$

in which ρ is the density of the medium and B_S is the isentropic bulk modulus. This equation forms the basis for determining κ_S or B_S from measurements of the velocity of sound.

The isothermal compressibility κ_T, the constant-volume heat capacity C_V, and the ratio of the heat capacities $\gamma_h (\equiv C_P/C_V)$ can be calculated through the thermodynamic relations:

$$\frac{\kappa_T}{\kappa_S} = \frac{C_P}{C_V} \tag{5.2}$$

$$\frac{C_P}{C_V} = 1 + \frac{\alpha^2 VT}{\kappa_S C_P} \tag{5.3}$$

where α is the volume expansivity, or the thermal expansivity. It is difficult to determine values for the isothermal compressibility and constant-volume heat capacity experimentally, but their values can be computed using these formulae together with sound velocity measurements.

Incidentally, the isothermal compressibility can be expressed in terms of the pair distribution function. By working with a grand canonical ensemble and considering the fluctuations in the number of atoms in a given volume, the isothermal compressibility can be formulated, in the form

$$\kappa_T = \frac{1}{n_0 kT}\left[1 + n_0 \int_0^\infty \{g(r) - 1\}\, 4\pi r^2 dr\right] \tag{5.4}$$

Equation (5.4) shows that κ_T is expressed in terms of only n_0 and $g(r)$ (i.e. no data for the pair potential $\phi(r)$ is needed). The isothermal compressibility is also related to the structure factor $S(0)$ (i.e. the structure factor $S(Q)$ as $Q \to 0$, where $Q = 4\pi \sin\theta/\lambda$) as follows:

$$\kappa_T = \frac{1}{n_0 kT} S(0) \tag{5.5a}$$

or

$$S(0) = n_0 kT \kappa_T \tag{5.5b}$$

At present, the accuracy of experimental data in the low Q region is not necessarily satisfactory for researchers in materials process science.

5.3 Theoretical Equations for the Velocity of Sound in Liquid Metallic Elements

5.3.1 The Jellium Model

According to Bohm and Staver's considerations of the jellium model, which assumes the presence of a free non-interacting electron gas, the velocity of sound in condensed metal phases can be expressed by [2]

$$U = \left(\frac{2zE_F}{3m}\right)^{1/2} \tag{5.6}$$

where z is the number of valence electrons per atom, E_F is the free electron value of the Fermi energy, and m is the atomic mass (i.e. the mass of an atom, or an ion).

Computed and measured values for the velocity of sound in liquid metallic elements at their melting points, and the corresponding δ_i, Δ, and S values needed for statistical assessment of the model are given in Tables 5.1 and 5.2, respectively.

5.3.2 The Hard-Sphere Model

Ascarelli [3] proposed an equation for the velocity of sound in liquid metallic elements based on a model of hard spheres immersed in a uniform background potential. His final expression for the velocity of sound U is:

$$U = \left[\frac{1}{m}\frac{C_P}{C_V}kT\left\{\frac{(1+2\eta)^2}{(1-\eta)^4} + \frac{2}{3}\frac{zE_F}{kT} - A\left(\frac{V_m}{V}\right)^{1/3}\frac{4}{3}\frac{kT_m}{kT}\right\}\right]^{1/2} \tag{5.7}$$

$$A \equiv 10 + \frac{2}{5}zE_F(T_m)/kT_m$$

where η is the packing fraction and subscript m denotes the melting point. In calculating melting point sound velocities, Ascarelli used a value $\eta_m = 0.45$ for all liquid metallic elements. At the melting point of a liquid metallic element (when $\eta_m = 0.45$), Eq. (5.7) becomes

$$U_m = \left[\frac{1}{m}\frac{C_P}{C_V}kT_m\left\{27 + \frac{1}{5}\left(\frac{2}{3}\frac{zE_F}{kT_m}\right)\right\}\right]^{1/2} \tag{5.8}$$

A comparison of measured values for the velocity of sound in liquid metallic elements with those calculated using the Ascarelli model represented by Eq. (5.8), and the corresponding δ_i, Δ, and S values, are also given in Tables 5.1 and 5.2, respectively. It is evident from Tables 5.1 and 5.2 that Ascarelli's results represent a remarkable improvement over those of the Bohm and Staver model.

Ascarelli has also provided calculated values for the temperature dependence of the velocity of sound. These are: for rubidium, −0.3; for zinc, −0.20 (low-temperature value), −0.6 (high-temperature value); for indium, −0.14; for tin, −0.15 m s⁻¹ K⁻¹,

Table 5.1 *Comparison of experimental and Ascarelli's calculated values for the velocity of sound in liquid metallic elements at their melting points.*

Element		$(U_m)_{exp}$ m s^{-1}	$(U_m)_{cal}$, m s^{-1}	
			BS[†]	A[‡]
Aluminium	Al	4680	8750	4900
Antimony	Sb	1910	5340	3150
Bismuth	Bi	1640	3940	2080
Cadmium	Cd	2237	2850	1840
Caesium	Cs	983	880	890
Copper	Cu	3440	2580	2700
Gallium	Ga	2873	5430	2850
Indium	In	2320	3760	2041
Lead	Pb	1821	3350	1900
Mercury	Hg	1511	2100	1220
Potassium	K	1876	1810	1720
Rubidium	Rb	1251	1140	1103
Silver	Ag	2790	1740	1920
Sodium	Na	2526	2960	2500
Thallium	Tl	1650	2760	1580
Tin	Sn	2464	4630	2440
Zinc	Zn	2850	4180	2610

[†] Bohm and Staver's model represented by Eq. (5.6).
[‡] Ascarelli's model represented by Eq. (5.8); C_P/C_V has been taken as equal to 1.15 for all metallic elements [3].

respectively. Considering the simplicity of the model, agreement with experimental data is fairly good (see Table 17.7).

Using a hard-sphere model, Rosenfeld [4] has discussed the velocity of sound and its temperature dependence. Rosenfeld pointed out that the sound velocity U divided by the thermal velocity, i.e. $(kT/m)^{1/2}$, has about the same value for many liquid metallic elements near their melting points. Typically,

$$U \left/ \left(\frac{kT}{m}\right)^{1/2} \right|_{T=T_m} \approx 10 \qquad (5.9)$$

Table 5.2 *Values of δ_i, Δ, and S obtained from four models for the velocity of sound in liquid metallic elements at their melting points.*

Element		BS[†]	A[‡]	Eq. (5.22)	Eq. (5.40)
				δ_i, %	
Aluminium	Al	−46.5	−4.5	−11.3	14.3
Antimony	Sb	−64.2	−39.4	9.3	−12.3
Bismuth	Bi	−58.4	−21.2	17.1	−9.4
Cadmium	Cd	−21.5	21.6	3.5	−13.1
Caesium	Cs	11.7	10.4	−13.0	−9.6
Copper	Cu	33.3	27.4	0.4	−12.8
Gallium	Ga	−47.1	0.8	5.1	−10.7
Indium	In	−38.3	13.7	10.0	−2.3
Lead	Pb	−45.6	−4.2	21.1	−3.0
Mercury	Hg	−28.0	23.9	4.9	−30.6
Potassium	K	3.6	9.1	−17.9	2.9
Rubidium	Rb	9.7	13.4	−16.1	−7.9
Silver	Ag	60.3	45.3	11.6	−2.4
Sodium	Na	−14.7	1.0	−21.5	−12.2
Thallium	Tl	−40.2	4.4	11.5	−15.6
Tin	Sn	−46.8	1.0	17.6	6.9
Zinc	Zn	−31.8	9.2	1.3	−18.5
	$\Delta(17)$ %	35.4	14.7	11.4	10.9
	S (17)	0.397	0.196	0.131	0.128

[†] Bohm and Staver's model represented by Eq. (5.6).
[‡] Ascarelli's model represented by Eq. (5.8).

with a spread in values between ∼6 and ∼12. Except for a few anomalous cases like antimony, tellurium, and cerium, the sound velocity decreases very slowly with temperature, and typically,

$$\left.\frac{\partial \ln U}{\partial \ln T}\right|_{T \approx T_m} = \left.\frac{T \partial U}{U \partial T}\right|_{T \approx T_m} \approx -0.2 \qquad (5.10)$$

with a spread in values between ∼ −0.1 and ∼ −0.3.

On the basis of the hard-sphere model, Rosenfeld has given a reasonable explanation for the relations represented by Eqs. (5.9) and (5.10).

Yokoyama [5, 6] has developed Ascarelli's and Rosenfeld's approaches, and has discussed the sound velocity and its temperature dependence for the liquid alkali and lanthanoid (lanthanum, cerium, and praseodymium) metals.

5.3.3   Statistical Mechanical Considerations (the Gitis–Mikhailov Model)

Using arguments involving the statistical mechanical theory of liquids, Gitis and Mikhailov [7] have derived another, simple expression for the velocity of sound, in terms of the molar cohesive energy, E_c, and the molar mass, M, as follows:

$$U = \left(\frac{2E_c}{M}\right)^{1/2} \tag{5.11}$$

where E_c is given by (see Subsections 2.2.5 and 4.1.2)

$$E_c = \frac{2\pi N_A^2}{V} \int_0^\infty g(r)\phi(r)r^2 dr \tag{5.12}$$

in which V is the molar volume.

We repeat here that the molar cohesive energy is approximately equal to the molar evaporation enthalpy, $\Delta_l^g H$, namely

$$\Delta_l^g H = RT - \frac{2\pi N_A^2}{V} \int_0^\infty g(r)\phi(r)r^2 dr = RT - E_c \approx -E_c, \text{ since } RT \ll |E_c| \tag{5.13}$$

To a good approximation, E_c is numerically equal to $\Delta_l^g H$. Thus, Eq. (5.11) may be written as

$$U = \left(\frac{2\Delta_l^g H}{M}\right)^{1/2} \tag{5.14}$$

Gitis and Mikhailov [8] have proposed the following correlation between the velocity of sound and a liquid metal's electrical conductivity σ_e:

$$\frac{1}{\sigma_e}\frac{d\sigma_e}{dT} = \frac{2}{U}\frac{dU}{dT} + \frac{2}{3\rho}\frac{d\rho}{dT} \tag{5.15}$$

On the basis of measured values for U and ρ, they determined temperature coefficients of electrical conductivity $(1/\sigma_e)(d\sigma_e/dT)$ for eight liquid metals using Eq. (5.15). In most cases, their calculations yielded results which were about 15 to 20 per cent lower than experimental values.

5.4 Semi-Empirical Models for the Velocity of Sound in Liquid Metallic Elements

5.4.1 The Einstein–Lindemann Model

According to Einstein [9], a simple relationship exists between the velocity of sound in solids and the mean frequency of atomic vibration ν. His relation can be written as

$$U \propto \nu V^{1/3} \tag{5.16}$$

in which V is the molar volume. By using Lindemann's formula, represented by Eq. (1.16), for the mean atomic frequency, we have

$$U_{\mathrm{m}} \propto \left(\frac{RT_{\mathrm{m}}}{M}\right)^{1/2} \tag{5.17}$$

where R is the molar gas constant. This relationship shows that the velocity of sound in solids at their melting point temperatures U_{m} is proportional to the square root of the metallic element's melting point and inversely proportional to the square root of the metallic element's molar mass. This expression has been known for a long time. Equation (5.17) shall be called the Einstein–Lindemann model (for the velocity of sound). Incidentally, as already mentioned, the relation represented by Eq. (5.17) can also be interpreted successfully on the basis of the hard-sphere model (i.e. Eq. (5.9))

Equation (5.17) can be expected to be roughly valid for liquid metallic elements, owing to the similarity of their sound velocities with metallic elements in their solid state. Figure 5.1 gives the relation represented by Eq. (5.17) for 41 liquid metallic elements, plus sulphur. (At present, experimental sound velocity data for liquid metallic elements only appear to be available for these 41 simple substances.) As can be seen, Figure 5.1 shows a reasonably linear correlation between the two variables, i.e. U_{m} and $(RT_{\mathrm{m}}/M)^{1/2}$; this can be expressed as

$$U_{\mathrm{m}} = 9.197 \left(\frac{RT_{\mathrm{m}}}{M}\right)^{1/2} \tag{5.18}$$

in which the proportionality factor of 9.197 (dimensionless) was determined so as to give the minimum S values for the 41 liquid metallic elements plotted in Figure 5.1. Table 5.3 lists a comparison between the experimental and calculated values based on Eq. (5.18) for the velocity of sound in liquid metallic elements at their melting point temperatures, together with δ_i, Δ, and S values. The Einstein–Lindemann model performs reasonably with $\Delta(41)$ and $S(41)$ values of 24.8 per cent and 0.293, respectively.

5.4.2 Modified Einstein–Lindemann Model

As already mentioned, Lindemann's melting model given by Eq. (1.16) provides only rough values for the average vibrational frequency of atoms in the solid state at the

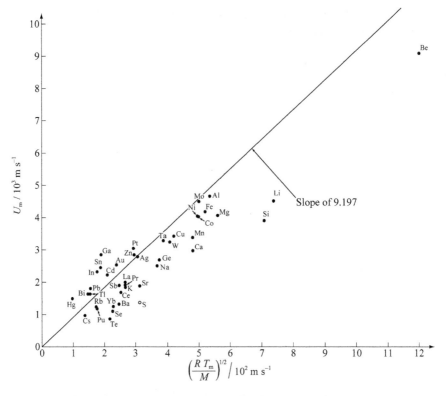

Figure 5.1 *The velocity of sound in liquid metallic elements at their melting point temperatures as a function of $(RT_m/M)^{1/2}$, i.e. the Einstein–Lindemann model.*

melting point. In the semi-empirical treatment of Iida and Guthrie [10] for the surface tension of liquid metals, it was proposed that Lindemann's formula needed correcting for the mean frequency of atoms in the liquid state. At the melting point, the mean atomic frequency v_l of liquid metallic elements can be expressed in the form (see Subsection 6.3.2)

$$v_l = \beta v_L = 3.1 \times 10^8 \beta \left(\frac{RT_m}{MV_m^{2/3}} \right)^{1/2} \tag{6.21}$$

where v_L is the atomic frequency which is calculated from Lindemann's formula by Eq. (1.16), and the correction factor β is given by (see Subsection 6.3.2)

$$\beta = \frac{1.13 \times 10^3 V_m^{1/3}}{\alpha} \left(\frac{\gamma_m}{RT_m} \right)^{1/2} \tag{6.35}$$

Table 5.3 *Comparison of experimental values for the velocity of sound in liquid metallic elements at their melting points with those calculated using the Einstein–Lindemann model, together with δ_i, Δ, and S values.*

Element		(U_m), m s^{-1}		δ_i %
		Experimental	Calculated	
Aluminium	Al	4680	4932	−5.1
Antimony	Sb	1910	2284	−16.4
Barium	Ba	1331	2262	−40.2
Beryllium	Be	9104	11034	−17.5
Bismuth	Bi	1640	1354	21.1
Cadmium	Cd	2237	1928	16.0
Caesium	Cs	983	1264	−22.2
Calcium	Ca	2978	4423	−32.7
Cerium	Ce	1693	2319	−27.0
Cobalt	Co	4031	4593	−12.2
Copper	Cu	3440	3877	−11.3
Gallium	Ga	2873	1748	64.4
Germanium	Ge	2693	3428	−21.4
Gold	Au	2568	2185	17.5
Indium	In	2320	1622	43.0
Iron	Fe	4200	4775	−12.0
Lanthanum	La	2022	2455	−17.6
Lead	Pb	1821	1427	27.6
Lithium	Li	4554	6780	−32.8
Magnesium	Mg	4065	5167	−21.3
Manganese	Mn	3381	4409	−23.3
Mercury	Hg	1511	906	66.8
Molybdenum	Mo	4502	4608	−2.3
Nickel	Ni	4047	4551	−11.1
Platinum	Pt	3053	2713	12.5

continued

Table 5.3 *(continued)*

Element		(U_m), m s^{-1}		δ_i %
		Experimental	Calculated	
Plutonium	Pu	1195	1622	−26.3
Potassium	K	1876	2460	−23.7
Praseodymium	Pr	1925	2451	−21.5
Rubidium	Rb	1251	1604	−22.0
Selenium	Se	1100	2097	−47.5
Silicon	Si	3920	6495	−39.6
Silver	Ag	2790	2837	−1.7
Sodium	Na	2526	3368	−25.0
Strontium	Sr	1902	2903	−34.5
Tantalum	Ta	3303	3576	−7.6
Tellurium	Te	889	1996	−55.5
Thallium	Tl	1650	1409	17.1
Tin	Sn	2464	1730	42.4
Tungsten	W	3279	3760	−12.8
Ytterbium	Yb	1274	2107	−39.5
Zinc	Zn	2850	2730	4.4
			$\Delta(41)$ %	24.8
			$S(41)$	0.293

where

$$\underline{\alpha} = \left(\frac{\rho_m}{\rho_c}\right)^{1/3} - 1 = \left(\frac{V_c}{V_m}\right)^{1/3} - 1 \approx 1.97\eta_m^{1/3} - 1$$

and γ_m is the surface tension of the liquid metallic element at its melting point. The values of ρ_c or V_c have been determined experimentally for only a few metals. According to Young and Alder [11, 12], the atomic volume at the critical point V_c can be expressed in terms of the effective hard-sphere diameter σ, based on the hard-sphere model:

$$V_c = 2.417 \times 10^{24}\sigma^3 \tag{5.19}$$

Combination of Eqs. (1.19)[1] and (5.19) gives $(V_c/V_m)^{1/3} \approx 1.97\eta_m^{1/3}$. From Young and Alder's consideration, which was based on the van der Waals model (though this model is not suitable for metals), α is approximately equal to $(1.97\eta_m^{1/3} - 1)$. The packing fraction at the melting point $\eta_m = 0.463$ is a good approximation for all liquid metals [5, 6]. Combining Eq. (6.21) for the mean frequency of atomic vibrations in the liquid state with Eq. (5.16), at the melting point we have

$$U_m \propto \beta \left(\frac{RT_m}{M} \right)^{1/2} \tag{5.20}$$

Substituting Eq. (6.35) into Eq. (5.20), we obtain[2]

$$U_m \propto \left(\frac{\gamma_m}{M} \right)^{1/2} V_m^{1/3} \tag{5.21}$$

Equation (5.20) or (5.21) shall be called a modified Einstein–Lindemann model.

Figure 5.2 shows a plot of U_m against $(\gamma_m/M)^{1/2} V_m^{1/3}$ for 41 metallic elements, and shows a good linear relationship between the two variables; the slope (or numerical factor) of 3.768×10^4 mol$^{-1/6}$ was determined so as to give the minimum S value for the 41 liquid metallic elements considered. Thus, we have

$$U_m = 3.768 \times 10^4 \left(\frac{\gamma_m}{M} \right)^{1/2} V_m^{1/3} \tag{5.22}$$

From thermodynamic relations represented by Eqs. (5.1), (5.2), and (5.5), a similar correlation to Eq. (5.20) is obtained in liquid metallic elements at their melting point temperatures, as follows:

$$U_m = \left\{ \frac{\gamma_h}{S(0)} \right\}^{1/2} \left(\frac{RT_m}{M} \right)^{1/2} \tag{5.23}$$

Comparing Eqs. (5.20) and (5.23), we obtain

$$\beta \propto \left\{ \frac{\gamma_h}{S(0)} \right\}^{1/2} \tag{5.24}$$

Consequently, we expect that, to a good approximation, Eq. (5.22) will hold for the velocity of sound U_m in liquid metallic elements.

---

[1] At the melting point, Eq. (1.19) becomes

$$\eta_m = \frac{4}{3} \pi \left(\frac{\sigma}{2} \right)^3 \frac{N_A}{V_m}$$

where V is the molar volume, N_A is the Avogadro constant, and subscript m denotes the melting point.
[2] $\alpha \approx 0.524$ for all liquid metals.

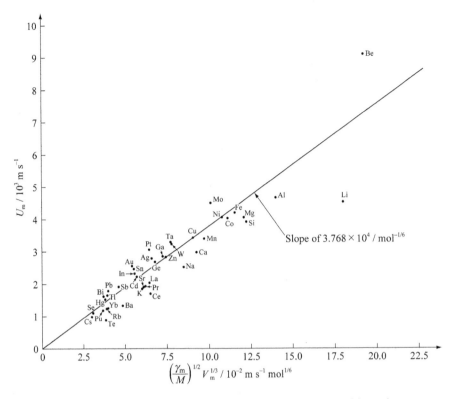

Figure 5.2 *The velocity of sound in liquid metallic elements at their melting point temperatures as a function of $(\gamma_m/M)^{1/2}V_m^{1/3}$, i.e. the modified Einstein–Lindemann model.*

5.5 Equations for the Velocity of Sound in Terms of New Dimensionless Parameters ξ_E and ξ_T

5.5.1 The Gitis–Mikhailov Model in Terms of a New Dimensionless Parameter ξ_E (a Corrected Gitis–Mikhailov Model)

Let us now assess the performance of the Gitis–Mikhailov model. Since the temperature dependence of the evaporation enthalpies of liquid metallic elements is negligibly small, Eq. (5.14) can be written as

$$U_m = \left(\frac{2\Delta_l^g H_b}{M} \right)^{1/2} \tag{5.25}$$

where $\Delta_l^g H_b$ is the evaporation enthalpy at the boiling point temperature.

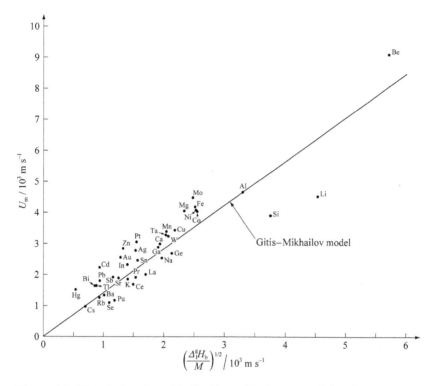

Figure 5.3 *The velocity of sound in liquid metallic elements at their melting point temperatures as a function of* $(\Delta_l^g H_b/M)^{1/2}$*, i.e. the Gitis–Mikhailov model.*

In Figure 5.3, experimental sound velocities in liquid metallic elements at their melting point temperatures are shown as a function of $(\Delta_l^g H_b/M)^{1/2}$ for 39 metallic simple substances. We see that the Gitis–Mikhailov model is approximately true for various liquid metallic elements. However, the periodic Group IIB metals (i.e. zinc group metals) show large deviations from the line of proportionality. Table 5.4 compares experimental values for the sound velocity in 39 liquid metallic elements for which experimental data on both U_m and $\Delta_l^g H_b$ are available, with those calculated from the Gitis–Mikhailov model represented by Eq. (5.25), as well as corresponding data for δ_i, Δ, and S. As is obvious from this table, the Gitis–Mikhailov model performs reasonably with $\Delta(39)$ and $S(39)$ values of 24.7 per cent and 0.291, respectively; these values are approximately the same as those calculated from the Einstein–Lindemann model.

From the microscopic point of view, sound velocity and compressibility are physical properties which are related to the curvature of the interatomic potential energy curve. Consequently, the velocity of sound in metallic liquids should be linked to both repulsive and attractive energies between metallic atoms.[3] Iida et al. [13] have, in this way, made a

---

[3] Sound in fluids, i.e. liquids and gases, travels only as longitudinal waves, or compressional waves. The longitudinal wave is the wave in which the propagation vector and the displacement of the individual oscillators are parallel to each other.

Table 5.4 *Comparison of experimental values for the velocity of sound in liquid metallic elements at their melting points with those calculated using the Gitis–Mikhailov model, together with δ_i, Δ, and S values.*

Element		(U_m), m s^{-1}		δ_i %
		Experimental	Calculated	
Aluminium	Al	4680	4668	0.3
Antimony	Sb	1910	1646	16.0
Barium	Ba	1331	1483	−10.2
Beryllium	Be	9104	8078	12.7
Bismuth	Bi	1640	1206	36.0
Cadmium	Cd	2237	1334	67.7
Caesium	Cs	983	996	−1.3
Calcium	Ca	2978	2736	8.8
Cerium	Ce	1693	2117	−20.0
Cobalt	Co	4031	3605	11.8
Copper	Cu	3440	3093	11.2
Gallium	Ga	2873	2699	6.4
Germanium	Ge	2693	3029	−11.1
Gold	Au	2568	1817	41.3
Indium	In	2320	1984	16.9
Iron	Fe	4200	3560	18.0
Lanthanum	La	2022	2400	−15.8
Lead	Pb	1821	1318	38.2
Lithium	Li	4554	6397	−28.8
Magnesium	Mg	4065	3295	23.4
Manganese	Mn	3381	2894	16.8
Mercury	Hg	1511	755	100.1
Molybdenum	Mo	4502	3507	28.4
Nickel	Ni	4047	3599	12.4
Platinum	Pt	3053	2193	39.2

Table 5.4 *(continued)*

Element		(U_m), m s^{-1}		δ_i %
		Experimental	Calculated	
Plutonium	Pu	1195	1679	−28.8
Potassium	K	1876	1990	−5.7
Praseodymium	Pr	1925	2174	−11.5
Rubidium	Rb	1251	1327	−5.7
Selenium	Se	1100	1500	−26.7
Silicon	Si	3920	5303	−26.1
Silver	Ag	2790	2166	28.8
Sodium	Na	2526	2784	−9.3
Strontium	Sr	1902	1781	6.8
Tantalum	Ta	3303	2885	14.5
Thallium	Tl	1650	1259	31.1
Tin	Sn	2464	2214	11.3
Tungsten	W	3279	2950	11.2
Zinc	Zn	2850	1875	52.0
			$\Delta(39)$ %	24.7
			$S(39)$	0.291

correction to the Gitis–Mikhailov model; at the melting point temperature, the velocity of sound U_m is

$$U_m = \left(\frac{2\xi_E \Delta_l^g H_b}{M} \right)^{1/2} \tag{5.26a}$$

where ξ_E is the common parameter, or the correction factor, introduced by Iida et al. Values for $\xi_E^{1/2}$ can be calculated on the basis of Eq. (5.26a) using experimental sound velocity data, namely

$$\xi_E^{1/2} = \left(\frac{M}{2\Delta_g^l H_b} \right)^{1/2} U_m \tag{5.26b}$$

Calculated values for $\xi_E^{1/2}$ at the melting points of liquid metallic elements are listed in Table 5.5. Their experimental data for the enthalpy of evaporation and the velocity of sound are given in Tables 17.3 and 17.7, respectively.

Table 5.5 *Values of the dimensionless parameters $\xi_E^{1/2}$ and $\xi_T^{1/2}$ of liquid metallic elements at their melting points, together with the repulsive exponent n determined by Matsuda and Hiwatari [12, 21].*

Element		$\xi_E^{1/2}$	$\xi_T^{1/2}$	n
Aluminium	Al	1.002	0.949	6.74
Antimony	Sb	1.160	0.836	–
Barium	Ba	0.898	0.588	–
Beryllium	Be	1.127	0.825	–
Bismuth	Bi	1.360	1.211	–
Cadmium	Cd	1.677	1.160	16.2
Caesium	Cs	0.987	0.778	5.04
Calcium	Ca	1.088	0.673	6.78
Cerium	Ce	0.800	0.730	–
Cobalt	Co	1.118	0.878	8.65
Copper	Cu	1.112	0.887	8.27
Gallium	Ga	1.064	1.644	7.45
Germanium	Ge	0.889	0.787	8.51
Gold	Au	1.414	1.175	14.5
Indium	In	1.169	1.430	8.08
Iron	Fe	1.180	0.880	8.60
Lanthanum	La	0.843	0.823	–
Lead	Pb	1.382	1.275	12.0
Lithium	Li	0.712	0.672	2.83
Magnesium	Mg	1.234	0.787	10.0
Manganese[†]	Mn	1.168	0.767	4.70
Mercury	Hg	2.001	1.667	18.5
Molybdenum	Mo	1.284	0.977	–
Nickel	Ni	1.125	0.889	8.60
Platinum	Pt	1.392	1.125	–
Plutonium	Pu	0.712	0.737	–
Potassium	K	0.943	0.763	4.50
Praseodymium	Pr	0.885	0.785	–
Rubidium	Rb	0.943	0.780	5.90
Selenium	Se	0.710	0.525	–

Table 5.5 *(continued)*

Element		$\xi_E^{1/2}$	$\xi_T^{1/2}$	n
Silicon	Si	0.739	0.604	–
Silver	Ag	1.288	0.983	10.8
Sodium	Na	0.907	0.750	4.33
Strontium	Sr	1.068	0.655	7.19
Sulphur[‡]	S	–	0.467	–
Tantalum	Ta	1.145	0.924	–
Tellurium	Te	–	0.445	–
Thallium	Tl	1.310	1.171	10.2
Tin	Sn	1.113	1.424	8.80(w)
Tungsten	W	1.111	0.872	–
Ytterbium	Yb	–	0.605	–
Zinc	Zn	1.520	1.044	12.7

[†] An experimental U_m value of 3381 m s^{-1} due to Blairs [1] has been adopted.
[‡] An experimental U_m value of 1360 m s^{-1} has been reported [10] (see also Figure 5.11).

About a quarter of a century ago, Iida and Guthrie [10] pointed out that a simple linear relationship exists between the correction factor ξ_E of the Gitis–Mikhailov model for the velocity of sound in liquid metals and the repulsive exponent n of the pair interaction potential for a classical particle system, i.e. $\xi_E \approx 0.15n$. At the beginning of this century, this issue was discussed further by Iida et al. [13–20].

As already mentioned, on the basis of the pair theory of liquids, their thermophysical properties can be formulated in terms of the pair distribution function and the pair potential. Toda et al. [12] and Matsuda and Hiwatari [21] chose the following pair potential $\phi(r)$:

$$\phi(r) = \varepsilon \left(\frac{a}{r}\right)^n - b\delta^3 \exp(-\delta r) \tag{5.27}$$

in which r is the radial distance between any two atoms, and $\varepsilon > 0$, $a > 0$, $n > 3$, $b \geq 0$, $\delta > 0$. The first and second terms of the right-hand side of Eq. (5.27) are called the inverse power and Kac potentials, respectively. If the repulsive exponent n tends to infinity, the inverse power potential becomes that of the hard sphere. Matsuda and Hiwatari calculated values for the repulsive exponent n through the use of experimental data for the bulk modulus (i.e. the reciprocal of compressibility) at 0 K. The results of their calculations for various metallic elements are also given in Table 5.5. Figure 5.4 shows a plot of $\xi_E^{1/2}$ against n for 25 metallic elements. As expected, a good linear relationship exists between the two variables. Incidentally, the accuracy of the experimental sound velocity data is relatively good compared with accuracies for the other thermodynamic

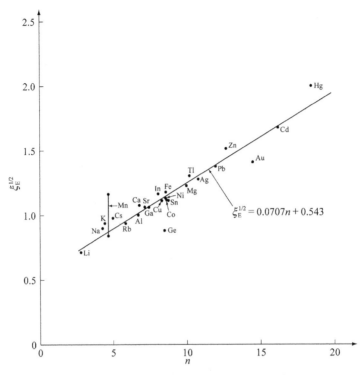

Figure 5.4 *Relationship between the dimensionless parameter, or the correction factor, $\xi_E^{1/2}$ (appearing in Eq. (5.26)) and the repulsive exponent n in the pair potential (appearing in Eq. (5.27)) for various metallic elements.*

and physical properties of liquid metallic elements. It appears reasonable to estimate that the uncertainties in the experimental sound velocity values lie approximately in the range of 2 to 10 per cent. Nevertheless, two very different experimental sound velocity values of 2442 and 3381 m s^{-1} have been reported for liquid manganese [1]. In Figure 5.4, the points linked by a vertical line represent values calculated using the two different sound velocities obtained for liquid manganese. The method of least squares for the other 24 data points, i.e. liquid metallic elements, excluding manganese, gives

$$\xi_E^{1/2} = 0.0707n + 0.543, \quad (n > 3) \tag{5.28}$$

Table 5.6 lists values for $\xi_E^{1/2}$ calculated from Eqs. (5.26b) and (5.28), and shows that Eq. (5.28) works extremely well with $\Delta(24)$ and $S(24)$ values of 4.8 per cent and 0.067, respectively, for the various liquid metallic elements considered. As can be seen from Figure 5.4 and Table 5.6, however, germanium (a semiconductor) has a large negative δ value.

Table 5.6 *Comparison of $\xi_E^{1/2}$ values calculated from Eqs. (5.26b) and (5.28), and of $\xi_T^{1/2}$ values from Eqs. (5.32b) and (5.33), together with δ_i, Δ, and S values.*

Element		$\xi_E^{1/2}$			$\xi_T^{1/2}$		
		Eq. (5.26b)	Eq. (5.28)	δ_i, pct	Eq. (5.32b)	Eq. (5.33)	δ_i, %
Aluminium	Al	1.002	1.020	−1.8	0.949	0.808	17.5
Cadmium	Cd	1.677	1.688	−0.7	1.160	1.313	−11.7
Caesium	Cs	0.987	0.899	9.8	0.778	0.717	8.5
Calcium	Ca	1.088	1.022	6.5	0.673	0.810	−16.9
Cobalt	Co	1.118	1.155	−3.2	0.878	0.910	−3.5
Copper	Cu	1.112	1.128	−1.4	0.887	0.890	−0.3
Gallium	Ga	1.064	1.070	−0.6	1.644	−	−
Germanium	Ge	0.889	1.145	−22.4	0.787	0.902	−12.7
Gold	Au	1.414	1.568	−9.8	1.175	1.222	−3.8
Indium	In	1.169	1.114	4.9	1.430	−	−
Iron	Fe	1.180	1.151	2.5	0.880	0.907	−3.0
Lead	Pb	1.382	1.391	−0.6	1.275	1.089	17.1
Lithium	Li	0.712	0.743	−4.2	0.672	0.599	12.2
Magnesium	Mg	1.234	1.250	−1.3	0.787	0.982	−19.9
Mercury	Hg	2.001	1.851	8.1	1.667	1.436	16.1
Nickel	Ni	1.125	1.151	−2.3	0.889	0.907	−2.0
Potassium	K	0.943	0.861	9.5	0.763	0.688	10.9
Rubidium	Rb	0.943	0.960	−1.8	0.780	0.763	2.2
Silver	Ag	1.288	1.307	−1.5	0.983	1.025	−4.1
Sodium	Na	0.907	0.849	6.8	0.750	0.679	10.5
Strontium	Sr	1.068	1.051	1.6	0.655	0.832	−21.3
Thallium	Tl	1.310	1.264	3.6	1.171	0.993	17.9
Tin	Sn	1.113	1.165	−4.5	1.424	−	−
Zinc	Zn	1.520	1.441	5.5	1.044	1.126	−7.3
			$\Delta(24)$ %	4.8		$\Delta(21)$ %	10.4
			$S(24)$	0.067		$S(21)$	0.123

The repulsive exponent n is also linked to the Grüneisen constant $\gamma_{G,E}$ or $\gamma_{G,T}$ (see Subsection 5.5.2), according to the relation [12].

$$n = 6\gamma_{G,E} - 2 \tag{5.29a}$$

$$n = 6\gamma_{G,T} - 2 \tag{5.29b}$$

Substituting Eq. (5.29a) into Eq. (5.28), we have

$$\xi_E^{1/2} = 0.424\gamma_{G,E} + 0.402 \tag{5.30}$$

or

$$\gamma_{G,E} = 2.36\xi_E^{1/2} - 0.947 \tag{5.31}$$

The Grüneisen constant $\gamma_{G,E}$ is given in terms of only one dimensionless number $\xi_E^{1/2}$, which can be revealed through data for the velocity of sound.

In Figure 5.5, values of $\xi_E^{1/2}$ are plotted for the 39 liquid metallic elements shown in Table 5.4 as a function of their respective atomic numbers. The figure shows that the values $\xi_E^{1/2}$ vary periodically with atomic number. The periodic Group IIB, i.e. zinc group metals, and magnesium (one of the Group IIA metals) occupy the peaks of the curve. Such a simple variation in the values of a quantity or parameter (e.g. $\xi_E^{1/2}$) with

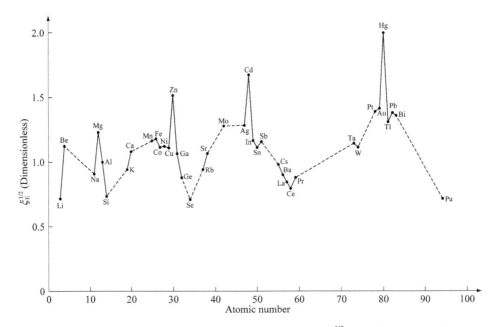

Figure 5.5 *The dimensionless parameter, or the correction factor, $\xi_E^{1/2}$ plotted against atomic number.*

atomic number allows us to estimate (through interpolation) values of the parameter for those not yet available [13, 15]. Additionally, the periodic variation with atomic number for such a parameter should give us an important clue in developing rigorous theories for the thermophysical properties of liquid metallic elements.

5.5.2 The Einstein–Lindemann Model in Terms of a Dimensionless Parameter ξ_T (a Corrected Einstein–Lindemann Model)

Iida et al. [14, 15] have also introduced a correction factor ξ_T into Eq. (5.18), as follows:

$$U_m = 9.197 \left(\frac{\xi_T R T_m}{M} \right)^{1/2} \tag{5.32a}$$

Values for ξ_T can be calculated from Eq. (5.32b) using experimental data on the velocity of sound:

$$\xi_T^{1/2} = \frac{1}{9.197} \left(\frac{M}{R T_m} \right)^{1/2} U_m \tag{5.32b}$$

Calculated values for $\xi_T^{1/2}$ are also given in Table 5.5. Figure 5.6 shows a plot of $\xi_T^{1/2}$ against n; as seen, a linear relationship exists between the two variables, apart from gallium, indium, and tin. With the exception of these metals and manganese data, treatment through the least-squares methods leads to:

$$\xi_T^{1/2} = 0.0534n + 0.448, \quad (n > 3) \tag{5.33}$$

As shown in Table 5.6, Eq. (5.33) performs well with $\Delta(21)$ and $S(21)$ values of 10.4 per cent and 0.123, respectively, for liquid metallic elements. Using the Grüneisen constant $\gamma_{G,T}$ (i.e. from Eqs. (5.29b) and (5.33)), Eq. (5.33) becomes

$$\xi_T^{1/2} = 0.320\gamma_{G,T} + 0.341 \tag{5.34}$$

or

$$\gamma_{G,T} = 3.12\xi_T^{1/2} - 1.06 \tag{5.35}$$

Figure 5.7 shows the periodic variation of $\xi_T^{1/2}$ with increase in the atomic number of the metallic elements; the Group IIIA (except for thallium) occupy the peaks, whereas the Group VIA elements, or the Group 16 elements in the IUPAC system (i.e. sulphur, selenium, tellurium, and ytterbium), occupy the valleys of the curve.

Both new parameters (i.e. $\xi_E^{1/2}$ and $\xi_T^{1/2}$), or the new dimensionless numbers, give a general indication of the extent of an atom's hardness or softness (to be exact, the hardness or softness of the pair interaction potential). These new parameters are also

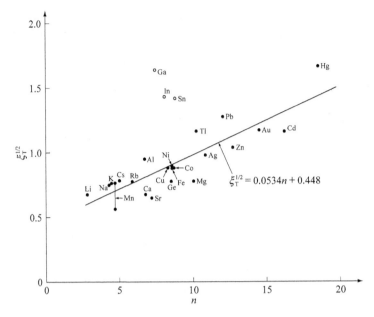

Figure 5.6 *Relationship between the dimensionless parameters, or the correction factor, $\xi_T^{1/2}$ (appearing in Eq. (5.32)) and the repulsive exponent n in the pair potential (appearing in Eq. (5.27)) for various metallic elements.*

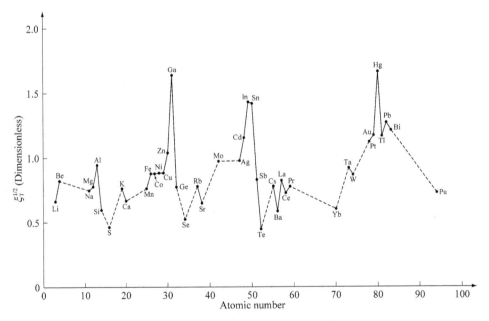

Figure 5.7 *The dimensionless parameter, or the correction factor, $\xi_T^{1/2}$ plotted against atomic number.*

useful in discussing the anharmonic effects of atomic motions in the liquid state. For instance, the volume expansion of a condensed matter is due to anharmonic effects (note that $\gamma_G = 0$ for a harmonic oscillator[4]). Especially in the liquid state, the effects of anharmonic motions of atoms are crucial importance in discussing its thermophysical properties. In addition, we can signify that the dimensionless parameters are closely related to one of the most fundamental physical quantities of liquids. Comparing Eqs. (5.23) and (5.32), we obtain

$$\xi_T = \frac{1}{9.197^2} \left\{ \frac{\gamma_h}{S(0)} \right\} = 1.182 \times 10^{-2} \left\{ \frac{\gamma_h}{S(0)} \right\} \tag{5.36}$$

From Eqs. (5.23) and (5.26), we have

$$\xi_E = \frac{1}{2} \left\{ \frac{\gamma_h}{S(0)} \right\} \frac{RT_m}{\Delta_l^g H_b} = 4.157 \left\{ \frac{\gamma_h}{S(0)} \right\} \frac{T_m}{\Delta_l^g H_b} \tag{5.37}$$

Equations (5.36) and (5.37) indicate that the two parameters (or the two correction factors), ξ_T and ξ_E, are dimensionless common parameters which reflect the structures of metallic liquids (see Subsection 2.2.3).

Substitution of Eq. (5.5b) into Eq. (5.36), or into Eq. (5.37), leads to

$$\xi_T = 1.182 \times 10^{-2} \frac{\gamma_h V}{RT \kappa_T} = 1.422 \times 10^{-3} \left. \frac{\gamma_h V}{T \kappa_T} \right|_{T=T_m} \tag{5.38a}$$

or

$$\xi_E = \left. \frac{\gamma_h V}{2\Delta_l^g H_b \kappa_T} \right|_{T=T_m} \tag{5.39a}$$

where R is the molar gas constant (8.314 J mol^{-1} K^{-1}) and V is the molar volume. Equations (5.38a) and (5.39a) can be rewritten, respectively, as

$$\kappa_T = 1.182 \times 10^{-2} \frac{\gamma_h V}{RT \xi_T} = 1.422 \times 10^{-3} \left. \frac{\gamma_h V}{T \xi_T} \right|_{T=T_m} \tag{5.38b}$$

or

$$\kappa_T = \left. \frac{\gamma_h V}{2\Delta_l^g H_b \xi_E} \right|_{T=T_m} \tag{5.39b}$$

On the basis of Eq. (5.26), we can obtain a sound velocity equation in terms of molar mass and molar volume. According to Figure 4.7, the cohesive energy, or the evaporation enthalpy, is roughly proportional to the reciprocal of a liquid metallic element's molar

---

[4] $\gamma_{G,E} = \gamma_{G,T} = 0$.

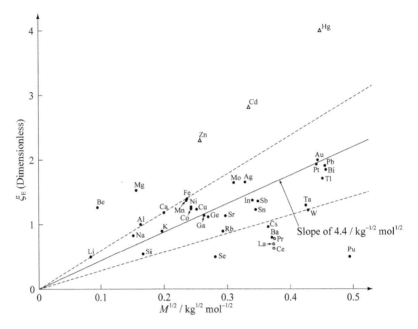

Figure 5.8 *The dimensionless parameter, or the correction factor,* ξ_E *plotted against the square root of molar mass,* $M^{1/2}$. $\triangle$, *Group IIB (zinc group) elements;* $\bigcirc$, *lanthanoid elements;* $\bullet$, *other elements. The broken lines denote the* ± 35 *per cent error band.*

volume V_m. On the other hand, the parameter ξ_E is roughly proportional to the square root of a liquid metallic element's molar mass, as shown in Figure 5.8. Substituting these correlations into Eq. (5.26), we have, therefore

$$U_m = \frac{5.579}{M^{1/4} V_m^{1/2}} \tag{5.40}$$

The numerical factor of 5.579 kg$^{1/4}$ m$^{5/2}$ s^{-1} mol$^{-3/4}$ was determined so as to give the minimum S value for the 41 liquid metallic elements plotted in Figure 5.9.

5.6   Assessment of Sound Velocity Models

Let us now evaluate the performance of the two models for the velocity of sound represented by Eqs. (5.22) and (5.40). Table 5.7 lists experimental and calculated values for the velocity of sound in 41 liquid metallic elements, and gives the corresponding δ_i, Δ, and S values needed for statistical assessment. As is clear from the table, Eq. (5.22) performs well with $\Delta(41)$ and $S(41)$ values of 14.5 per cent and 0.170, respectively. These results prove that Eq. (5.22), i.e. the modified Einstein–Lindemann

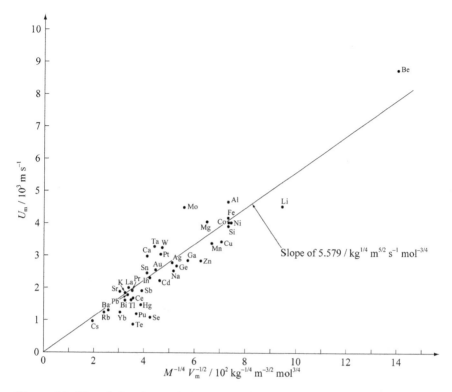

Figure 5.9 *The velocity of sound in liquid metallic elements at their melting point temperatures as a function of* $(M^{-1/4} V_m^{-1/2})$.

model, makes a remarkable improvement over those of the Einstein–Lindemann model, which gives $\Delta(41)$ and $S(41)$ values of 24.8 per cent and 0.293, respectively. Incidentally, for the liquid metallic elements listed in Table 5.1, Eq. (5.22) gives $\Delta(17)$ and $S(17)$ values of 11.4 per cent and 0.131, respectively (see Table 5.2). In spite of a simple equation in terms of two well-known physical quantities, the empirical relation represented by Eq. (5.40), against our expectation, performs well with $\Delta(41)$ and $S(41)$ values of 15.4 per cent and 0.206, respectively. As is evident from Table 5.7, however, the mean value of the sound velocities calculated from Eqs. (5.22) and (5.40) gives a better agreement with experimental values.

5.7 Experimental Sound Velocity Data

At present, experimentally derived data for the velocity of sound are available for some 41 liquid metallic elements. As mentioned earlier, the velocity of sound in liquid metallic elements is an indispensable basic quantity in discussing their thermophysical properties

Table 5.7 *Comparison of experimental values for the velocity of sound in liquid metallic elements at their melting points with those calculated from Eqs. (5.22) and (5.40), together with δ_i, Δ, and S values.*

Element		$(U_m)_{exp}$ m s^{-1}	$(U_m)_{cal}$, m s^{-1}			δ_i, %		
			Eq. (5.22)	Eq. (5.40)	Mean[†]	Eq. (5.22)	Eq. (5.40)	Mean[†]
Aluminium	Al	4680	5277	4093	4685	−11.3	14.3	−0.1
Antimony	Sb	1910	1748	2179	1964	9.3	−12.3	−2.7
Barium	Ba	1331	1837	1425	1631	−27.5	−6.6	−18.4
Beryllium	Be	9104	7273	7841	7557	25.2	16.1	20.5
Bismuth	Bi	1640	1401	1810	1606	17.1	−9.4	2.1
Cadmium	Cd	2237	2162	2575	2369	3.5	−13.1	−5.6
Caesium	Cs	983	1130	1087	1109	−13.0	−9.6	−11.4
Calcium	Ca	2978	3498	2301	2900	−14.9	29.4	2.7
Cerium	Ce	1693	2473	1992	2233	−31.5	−15.0	−24.2
Cobalt	Co	4031	4204	4110	4157	−4.1	−1.9	−3.0
Copper	Cu	3440	3426	3943	3685	0.4	−12.8	−6.6
Gallium	Ga	2873	2733	3216	2975	5.1	−10.7	−3.4
Germanium	Ge	2693	2574	2959	2767	4.6	−9.0	−2.7
Gold	Au	2568	2046	2485	2266	25.5	3.3	13.3
Indium	In	2320	2109	2374	2242	10.0	−2.3	3.5
Iron	Fe	4200	4361	4073	4217	−3.7	3.1	−0.4
Lanthanum	La	2022	2464	1892	2178	−17.9	6.9	−7.2
Lead	Pb	1821	1504	1877	1691	21.1	−3.0	7.7
Lithium	Li	4554	6785	5281	6033	−32.9	−13.8	−24.5
Magnesium	Mg	4065	4556	3614	4085	−10.8	12.5	−0.5
Manganese	Mn	3381	3659	3731	3695	−7.6	−9.4	−8.5
Mercury	Hg	1511	1440	2178	1809	4.9	−30.6	−16.5
Molybdenum	Mo	4502	3817	3124	3471	17.9	44.1	29.7
Nickel	Ni	4047	4066	4160	4113	−0.5	−2.7	−1.6
Platinum	Pt	3053	2454	2613	2534	24.4	16.8	20.5
Plutonium	Pu	1195	1384	2073	1729	−13.7	−42.4	−30.9
Potassium	K	1876	2285	1824	2055	−17.9	2.9	−8.7
Praseodymium	Pr	1925	2355	1973	2164	−18.3	−2.4	−11.0

Table 5.7 *(continued)*

Element		$(U_m)_{exp}$ m s^{-1}	$(U_m)_{cal}$, m s^{-1}			δ_i, %		
			Eq. (5.22)	Eq. (5.40)	Mean†	Eq. (5.22)	Eq. (5.40)	Mean†
Rubidium	Rb	1251	1491	1358	1425	−16.1	−7.9	−12.2
Selenium	Se	1100	1162	2371	1767	−5.3	−53.6	−37.7
Silicon	Si	3920	4624	4090	4357	−15.2	−4.2	−10.0
Silver	Ag	2790	2499	2858	2679	11.6	−2.4	4.1
Sodium	Na	2526	3216	2877	3047	−21.5	−12.2	−17.1
Strontium	Sr	1902	2308	1685	1997	−17.6	12.9	−4.8
Tantalum	Ta	3303	2912	2459	2686	13.4	34.3	23.0
Tellurium	Te	889	1445	1990	1718	−38.5	−55.3	−48.3
Thallium	Tl	1650	1480	1955	1718	11.5	−15.6	−4.0
Tin	Sn	2464	2095	2305	2200	17.6	6.9	12.0
Tungsten	W	3279	2916	2642	2779	12.4	24.1	18.0
Ytterbium	Yb	1274	1514	1703	1609	−15.9	−25.2	−20.8
Zinc	Zn	2850	2814	3498	3156	1.3	−18.5	−9.7
					$\Delta(41)$ %	14.5	15.4	12.4
					S (41)	0.170	0.206	0.165

† The mean value of sound velocities calculated from Eqs. (5.22) and (5.40).

and behaviour. Consequently, it is to be hoped that 50 or more experimental data will become available, especially for reactive (e.g. rare earth metals) and high melting point transition metals. The accuracy of the sound velocity data is comparatively good. Nevertheless, in order to evaluate other properties of metallic liquids, for example, compressibility, using measured sound velocities, even more accurate data are needed since $\kappa_s = \rho\, U^2$ and any uncertainties or errors are therefore approximately doubled.

In general, the velocity of sound in liquid metallic elements decreases linearly as temperature rises, i.e. the temperature dependence dU/dT is negative, with dU/dT varying from $\sim$0.1 to $\sim$0.6 m s^{-1} K^{-1}. However, there are fairly large discrepancies between some of the experimental values of temperature coefficient of a liquid metallic element's sound velocity. Several liquid metallic elements exhibit nonlinear or positive variations in their velocity of sound with respect to temperature. A few examples of these anomalous liquid metallic elements are given below. As is apparent from Figure 5.10, the velocity of sound in liquid tellurium first increases very rapidly and linearly with $dU/dT \approx 0.8$ m^{-1}K^{-1}. It then slows markedly. Figure 5.11 shows experimental data for the temperature dependence of the velocity of sound in liquid sulphur and selenium, respectively. These dU/dT values cease to be constant with increasing temperature.

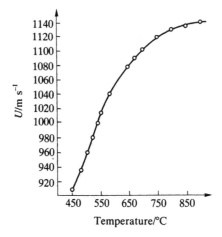

Figure 5.10 *Temperature dependence of the velocity of sound in liquid tellurium (after Gitis and Mikhailov [8]).*

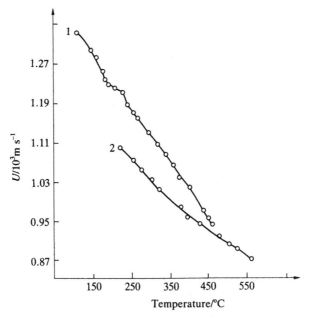

Figure 5.11 *Temperature dependence of the velocity of sound in liquid sulphur (curve 1) and selenium (curve 2) (after Gitis and Mikhailov [22]).*

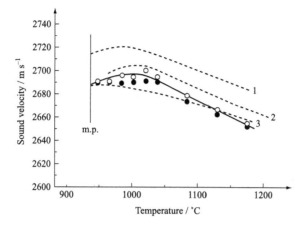

Figure 5.12 *Temperature dependence of the velocity of sound in liquid germanium (re-plotted from Hayashi et al. [23]). Open and closed circles represent two different runs (after Hayashi et al. [23]): 1, Baidov and Gitis; 2, Glazov et al.; 3, Yoshimoto et al.*

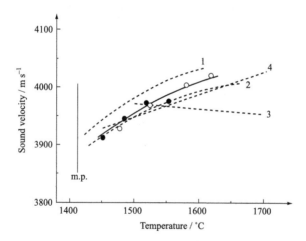

Figure 5.13 *Temperature dependence of the velocity of sound in liquid silicon (re-plotted from Hayashi et al. [23]). Open and closed circles represent two different runs (after Hayashi et al. [23]): 1, Yoshimoto et al.; 2, Glazov et al.; 3, Sokolov et al.; 4, Keita and Steinemann.*

Liquid germanium and silicon (semiconductors) also show anomalous variations in their velocity of sound with respect to temperature, as illustrated in Figures 5.12 and 5.13 [23], respectively. As can be seen from Figure 5.12, the velocity of sound in liquid germanium increases as temperature rises to around 1000 °C, above which it decreases linearly, excluding the result obtained by Yoshimoto et al. Figure 5.13 shows that the velocity of sound in liquid silicon increases with increasing temperature, in temperature ranges measured, excluding the result of Sokolov et al. Such anomalous behaviour in sound velocity, i.e. nonlinear or positive dependence on temperature, is interpreted as being based on the structural rearrangement, or microscopic structural changes, in the liquids. However, to have more detailed discussions on this subject, experimental re-examinations for the sound velocities in these liquid semiconductors may still be needed.

Experimental values for the velocity of sound in liquid metallic elements at their melting point temperatures, together with equations for their variations in temperature are listed in Table 17.7(a) and (b).

....

REFERENCES

1. S. Blairs, *Int. Mater. Rev.*, **52** (2007), 321; see also S. Blairs, *Phys. Chem. Liq.*, **45** (2007), 339.
2. D. Bohm and T. Staver, *Phys. Rev.*, **84** (1951), 836.
3. P. Ascarelli, *Phys. Rev.*, **173** (1968), 271.
4. Y. Rosenfeld, *J. Phys.: Condens. Matter*, **11** (1999), L71.
5. I. Yokoyama, *Physica B*, **293** (2001), 338.
6. I. Yokoyama and Y. Waseda, *High Temp. Mater. Process*, **24** (2005), 213.
7. M.B. Gitis and I.G. Mikhailov, *Sov. Phys. Acoust.*, **13** (1968), 473.
8. M.B. Gitis and I.G. Mikhailov, *Sov. Phys. Acoust.*, **12** (1966), 14.
9. A. Einstein, *Ann. Phys.*, **34** (1911), 170.
10. T. Iida and R.I.L. Guthrie, *The Physical Properties of Liquid Metals*, Clarendon Press, Oxford, 1988.
11. D.A. Young and B.J. Alder, *Phys. Rev. A*, **3** (1971), 364.
12. M. Toda, H. Matsuda, Y. Hiwatari, and M. Wadatsu, *The Structure and Physical Properties of Liquids*, Iwanami Shoten Publishers, Tokyo, 1976.
13. T. Iida, R.I.L Guthrie, and M. Isac, in *ICS Proceedings of the 3rd International Congress on Science and Technology of Steelmaking*, Association for Iron and Steel Technology, Charlotte, NC, 2005, p.3.
14. T. Iida, R.I.L Guthrie, and M. Isac, in *ICS Proceedings of the 3rd International Congress on Science and Technology of Steelmaking*, Association for Iron and Steel Technology, Charlotte, NC, 2005, p.57.
15. T. Iida, R. Guthrie, M. Isac, and N. Tripathi, *Metall. Mater. Trans. B*, **37** (2006), 403.
16. T. Iida, R. Guthrie, and N. Tripathi, *Metall. Mater. Trans B*, **37** (2006), 559.
17. T. Iida and R. Guthrie, *Metall. Mater. Trans B*, **40** (2009), 949.
18. T. Iida and R. Guthrie, *Metall. Mater. Trans B*, **40** (2009), 959.
19. T. Iida and R. Guthrie, *Metall. Mater. Trans B*, **40** (2009), 967.
20. T. Iida and R. Guthrie, *Metall. Mater. Trans B*, **41** (2010), 437.
21. H. Matsuda and Y. Hiwatari, in *Cooperative Phenomena*, edited by H. Haken and M. Wagner, Springer-Verlag, 1973; see also H. Matsuda, *Prog. Theor. Phys. (Kyoto)*, **42** (1969), 414; Y. Hiwatari and H. Matuda, *Prog. Theor. Phys. (Kyoto)*, **47** (1972), 741.
22. M.B. Gitis and I.G. Mikhailov, *Sov. Phys. Acoust.*, **13** (1967), 251.
23. M. Hayashi, H. Yamada, N. Nabeshima, and K. Nagata, *Int. J. Thermophys.*, **28** (2007), 83.

6

Surface Tension

6.1 Introduction

Knowledge of the surface tension of metallic liquids is essential for understanding various processing concepts. For example, in smelting and refining operations, the surface tension or interfacial tension is a dominating factor for phenomena such as gas absorption, nucleation of gas bubbles, nucleation and growth of non-metallic inclusions, and slag/metal reactions. Figure 6.1 gives an example of the important role surface tension can play in metallurgical mass transport phenomena [1]. As can be seen, there is a simple relation between the surface tension of metallic liquids and apparent mass transfer coefficients. Similarly, it will be appreciated that many other materials technologies such as crystal growth, casting, welding, brazing, melt spinning and extraction, zone melting, sintering, and spraying, are greatly influenced by the role played by the surface tension of a metallic liquid. To elaborate on this, as an example, we need a clear understanding of Marangoni flows which are caused by the liquid's surface tension gradient. This is of utmost importance for the manufacture of highly value-added metallic materials such as semiconductor crystals. Therefore, accurate and reliable data on the surface tensions of all liquid metallic elements, or liquid metallic 'simple substances', i.e. liquid metals, semimetals, and semiconductors, are needed in the field of materials process science.

Unfortunately, experimental and theoretical investigations on the nature and behaviour of surfaces or interfaces between phases are generally unsatisfactory. This situation results from the difficulty of precise experimental determinations of metallic liquids' surface tensions or interfacial tensions; the main problem is that it is extremely difficult to eradicate all impurities (especially surface active elements such as oxygen) from both the surface of the specimen and the atmosphere. Furthermore, from a more fundamental viewpoint, we have, as yet, no experimental methods available for determining the ionic and electronic structures of an interface, i.e. the inhomogeneous transition region between two phases. Therefore, some drastic, oversimplified assumptions are necessarily involved in any theoretical treatment of surface tension or interfacial tension.

Incidentally, the surface tension is thermodynamically defined as the surface free energy per unit area. In a dynamic sense, surface tension represents the work or energy required to create one unit of additional surface area at constant temperature.

The Thermophysical Properties of Metallic Liquids: Volume 1 – Fundamentals. First Edition.
Takamichi Iida and Roderick I. L. Guthrie. © Takamichi Iida and Roderick I. L. Guthrie 2015.
Published in 2015 by Oxford University Press.

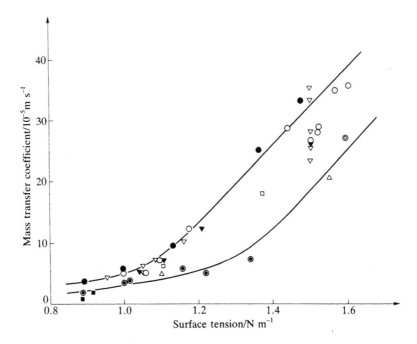

Figure 6.1 *Relation between mass transfer coefficient and surface tension in iron–sulphur and iron–oxygen melts (after Inouye and Choh [1]): Inouye and Choh: ○, Fe–O 1600 °C, ●, Fe–S 1600 °C, ⊙, Fe–S 1550 °C, ◎, pure iron (0.007 per cent O) 1550 °C; Eliott and Pehlke: ▽, Fe–O 1606 °C, ▼, Fe–S 1606 °C, △, Fe–O 1556–1561 °C; Schenck et al.: □, Fe–O 1560 °C, ■, Fe–S 1560 °C.*

These definitions are the starting point of the theoretical or semi-theoretical (semi-empirical) formulations for the surface tension of a liquid.

6.2   Theoretical Equations and Models for the Surface Tension of Liquids

Let us consider a one-component system in which two different bulk phases, a liquid and a gas phase, coexist. The presence of a gravitational field will provide a plane liquid–gas (vapour) interface. At the interface between the liquid and the gas phase, a non-uniform equilibrium phase, which is called a transition zone, or an interface zone, will exist, as illustrated in Figure 6.2(a). One of the basic quantities characterizing the transition zone is the number density or the distribution of atoms. The distribution function must vary continuously with z in the transition zone, which will exhibit a certain thickness (see Figure 6.2(a)). However, we have no means of experimentally determining either this variation of $g(r, z)$, where r denotes a vector in the (x, y) plane, or the thickness of the associated interfacial transition zone.

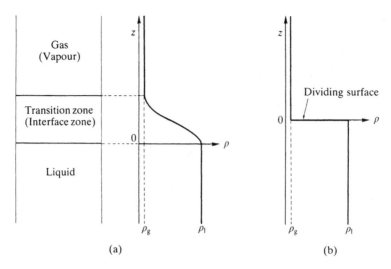

Figure 6.2 *Density variations at the interface between a liquid phase and a gas (vapour) phase: (a) liquid–gas interface region; (b) Fowler's approximation, $g(r, z) = g(r)$ for $z < 0$, $g(r, z) = 0$ for $z > 0$.*

6.2.1 An Equation in Terms of the Pair Distribution Function $g(r)$ and the Pair Potential $\phi(r)$

A statistical mechanical analysis of the relation between surface tension and intermolecular (interatomic) forces acting at the interface between a liquid and a gas phase was first made by Fowler [2]. In the derivation of an expression for the surface tension, Fowler introduced the drastic but simple approximation that there is a mathematical surface at which a discontinuity in density exists between the liquid and the gas phase of negligible density, as indicated in Figure 6.2(b). Fowler's expression for the surface tension γ of the liquid is given by

$$\gamma = \frac{\pi n_0^2}{8} \int_0^\infty g(r) \frac{\partial \phi(r)}{\partial r} r^4 \mathrm{d}r \qquad (6.1)$$

Since then, several researchers have provided further contributions towards this approach. Kirkwood and Buff [3] and Harashima [4] have both presented rigorous expressions for surface tension in terms of the intermolecular potentials and molecular distribution functions, using a general statistical mechanical theory of interfacial phenomena. However, in their calculations of surface tension values, the approximation of a discontinuous surface between the two phases had to be introduced because of the present lack of information on the spatial arrangement of molecules, i.e. $g(r, z)$, in the transition zone.

Johnson et al. [5] and others (e.g. Waseda and Suzuki [6]) have made calculations of the surface tension values of liquid metals using experimental data available for $g(r)$

and long-range oscillatory pair potential $\phi(r)$ (see Subsection 2.2.4(vi)) through the use of Fowler's formula (Eq. (6.1)). Calculated values for the surface tensions of liquid metals are in quite good agreement with experimental data. However, a close examination of the accuracy and validity of the pair potentials and Fowler's approximation is needed.

Incidentally, it is interesting that the integral appearing in Eq. (6.1) is exactly equal to that in Born–Green viscosity formula represented by Eq. (7.1).

6.2.2   An Equation Based on Fluctuation Theory

Bhatia and March [7] and other researchers have developed a theory of surface tension based on (statistical) density fluctuations in non-uniform systems. According to their fluctuation theory, the surface tension can be written as the sum of two terms:

$$\gamma = \gamma_1 + \gamma_2 = \frac{l(\Delta n)^2}{2n^2\kappa_T} + Bl\left(\frac{\Delta n}{l}\right)^2 \tag{6.2}$$

where l is the effective thickness of the interface, Δn is the fluctuation in number density, and B is a constant. In Eq. (6.2), the first term arises from treating a surface inhomogeneity as an accidental fluctuation in number density, while the second term evidently originates from the density gradient. Minimizing γ with respect to l (to determine the effective surface thickness), Bhatia and March [7] obtained

$$\gamma = \frac{l(\Delta n)^2}{n^2\kappa_T} \tag{6.3}$$

or

$$\gamma\,\kappa_T \approx l \tag{6.4}$$

Egelstaff and Widom [8] have presented an expression for the product of the isothermal compressibility and surface tension of a liquid near its triple point, based on Fisk and Widom's theoretical studies of surface tension in the region of the critical point, namely

$$\gamma\,\kappa_T \approx 0.07\,l \tag{6.5}$$

All these equations correlate the bulk property (i.e. isothermal compressibility κ_T of the liquid metal) with the surface property. Equation (6.5) was verified experimentally by Egelstaff and Widom [8] for liquids at, or near, their triple points. Values of $\gamma\,\kappa_T$ for various liquid metallic elements are listed in Table 6.1. For liquid metallic elements, the liquid metal–vapour interface thickness l ($\approx \gamma\,\kappa_T/0.07$) is in the range 2×10^{-10} to 7×10^{-10} m (0.2–0.7 nm), that is, in the order of atomic sizes. The result demonstrates that the density gradient at the interface of a liquid metallic element with a gas is very sharp, as might be expected.

March and Tosi [9] proved that an expression similar to Eqs. (6.4) and (6.5) can be derived from the behaviour of conduction electrons in metals.

Table 6.1 *Values of* $\gamma \kappa_T$ *for various liquid metallic elements at or near their triple points.*

Metal	γ $10^{-3}\,\mathrm{N\,m^{-1}}$	κ_T $10^{-11}\,\mathrm{m^2\,N^{-1}}$	$\gamma\kappa_T$ $10^{-10}\,\mathrm{m}$
Alkali metals			
Sodium	194	21	0.40
Potassium	113	40	0.45
Rubidium	95	49	0.46
Caesium	71	67	0.47
Other metals and a semimetal			
Iron	1790	1.43	0.25
Copper	1280	1.45	0.19
Silver	940	1.86	0.18
Zinc	785	2.4	0.19
Cadmium	666	3.2	0.21
Lead	470	3.5	0.17
Bismuth	395	4.3	0.17

Data from Egelstaff and Widom [8].

6.2.3 Equations Based on Hard-Sphere Models

An expression has also been derived for the surface tension of a hard-sphere liquid, which takes the form [9]

$$\gamma = \frac{9kT\eta^2(1+\eta)}{2\pi\sigma^2(1-\eta)^3} \tag{6.6}$$

where σ is an effective hard-sphere diameter and η is the packing fraction.

A correlation between surface tension and isothermal compressibility yields [8]

$$\gamma\,\kappa_T \approx \frac{\sigma(2-3\eta+\eta^3)}{4(1+2\eta)^2} \tag{6.7}$$

As already stated, the packing fraction η_m is approximately equal to 0.46 for many liquid metals, so that Eq. (6.7) becomes

$$\gamma\,\kappa_T \approx 0.05\sigma \tag{6.8}$$

6.2.4   An Equation Based on a Free Electron Model

Several workers have developed a theory of surface tension, based on a free electron gas model of a liquid metal. Gogate and Kothari [10] considered the surface tension layer of liquid metals to be a two-dimensional electron gas which obeyed Fermi–Dirac statistics. Their resulting expression for the surface tension of liquid metals is that γ is inversely proportional to the liquid metal's molar volume according to

$$\gamma \propto V^{-4/3} \tag{6.9}$$

This correlation holds roughly for a large number of liquid metallic elements.

Stratton [11] has since refined the theoretical calculation of the surface energy of liquid metals by taking into account the phenomenon of charge conservation at the metal surface. Calculated surface energy values for the alkali metals by Stratton appear to reproduce experimental data with good agreement. However, for other metals, the agreement between calculation and experiment is generally poor, owing to the simplicity of the free electron model when applied to more complex electronic structures such as transition metals.

6.2.5   Equations Based on the Combination of the Theoretical Formulae

(1) By combining Eq. (6.1) with Born–Green's viscosity formula represented by Eq. (7.1), the integrals expressed in terms of $g(r)$ and $\phi(r)$ cancel. As a result, we have a simple relation between surface tension γ and viscosity μ of a pure liquid [12]:

$$\gamma = \frac{15}{16}\left(\frac{kT}{m}\right)^{1/2}\mu \tag{6.10}$$

where k is the Boltzmann constant, and m is the mass of an atom (the atomic mass). This equation shall be called the Fowler–Born–Green relation.

In discussing this relation, Egry et al. [13] introduced a parameter, denoted by Q, at the melting point

$$Q = \frac{16}{15}\frac{\gamma}{\mu}\left(\frac{m}{kT}\right)^{1/2}\Bigg|_{T=T_m} \tag{6.11}$$

Although many approximations are involved in these theories, the values of Q for six liquid metals, i.e. iron, cobalt, nickel, copper, silver, and gold, at their melting points are in the range 0.81–1.02. (If the Fowler–Born–Green relation holds perfectly, $Q = 1$.)

(2) Combination of Eq. (6.10) and the Stokes–Einstein formula represented by Eq. (8.24) gives [14]

$$\gamma = \frac{15}{32\pi} \left(\frac{kT}{m}\right)^{1/2} \frac{kT}{\sigma D_{HS}} \tag{6.12}$$

where D_{HS} is the self-diffusivity in a hard-sphere fluid, with σ representing the hard-sphere diameter.

Yokoyama [14] checked this relation for 23 liquid metallic elements near their melting point temperatures, on the basis of the hard-sphere model. According to Yokoyama, calculated values of γ are in excellent agreement with experimental values for liquid alkali metals. Even for the noble and polyvalent metals, the agreement between theory and experiment is reasonable, apart from mercury, gallium, indium, tin, and antimony. Incidentally, these latter elements are classified into anomalous metallic elements according to Wallace [15].

6.3 Semi-Empirical Equations for the Surface Tension of Liquid Metallic Elements

6.3.1 The Skapski Model

Skapski [16] and, shortly after, Oriani [17] proposed a semi-empirical model for the surface tension of liquids in terms of their enthalpies of sublimation at 0 K.

According to Skapski's treatment, the total molar surface energy of a liquid is determined by the amount of energy required to bring N_A atoms (where N_A is the Avogadro constant) from the bulk of the condensed phase to its free surface. The total molar surface energy G_s of the liquid is defined as

$$G_s = S_A \left\{ \gamma - T \left(\frac{d\gamma}{dT}\right) \right\} \tag{6.13}$$

where S_A is the surface area occupied by a monatomic layer of N_A atoms. Such a S_A is given by

$$S_A = f N_A^{1/3} \left(\frac{M}{\rho}\right)^{2/3} = f N_A \left(\frac{V}{N_A}\right)^{2/3} \tag{6.14}$$

where f is a surface-packing, or configuration, factor.[1] Skapski supposed that the total molar surface energy G_s of a liquid metal at its melting point is roughly a constant fraction of its molar enthalpy of sublimation (or molar heat of evaporation) $\Delta_s^g H_0$ at 0 K (i.e. $G_s \approx Q_0 \Delta_s^g H_0$, where Q_0 is a constant). Thus, the surface energy per unit area $(g_s)_m$ at the melting point can be written as

$$(g_s)_m = q_0 \frac{\Delta_s^g H_0}{V_m^{2/3}} \tag{6.15}$$

---

[1] For close-packed liquids, $f = 1.09$; for bcc liquids (coordination number 8), $f = 1.12$; for liquid mercury (coordination number 6), $f = 1.04$.

where q_0 is a constant. Incidentally, surface energy, g_s, is equal to the surface tension γ at 0 K, i.e. $g_s = \gamma - T(d\gamma / dT)$; at 0 K, $g_s = \gamma$. The relation represented by Eq. (6.15) is called the Skapski model. A similar relation to Eq. (6.15) holds between the surface tension γ_m at the melting point and the enthalpy of evaporation $\Delta_l^g H_b$ at the boiling point:

$$\gamma_m = q \frac{\Delta_l^g H_b}{V_m^{2/3}} \tag{6.16}$$

where q is a constant. This expression is also called the Skapski model. Skapski gave an explanation to the relations represented by Eqs. (6.13) and (6.15) on the basis of the quasi-chemical model.

Several research workers have demonstrated the validity of the relationship described by Eq. (6.15) for various metallic elements (e.g. Utigard [18]; cf. also Kaptay [19]).

Oriani [17] further developed the findings of Skapski using the excess binding energy. Oriani's expressions for the surface tension and the total molar surface energy are given, respectively, by

$$\gamma = \frac{1}{S_A} \left\{ \left(\frac{Z_i - Z_s}{Z_i} \right) \Delta_s^g H_0 - \frac{Z_s \phi}{2} \right\} + T \left(\frac{d\gamma}{dT} \right) \tag{6.17}$$

and

$$G_s = \left(\frac{Z_i - Z_s}{Z_i} \right) \Delta_s^g H_0 - \frac{Z_s \phi}{2} \tag{6.18}$$

in which Z_i is the coordination number within the interior (or bulk) of the liquid, Z_s is the equivalent coordination number at the surface of liquid, and ϕ is the excess binding energy. This is defined as the difference in the pairwise interaction energy among atoms within the bulk of the liquid u and that on the surface layer v (i.e. $\phi \equiv v - u$). By assuming that the pairwise binding energy of surface atoms is equal to that for atoms within the interior of the liquid, i.e. $\phi = v - u = 0$, Eqs. (6.17) and (6.18) become identical to Skapski's expression for γ and G_s.

Miedema and Boom [20] discussed the Skapski model on the basis of an idea of electron density; they introduced the electron density at the boundary of an atomic cell in a pure metal n_{ws} as a fundamental parameter, and calculated that both the surface energy and the heat of vaporization for pure metals should be linearly related to n_{ws}. According to Keene's examination [21] on this subject, the relationship between surface tension and the parameter (n_{ws}) would appear to be classified into two groups: transition metals and non-transition metals.

6.3.2 The Schytil Model

Schytil [22] presented a semi-empirical model for the surface tension of liquid metallic elements at their melting point temperatures in terms of well-known physical quantities.

The Schytil model is expressed by the equation

$$\gamma_{\mathrm{m}} = q_s \frac{RT_{\mathrm{m}}}{V_{\mathrm{m}}^{2/3}} \tag{6.19}$$

where q_s is a numerical factor that is approximately the same for liquid metallic elements. The Schytil-type equations can be derived from some different viewpoints (e.g. the Eötvös law given by Eq. (6.40), and the hard-sphere model represented by Eq. (6.6)). However, it is said that Schytil was the first researcher to report on the correlation between the surface tension and the factor $T_{\mathrm{m}} / V_{\mathrm{m}}^{2/3}$, the latter being an indirect characteristic of the binding energy in metals [23]. Iida et al. have also proposed the Schytil-type equation for the surface tension of liquid metallic elements at their melting point temperatures, based on a simple harmonic oscillator model. (These approaches are described below, as an exercise in liquid properties.)

By assuming a simple function for the interatomic potential, Iida et al. derived an expression for the surface tension of liquid metallic elements. For this, they considered the work required to separate atoms to a distance at which interatomic forces become negligible.

We first present an expression for the surface tension of liquid metallic elements at their melting points. To simplify the treatment, let us consider a homogeneous liquid metallic element (or simple substance) at its melting point, which consists of atoms of mass m, performing harmonic vibrations of the same frequency ν_1 (i.e. Einstein frequency) about their positions of equilibrium.

The Einstein frequency is given by

$$\nu_1 = \frac{1}{2\pi}\left(\frac{K_{\mathrm{f}}}{m}\right)^{1/2} \tag{6.20}$$

where K_{f} is the force constant. In order to represent the force constant in terms of well-known physical parameters, we employ a modified version of Lindemann's melting formula for atomic frequency, this being

$$\nu_1 = \beta\nu_{\mathrm{L}} = \beta c\left(\frac{RT_{\mathrm{m}}}{MV_{\mathrm{m}}^{2/3}}\right)^{1/2}, \quad (c \approx 3.1 \times 10^8 \text{ in SI units}) \tag{6.21}$$

in which β is a correction factor and ν_{L} is the atomic frequency based on Lindemann's equation (Eq. (1.16)). From Eqs. (6.20) and (6.21), we have

$$K_{\mathrm{f}} = 4\pi^2\beta^2c^2\frac{RT_{\mathrm{m}}}{N_{\mathrm{A}}V_{\mathrm{m}}^{2/3}} \tag{6.22}$$

The interatomic force, $f(s)$, acting between a pair of atoms can be expressed by

$$f(s) = -k\,s$$

$$k = K_{\mathrm{f}}/Z_{\mathrm{i}} \tag{6.23}$$

In Eq. (6.23), Z_i is the nearest-neighbour coordination number of atoms in the bulk of the liquid and s is the distance of displacement of the central atom from its position of equilibrium.[2]

The next step is to select an arbitrary dividing surface in the interior of the metallic liquid and to define a rectangular coordinate system (x, y, z) in which the dividing surface lies in the (x, y) plane, as depicted in Figure 6.3. Consider an atom 1, situated at a point $z = 0$ on the dividing surface, i.e. the (x, y) plane, and an atom 2, situated below (x, y) plane at the equilibrium separation distance, or the average interatomic distance, a_m (where subscript m refers to the melting point), from atom 1. When we displace atom 1 in the direction of the z-axis by a distance z, the component of force f_z in the z direction acting between the two atoms separated by a distance $r_{1'2}$ is given by

$$f_z = -k\left(\sqrt{z^2 + 2a_m z \cos\theta + a_m^2} - a_m\right)\cos\phi \qquad (6.24)$$

where

$$\cos\phi = \frac{z + a_m \cos\theta}{\sqrt{z^2 + 2a_m z \cos\theta + a_m^2}}$$

For simplicity, we now introduce the drastic assumption that the potential of the interatomic force acting between a pair of atoms separated by a distance r may be expressed in the form illustrated in Figure 6.4. On the assumption that this potential energy function applies, the distance z_1 over which the interatomic force extends is

$$\sqrt{z_1^2 + 2a_m z_1 \cos\theta + a_m^2} = (1 + \alpha)a_m \qquad (6.25)$$

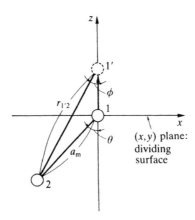

Figure 6.3 *Schematic diagram of the explanation of surface tension.*

[2] $\frac{1}{2}Z_i = \int_0^{r_1}\int_0^{\pi/2} 2\pi r^2 n_0 g(r) \sin\theta\, d\theta\, dr$, where r_1 is the maximum value of nearest-neighbour distance.

or

$$z_1 = a_m \left[\left\{ \cos^2\theta + \underline{\alpha}(\underline{\alpha} + 2) \right\}^{1/2} - \cos\theta \right] \tag{6.26}$$

Consequently, the work, or the energy, w_m necessary to the separate atom 1 to a distance beyond which the interatomic force between two atoms (1 and 2) are no longer felt (i.e. the force declines to zero) can be expressed as follows:

$$w_m = -\int_0^{z_1} f_z \, dz = \int_0^{z_1} k \left\{ \left(z^2 + 2a_m z \cos\theta + a_m^2 \right)^{1/2} - a_m \right\}$$

$$\times \frac{z + a_m \cos\theta}{(z^2 + 2a_m z \cos\theta + a_m^2)^{1/2}} \, dz = \frac{1}{2} k (\underline{\alpha} a_m)^2 \tag{6.27}$$

Combination of Eqs. (6.22), (6.23), and (6.27) gives

$$w_m = \frac{2(\pi \underline{\alpha} \beta c a_m)^2 R T_m}{N_A Z_i V_m^{2/3}} \tag{6.28}$$

If we neglect the work needed to change the distribution of atoms in the bulk of the liquid into their new distribution within the surface transition zone (in deriving Eq. (6.1), the same approximation was introduced by Fowler), we have for the surface tension γ_m of liquids at their melting point temperatures.

$$\gamma_m \approx \frac{Z_i w_m}{4 a_m^2} \tag{6.29}$$

In Eq. (6.29), $(1/a_m^2)$ and $(Z_i/2)$ are, respectively, the average number of atoms per unit area on the dividing surface and the number of equivalent atoms to atom 2 in the correlation between the atoms 1 and 2. (By dividing the bulk of liquid, two new surfaces can be produced.) By substituting Eq. (6.28) into Eq. (6.29), we obtain a Schytil-type expression, as follows:

$$\gamma_m = \frac{(\pi \underline{\alpha} \beta c)^2}{2 N_A} \frac{R T_m}{V_m^{2/3}} = 7.87 \times 10^{-7} (\underline{\alpha} \beta)^2 \frac{R T_m}{V_m^{2/3}} \tag{6.30}$$

A comparison of Eq. (6.30) with Eq. (6.19) implies that the product of $\underline{\alpha} \beta$ must be approximately the same for a large number of liquid metallic elements.

Now, if we assume that, at any temperature above the melting point, the isothermal work w needed to create unit area of free surface is reduced by the work of thermal expansion, w can be expressed by the relation (see Figure 6.5).

$$w = \frac{1}{2} k \{ (1 + \underline{\alpha}) a_m - a \}^2 \tag{6.31}$$

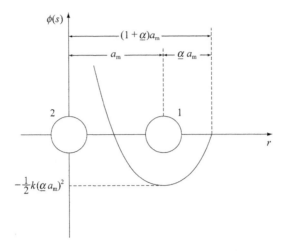

Figure 6.4 *Schematic diagram of the pair potential for deriving an expression for the surface tension of liquid metallic elements at their melting point temperatures.*

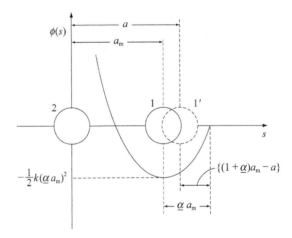

Figure 6.5 *Schematic diagram of the pair potential for the explanation of temperature variations of surface tension.*

Following equivalent procedures to those described earlier for the surface tension of liquid metallic elements at their melting point temperatures, the surface tension of liquid metallic elements at any temperature above their melting points can then be expressed by

$$\gamma \approx \frac{(\pi \beta c)^2 T_\mathrm{m}}{2N_\mathrm{A}} \frac{\left\{(1+\underline{\alpha}) V_\mathrm{m}^{1/3} - V^{1/3}\right\}^2}{V^{4/3}} \tag{6.32}$$

Table 6.2 *Values of β for various liquid metallic elements.*

Element		β
Aluminium	Al	0.56
Antimony	Sb	0.40
Barium	Ba	0.43
Beryllium	Be	0.35
Bismuth	Bi	0.55
Cadmium	Cd	0.59
Caesium	Cs	0.47
Calcium	Ca	0.42
Cerium	Ce	0.56
Cobalt	Co	0.48
Copper	Cu	0.47
Gallium	Ga	0.83
Germanium	Ge	0.40
Gold	Au	0.49
Indium	In	0.69
Iron	Fe	0.48
Lanthanum	La	0.53
Lead	Pb	0.56
Lithium	Li	0.53
Magnesium	Mg	0.47
Manganese	Mn	0.44
Mercury	Hg	0.84
Molybdenum	Mo	0.44
Nickel	Ni	0.47
Platinum	Pt	0.48
Plutonium	Pu	0.45
Potassium	K	0.49
Praseodymium	Pr	0.51
Rubidium	Rb	0.49
Selenium	Se	0.29

continued

Table 6.2 *(continued)*

Element		β
Silicon	Si	0.38
Silver	Ag	0.46
Sodium	Na	0.50
Strontium	Sr	0.42
Tantalum	Ta	0.43
Tellurium	Te	0.38
Thallium	Tl	0.55
Tin	Sn	0.64
Titanium	Ti	0.50
Tungsten	W	0.41
Ytterbium	Yb	0.38
Zinc	Zn	0.54

or

$$\gamma \approx \frac{(\pi\beta c)^2 T_{\mathrm{m}}}{2N_{\mathrm{A}} M^{2/3}} \left(\frac{\rho}{\rho_m}\right)^{2/3} \left\{(1 + \underline{\alpha})\rho^{1/3} - \rho_{\mathrm{m}}^{1/3}\right\}^2 \tag{6.33}$$

If Eq. (6.32), or Eq. (6.33), holds over the entire range of temperatures, we can obtain a value for the parameter $\underline{\alpha}$ since surface tension values reduce to zero at the critical point. From Eqs. (6.32) and (6.33), we then have (see Subsection 5.4.2)

$$\underline{\alpha} = \left(\frac{V_{\mathrm{c}}}{V_{\mathrm{m}}}\right)^{1/3} - 1 = \left(\frac{\rho_{\mathrm{m}}}{\rho_{\mathrm{c}}}\right)^{1/3} - 1 \approx 1.97\eta_{\mathrm{m}}^{1/3} - 1 \tag{6.34}$$

As already mentioned, the value of $\underline{\alpha}$ is about 0.524 for liquid metallic elements at or near the melting point temperatures. Consequently, on the basis of the relationship of Eq. (6.30), the correction factor β for Lindemann's melting formula can be evaluated from the expression:

$$\beta = \frac{1.13 \times 10^3 V_{\mathrm{m}}^{1/3}}{\underline{\alpha}} \left(\frac{\gamma_{\mathrm{m}}}{RT_{\mathrm{m}}}\right)^{1/2} = 2.16 \times 10^3 V_{\mathrm{m}}^{1/3} \left(\frac{\gamma_{\mathrm{m}}}{RT_{\mathrm{m}}}\right)^{1/2} \tag{6.35}$$

Substituting Eq. (6.35) into Eq. (6.21), we obtain

$$\nu_{\mathrm{l}} = \beta\nu_{\mathrm{L}} = 6.8 \times 10^{11} \left(\frac{\gamma_{\mathrm{m}}}{M}\right)^{1/2} \tag{6.36}$$

Values of β for various liquid metallic elements are given in Table 6.2.

6.4 Equations for the Surface Tension in Terms of New Dimensionless Parameters $\xi_E^{1/2}$ and $\xi_T^{1/2}$

It has been known for a long time that the thermophysical properties of liquid metallic elements at, or near, their melting point temperatures can be expressed using a few basic physical quantities. We can take Eqs. (5.17), (6.19), and (7.15) as examples:

$$\text{Sound velocity} \quad U_m \propto \left(\frac{T_m}{M} \right)^{1/2} \tag{5.17}$$

$$\text{Surface tension} \quad \gamma_m \propto \frac{T_m}{V_m^{2/3}} \tag{6.19}$$

$$\text{Viscosity} \quad \mu_m \propto \frac{(M T_m)^{1/2}}{V_m^{2/3}} \tag{7.15}$$

Combination of these three equations leads to the following important relation for surface tension, viscosity, and sound velocity at the melting point of a liquid metallic element [24–26]:

$$\gamma_m = k_0 \mu_m U_m \tag{6.37}$$

where k_0 is a dimensionless numerical factor. Figure 6.6 shows a plot of k_0 against the atomic number for 38 liquid metallic elements for which experimental data are available. As seen, values of k_0 for 35 metallic elements are roughly the same, with the exceptions of germanium, silicon, and selenium. A mean value of k_0 is 0.117 for these 35 elements excluding germanium, silicon, and selenium. In some detail, the k_0 values of the d-block transition metals in each period do not vary much, having lower values as a whole. Incidentally, the three exceptional elements are semiconductors. Of these elements, germanium and silicon have abnormally low values of the Andrade coefficient C_A, whereas selenium abnormally high value of C_A (see Subsection 7.3.1). Remarking parenthetically, from the viewpoint of theoretical physics [24], there has been great interest in understanding the significance of the inter-relationship expressed in Eq. (6.37), and in the corresponding dimensionless ratio $\gamma / \mu U$.

Substituting Eqs. (5.26) and (7.14) into Eq. (6.37), and using the relation given by Eq. (4.10) (i.e. $T_m = 2.844 \times 10^{-3} \xi^{-1} \Delta_l^g H_b$, where $\xi \equiv \xi_T/\xi_E$), we have

$$\gamma_m = \gamma_{0,E} \frac{\xi_E \Delta_l^g H_b}{\xi_T^{1/2} V_m^{2/3}} \tag{6.38}$$

$$\gamma_{0,E} \equiv 7.542 \times 10^{-2} \gamma_0, \quad (\gamma_0 \equiv k_0 C_A)$$

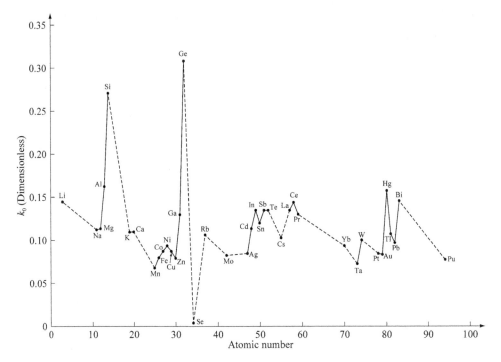

Figure 6.6 *The numerical factor k_0 appearing in Eq. (6.37) plotted against atomic number.*

Similarly, substituting Eqs. (5.32) and (7.14) into Eq. (6.37), we obtain

$$\gamma_m = \gamma_{0,\mathrm{T}} \frac{\xi_{\mathrm{T}}^{1/2} R T_m}{V_m^{2/3}} \tag{6.39}$$

$$\gamma_{0,\mathrm{T}} \equiv 3.190 \gamma_0$$

where $\gamma_{0,\mathrm{T}}$ or $\gamma_{0,\mathrm{E}}$ has roughly the same value for various liquid metallic elements. We see that the melting point surface tension values γ_m of liquid metallic elements depend on not only those of the factor $\Delta_l^g H_b / V_m^{2/3}$ or $R T_m / V_m^{2/3}$, but also on those of the parameter $\xi_{\mathrm{E}}^{1/2}$ or $\xi_{\mathrm{T}}^{1/2}$ and γ_0 (i.e. the product of the dimensionless numerical factor k_0 and the Andrade coefficient C_{A} of viscosity); in other words, if accurate values of these physical parameters, i.e. $\Delta_l^g H_b$ or T_m, $V_m (= M / \rho_m)$, $\xi_{\mathrm{E}}^{1/2}$ or $\xi_{\mathrm{T}}^{1/2}$ (sound velocity), and γ_0 (the product of k_0 and C_{A}), are already-known, an accurate surface tension value can be obtained by a simple numerical calculation.

Figure 6.7 shows γ_0 values for 41 liquid metallic elements plotted against their atomic number Z. It can be seen that there is a periodic variation in γ_0 values, the periodic Group IIIA metals (excluding thallium) occupying the peaks of the curve. In addition, Figure 6.7 has the following characteristic features. First, almost all the anomalous metallic elements (i.e. liquid gallium, cerium, mercury, silicon, selenium, and antimony, which are classified into anomalous metallic elements on the basis of the behaviour of

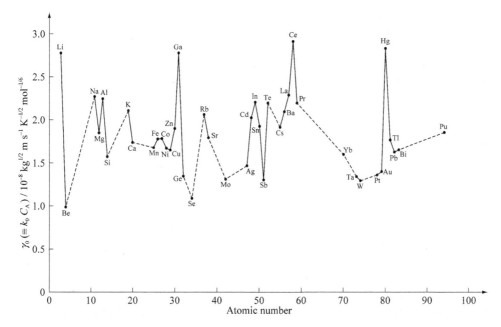

Figure 6.7 *The numerical coefficient $\gamma_0 (\equiv k_0 C_A)$ appearing in Eqs. (6.38) and (6.39) plotted against atomic number.*

$$\gamma_0 = \gamma_m V_m^{2/3} \Big/ 3.190 \xi_T^{1/2} RT_m \qquad (6.39')$$

γ_0 *values have been calculated from Eq. (6.39'). Data, except for $\xi_T^{1/2}$, are given in Chapter 17; $\xi_T^{1/2}$ in Table 5.5.*

entropy [14,15], sound velocity [27–29], and viscosity [26,30] in the liquid state), and further, normal metals, such as lithium and beryllium, obeying diagonal relationships, lie on the conspicuous lofty peaks or deep valleys of the curve. Second, although the γ_0 values of the d-block transition metals change very little, the relationship of γ_0 values against Z is slightly different from that of k_0 against Z: the d-block transition metals in Period 4 have γ_0 values of $(1.72 \pm 0.06) \times 10^{-8} \mathrm{kg}^{1/2} \mathrm{m\,s^{-1}K^{-1/2}mol^{-1/6}}$, while the d-block transition metals in Periods 5 and 6 have those of $(1.34 \pm 0.06) \times 10^{-8} \mathrm{kg}^{1/2}$ m s^{-1} K$^{-1/2}$ mol$^{-1/6}$.

6.5 Temperature Coefficients of the Surface Tension of Liquid Metallic Elements

Since the surface between the liquid phase and the gas phase disappears at the critical temperature T_c, it follows the surface tensions of liquids are reduced to zero at T_c. As a consequence, the surface tensions of liquids must decrease with rising temperature.

From a practical standpoint, the temperature coefficient $d\gamma/dT$, or the temperature dependence, of metallic liquids plays an important part in many fields of materials processing operations involving free liquid surfaces (e.g. in melt growth of semicon-ductor crystals, or in weld pools where Marangoni flows are produced by surface tension gradients within the melt).

A well-known relationship between surface tension and absolute temperature is the Eötvös law, which can be expressed in the form

$$\gamma = \frac{k_\gamma}{V^{2/3}}(T_c - T) \tag{6.40}$$

where k_γ is roughly equal to 6.4×10^{-8} J mol$^{-2/3}$ K^{-1} for liquid metallic elements. By dif-ferentiating Eq. (6.40) with respect to temperature T, we have for the temperature coefficient of surface tension $d\gamma/dT$ that

$$\frac{d\gamma}{dT} = \frac{k_\gamma}{V^{2/3}}\left\{\frac{2(T_c - T)}{3\rho}\frac{d\rho}{dT} - 1\right\} = \frac{\gamma}{T_c - T}\left\{\frac{2(T_c - T)}{3\rho}\frac{d\rho}{dT} - 1\right\} \tag{6.41}$$

Using this relationship, Allen [31] calculated values of $d\gamma/dT$ for high melting point metals for which no experimental data exist, in which estimated values of T_c by Grosse were employed (see Subsection 3.4.2).

On substituting the density–temperature correlation, i.e. $\rho = \rho_m + \Lambda T$, where $\Lambda (\equiv d\rho/dT)$, into Eq. (6.33) and differentiating γ with respect to T, we have

$$\frac{d\gamma}{dT} = \frac{(\pi\beta c)^2 \Lambda T_m}{3N_A M^{2/3}}\left\{2(1+\underline{\alpha})^2 \rho^{1/3} \rho_m^{-2/3} + \rho^{-1/3} - 3(1+\underline{\alpha})\rho_m^{-1/3}\right\} \tag{6.42}$$

On the basis of two different physical approaches, Miedema and Boom [20] and Eustathopoulos et al. [32] were led finally to propose the same expression for $d\gamma/dT$ in the form (see also Vinet et al. [33])

$$\frac{d\gamma}{dT} = \frac{b}{V_m^{2/3}} + \frac{2}{3}\frac{\gamma}{\rho}\frac{d\rho}{dT}, \quad \text{(Miedema and Boom)} \tag{6.43}$$

$$\frac{d\gamma}{dT} = -\frac{C'S_s}{V_m^{2/3}} + \frac{2}{3}\frac{\gamma}{\rho}\frac{d\rho}{dT} \quad \text{with} \quad C' = \frac{1}{f N_A^{1/3}}, \quad \text{(Eustathopoulos et al.)} \tag{6.44}$$

where b can be identified with entropy, S_s is the molar excess surface entropy, f is a factor of compactness, depending on the type of the 'quasicrystalline' lattice and on the surface plane. In the work of Miedema and Boom, values for b were determined from experimental data for γ, $d\gamma/dT$, V_m, and α (where α is the volume expansivity); on the other hand, in that of Eustathopoulos et al., values for $C'S_s$ were evaluated by applying the Skapski model.

Digilov [34] discussed the surface tension of pure liquid metals on the basis of the principle of corresponding states, and presented simple expressions for the surface tension and its temperature coefficient, as follows:

$$\gamma = 4.56\frac{RT_m}{(V_m^2 N_A)^{1/3}}\left[1 - 0.13\left(\frac{T}{T_m} - 1\right)\right]^{1.67} \tag{6.45}$$

$$\frac{d\gamma}{dT} \approx -0.217 \frac{\gamma_m}{T_m}\left[1 - 0.13\left(\frac{T}{T_m} - 1\right)\right]^{0.67} \tag{6.46}$$

At the melting point

$$\gamma_m \approx 4.56\frac{RT_m}{(V_m^2 N_A)^{1/3}} = 5.40 \times 10^{-8}\frac{RT_m}{V_m^{2/3}} \tag{6.47}$$

$$\left.\frac{d\gamma}{dT}\right|_{T=T_m} \approx -0.217\frac{\gamma_m}{T_m} \tag{6.48}$$

Substitution of $\alpha_m T_m \approx 0.09$ (i.e. Eq. (3.9)) into Eq. (6.48) yields another expression for $d\gamma/dT$ at the melting point

$$\left.\frac{d\gamma}{dT}\right|_{T=T_m} \approx -2.4\alpha_m\gamma_m \tag{6.49}$$

Lu et al. [35] investigated the surface tension of liquid silicon and germanium on the basis of a model for solid–vapour surface energy of elemental crystals. Kaptay [19] presented an expression for the surface tension of pure liquid metals in terms of constant-pressure heat capacity; at the melting point, his model becomes the same as the Schytil model.

Calculated values for $d\gamma/dT$ from these models appear to be in good agreement with experimental data. However, many approximations and/or assumptions are involved in all of these models; therefore, accurate and reliable experimental data for the temperature coefficients of liquid metallic elements are needed for building more accurate (predictive) models.

6.6 Assessment of Surface Tension Models

Let us now assess the performances of several of the surface tension models described in this Chapter.

Using the relation represented by Eq. (6.12), Yokoyama [14] calculated the surface tension values of 23 liquid metallic elements. In his calculations, the values of the self-diffusivity in hard-sphere fluids given by Eq. (8.16) were used for D_{HS} appearing in Eq. (6.12). Incidentally, the value of the packing fraction of 0.463, by which the hard-sphere structure factor can well describe experimental structure factor data for liquid metals near their melting point temperatures, was used for all 23 elements. On the basis of Yokoyama's calculated values for the surface tensions, $\Delta(23)$ and $S(23)$ values of 34.7 per cent and 0.576 were obtained for the liquid metallic elements near their melting point temperatures, as shown in Table 6.3. Experimentally determined surface tension data are taken from the review articles by Mills and Su [36] and Keene [21]. As is obvious from Table 6.3, with the exception of antimony, gallium, indium, mercury, and tin (these are classified into anomalous metallic elements [14,15]), the agreement between calculation and experiment is very good with $\Delta(18)$ and $S(18)$ values of 15.2 per cent and 0.178, respectively.

Table 6.3 *Comparison of experimental and Yokoyama's calculated values for the surface tension of liquid metallic elements near their melting point temperatures.*

Element		Temperature		Surface tension/N m^{-1}		δ_i %
		K	(°C)	Experimental[†]	Calculated	
Aluminium	Al	943	(670)	1.048	0.820	27.8
Antimony	Sb	933	(660)	0.370[a]	0.578	−36.0
Bismuth	Bi	573	(300)	0.380	0.332	14.5
Cadmium	Cd	623	(350)	0.633	0.469	35.0
Caesium	Cs	303	(30)	0.069[a]	0.076	−9.2
Cobalt	Co	1823	(1550)	1.881	2.058	−8.6
Copper	Cu	1423	(1150)	1.302	1.560	−16.5
Gallium	Ga	323	(50)	0.723	0.279	159
Gold	Au	1423	(1150)	1.137	1.228	−7.4
Indium	In	433	(160)	0.560	0.295	89.8
Iron	Fe	1833	(1560)	1.871	2.015	−7.1
Lead	Pb	613	(340)	0.456	0.365	24.9
Lithium	Li	463	(190)	0.398	0.358	11.2
Magnesium	Mg	953	(680)	0.569	0.676	−15.8
Mercury	Hg	238	(−35)	0.488	0.174	180
Nickel	Ni	1773	(1500)	1.780	2.010	−11.4
Potassium	K	343	(70)	0.110	0.155	−29.0
Rubidium	Rb	313	(40)	0.090	0.092	−2.2
Silver	Ag	1273	(1000)	0.917	1.090	−15.9
Sodium	Na	378	(105)	0.196	0.195	0.5
Thallium	Tl	588	(315)	0.458	0.373	22.8
Tin	Sn	523	(250)	0.554	0.347	59.7
Zinc	Zn	723	(450)	0.783	0.684	14.5
					$\Delta(23)$ %	34.7
					$S(23)$	0.576

[†] Data, except for those bearing the superscript a, are taken from Mills and Su [36].
[a] Data from Keene [21].

Figure 6.8 shows a plot of γ_m vs. $\Delta_g^l H_b / V_m^{2/3}$, i.e. the Skapski model, for 39 liquid metallic elements for which experimental data on surface tension γ_m, evaporation enthalpy $\Delta_g^l H_b$, and sound velocity U_m are available (in order to compare with performances of Eqs. (6.50), (6.51), and (6.52), or Eq. (6.53), expressed in terms of the new dimensionless parameter, $\xi_E^{1/2}$ or $\xi_T^{1/2}$, which is obtained from experimental data for the velocity of sound). Surface tension data are taken from the review articles mentioned previously. Experimentally derived data for γ_m, $\Delta_l^g H_b$, and U_m are collected in Chapter 17. As can be seen from Figure 6.8, there is a reasonably linear correlation with Eq. (6.16). The slope of 2.191×10^{-9} mol$^{1/3}$ was determined so as to give the minimum value for a relative standard deviation S for the 39 liquid metallic elements plotted in Figure 6.8. This minimization approach is also used in the evaluation of the numerical constants in Eqs. (6.51) to (6.53). As such, the Skapski model is written as

$$\gamma_m = 2.191 \times 10^{-9} \frac{\Delta_l^g H_b}{V_m^{2/3}} \qquad (6.50)$$

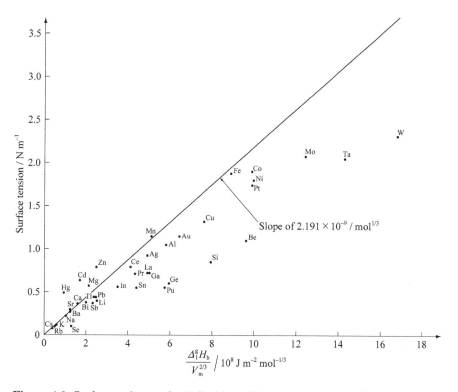

Figure 6.8 *Surface tensions γ_m for 39 liquid metallic elements at their melting point temperatures vs. $\Delta_l^g H_b / V_m^{2/3}$, i.e. the Skapski model.*

Figures 6.9 and 6.10 show plots of γ_m against $RT_m / V_m^{2/3}$, i.e. the Schytil model, and $\xi_T^{1/2} RT_m / V_m^{2/3}$, i.e. the modified Schytil model, for 41 liquid metallic elements for which experimental data on the velocity of sound are available (because of a comparison with the performances between Eqs. (6.51) and (6.52)). These figures show linear correlations between the two variables, respectively. As seen, however, the latter provides a much better correlation. The slopes of 6.755×10^{-8} mol$^{1/3}$ and 6.247×10^{-8} mol$^{1/3}$ were obtained for the 41 elements illustrated in Figures 6.9 and 6.10, respectively; hence the relationships are expressed as follows:

$$\gamma_m = 6.755 \times 10^{-8} \frac{RT_m}{V_m^{2/3}} \tag{6.51}$$

and

$$\gamma_m = 6.247 \times 10^{-8} \frac{\xi_T^{1/2} RT_m}{V_m^{2/3}} \tag{6.52}$$

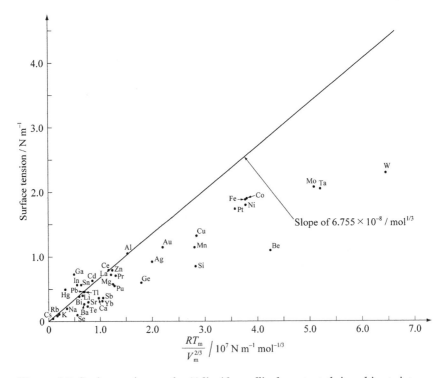

Figure 6.9 *Surface tensions γ_m for 41 liquid metallic elements at their melting point temperatures vs. $RT_m / V_m^{2/3}$, i.e. the Schytil model.*

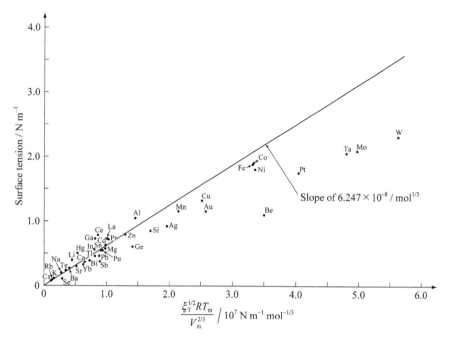

Figure 6.10 *Surface tensions γ_m for 41 liquid metallic elements at their melting point temperature vs. $\xi_T^{1/2} RT_m / V_m^{2/3}$, i.e. the modified Schytil model.*

For the 39 liquid metallic elements in Figure 6.8, Eq. (6.38) is expressed as

$$\gamma_m = 1.478 \times 10^{-9} \frac{\xi_E \Delta_l^g H_b}{\xi_T^{1/2} V_m^{2/3}} \tag{6.53}$$

Equations (6.52) and (6.53) give practically the same results from calculations. To be exact, if the numerical factors of Eq. (6.52) were determined (excluding tellurium and ytterbium), for the 39 liquid metallic elements (then, in this case, a value of 6.249×10^{-8} mol$^{1/3}$ is obtained for the numerical factor of Eq. (6.52) and we have a $\gamma_0 (\equiv k_0 C_A)$ value of 1.959×10^{-8} kg$^{1/2}$ m s^{-1} K$^{-1/2}$ mol$^{-1/6}$, using Eq. (6.38), or Eq. (6.39), both of these equations, i.e. Eqs. (6.52) and (6.53), would give exactly the same results in calculations; however, the difference between the numerical factors of 6.247×10^{-8} mol$^{1/3}$ and 6.249×10^{-8} mol$^{1/3}$ is negligibly small.

Incidentally, a comparison of Eqs. (6.51) and (6.30) gives

$$\underline{\alpha}\,\beta = 0.293 \tag{6.54}$$

and if the value of $\underline{\alpha}$ is 0.524, we have

$$\beta = 0.559 \tag{6.55}$$

Table 6.4 compares experimental values for the surface tension γ_m of the 39 or the 41 liquid metallic elements with those calculated from Eqs. (6.50), (6.51), and (6.52), or Eq. (6.53), together with the corresponding δ_i, Δ, and S values. It is evident from Table 6.4 that the model represented by Eq. (6.52), or Eq. (6.53), in terms of new parameter makes, statistically (i.e. in view of Δ and S), a noticeable improvement over both the original Skapski and Schytil models. This is particularly true over the Schytil model, giving $\Delta(41)$ and $S(41)$ values of 19.8 per cent and 0.241, respectively. This level of uncertainty is considerably closer to the uncertainties in the experimental measurements for the surface tensions of high melting point, toxic, and/or reactive liquid metallic elements. Thus, Eqs. (6.52) and (6.53), expressed in terms of new parameters $\xi_T^{1/2}$ and $\xi_E^{1/2}$, can be regarded as a modified Skapski and a modified Schytil model. (The physical significance of the parameters $\xi_E^{1/2}$ and $\xi_T^{1/2}$ is described in Section 5.5.)

Table 6.4 *Comparison of experimental values for the surface tension of liquid metallic elements at their melting point temperatures with those calculated from Eqs. (6.50), (6.51), and (6.52), together with values for δ_i, Δ, and S*

Element		$(\gamma_m)_{exp}$ $N\,m^{-1}$	$(\gamma_m)_{cal}$ / N m^{-1}			δi / %		
			Eq. (6.50)	Eq. (6.51)	Eq. (6.52)	Eq. (6.50)	Eq. (6.51)	Eq. (6.52)
Aluminium	Al	1.050	1.278	1.040	0.913	−7.8	1.0	15.0
Antimony	Sb	0.371	0.512	0.719	0.556	−27.5	−48.4	−33.3
Barium	Ba	0.273	0.277	0.469	0.256	−1.4	−41.8	6.6
Beryllium	Be	1.100	2.111	2.871	2.190	−47.9	−61.7	−49.8
Bismuth	Bi	0.382	0.440	0.405	0.453	−13.2	−5.7	−15.7
Cadmium	Cd	0.637	0.377	0.575	0.617	69.0	10.8	3.2
Caesium	Cs	0.069	0.083	0.098	0.071	−16.9	−29.6	−2.8
Calcium	Ca	0.363	0.345	0.658	0.410	5.2	−44.8	−11.5
Cerium	Ce	0.794	0.905	0.791	0.534	−12.3	0.4	48.7
Cobalt	Co	1.900	2.173	2.572	2.088	−12.6	−26.1	−9.0
Copper	Cu	1.320	1.673	1.916	1.572	−21.1	−31.1	−16.0
Gallium	Ga	0.724	1.099	0.336	0.510	−34.1	115	42.0
Germanium	Ge	0.607	1.306	1.217	0.885	−53.5	−50.1	−31.4
Gold	Au	1.150	1.410	1.487	1.616	−18.4	−22.7	−28.8
Indium	In	0.560	0.770	0.376	0.497	−27.3	48.9	12.7
Iron	Fe	1.880	1.949	2.556	2.078	−3.5	−26.4	−9.5
Lanthanum	La	0.728	1.073	0.819	0.623	−32.2	−11.1	16.9

Table 6.4 *(continued)*

Element		$(\gamma_m)_{exp}$ $N m^{-1}$	$(\gamma_m)_{cal}/N m^{-1}$			$\delta i / \%$		
			Eq. (6.50)	Eq. (6.51)	Eq. (6.52)	Eq. (6.50)	Eq. (6.51)	Eq. (6.52)
Lead	Pb	0.457	0.546	0.467	0.551	−16.3	−2.1	−17.1
Lithium	Li	0.399	0.551	0.452	0.281	−27.6	−11.7	42.0
Magnesium	Mg	0.577	0.470	0.842	0.613	22.8	−31.5	−5.9
Manganese	Mn	1.152	1.120	1.897	1.346	2.9	−39.3	−14.4
Mercury	Hg	0.489	0.209	0.220	0.339	134	122	44.2
Molybdenum	Mo	2.080	2.731	3.436	3.105	−23.8	−39.5	−33.0
Nickel	Ni	1.795	2.187	2.550	2.096	−17.9	−29.6	−14.4
Platinum	Pt	1.746	2.168	2.419	2.517	−19.5	−27.8	−30.6
Plutonium	Pu	0.550	1.258	0.856	0.583	−56.3	−35.7	−5.7
Potassium	K	0.110	0.130	0.145	0.102	−15.4	−24.1	7.8
Praseodymium	Pr	0.716	0.949	0.880	0.639	−24.6	−18.6	12.1
Rubidium	Rb	0.0896	0.110	0.118	0.0850	−18.5	−24.1	5.4
Selenium	Se	0.103	0.285	0.380	0.185	−63.9	−72.9	−44.3
Silicon	Si	0.850	1.739	1.902	1.063	−51.1	−55.3	−20.0
Silver	Ag	0.924	1.082	1.354	1.231	−14.4	−31.6	−24.8
Sodium	Na	0.197	0.230	0.245	0.170	−14.3	−19.6	15.9
Strontium	Sr	0.296	0.274	0.531	0.322	8.0	−44.3	−8.1
Tantalum	Ta	2.050	3.130	3.506	2.996	−34.5	−41.5	−31.6
Tellurium	Te	0.239	−	0.517	0.213	−	−53.8	12.2
Thallium	Tl	0.459	0.517	0.471	0.510	−11.2	−2.5	−10.0
Tin	Sn	0.555	0.964	0.429	0.565	−42.4	29.4	−1.8
Tungsten	W	2.310	3.678	4.356	3.513	−37.2	−47.0	−34.2
Ytterbium	Yb	0.320	−	0.703	0.393	−	−54.5	−18.6
Zinc	Zn	0.789	0.545	0.841	0.812	44.8	−6.2	−2.8
					$\Delta \%$	28.6[†]	35.1[‡]	19.8[‡]
					S	0.373[†]	0.436[‡]	0.241[‡]

[†] $\Delta(39)$, $S(39)$;
[‡] $\Delta(41)$, $S(41)$.

Table 6.5 *Comparison of experimental temperature coefficients of surface tension for liquid metallic elements with those calculated from Eqs. (6.42) and (6.44).*

Element		$-(d\gamma/dT)_{exp}$ $mN\,m^{-1}K^{-1}$	Range[†]	$-(d\gamma/dT)_{cal}/mN\,m^{-1}K^{-1}$	
				Eq. (6.42)	Eq. (6.44)[‡]
Aluminium	Al	0.25	0.12–0.34	0.26	–
Antimony	Sb	0.045	0.034–0.084	0.09	0.10
Barium	Ba	0.072	0.069–0.095	0.06	0.06
Bismuth	Bi	0.08	0.06–0.15	0.11	–
Cadmium	Cd	0.15	0.065–0.25	0.23	0.16
Caesium	Cs	0.047	0.0436–0.057	0.05	–
Calcium	Ca	0.10	0.068–0.112	0.14	–
Cobalt	Co	0.35	0.172–0.49	0.40	–
Copper	Cu	0.28	0.087–0.43	0.28	–
Gallium	Ga	0.07	0.037–0.13	0.18	–
Germanium	Ge	0.14	0.095–0.156	0.14	–
Gold	Au	0.15	0.10–0.52	0.27	–
Indium	In	0.09	0.0668–0.13	0.17	–
Iron	Fe	0.41	0.20–0.704	0.47	0.29
Lead	Pb	0.11	0.060–0.24	0.13	–
Lithium	Li	0.15	0.14–0.163	0.16	0.14
Magnesium	Mg	0.26	0.149–0.33	0.24	0.16
Mercury	Hg	0.23	0.19–0.30	0.22	–
Nickel	Ni	0.33	0.22–0.50	0.45	–
Potassium	K	0.07	0.060–0.11	0.07	–
Rubidium	Rb	0.056	0.052–0.0657	0.06	0.05
Silver	Ag	0.22	0.117–0.54	0.24	–
Sodium	Na	0.09	0.05–0.11	0.14	–
Thallium	Tl	0.106	0.08–0.16	0.13	–
Tin	Sn	0.07	0.0476–0.222	0.14	–
Zinc	Zn	0.21	0.090–0.25	0.25	0.22

[†] Sources of data: Keene [21] and Mills and Su [36].
[‡] Eustathopoulos et al.' s [32] calculated values.
See also Iida and Guthrie [54].

In Table 6.5, calculated values for the temperature coefficients $d\gamma / dT$ of surface tension based on Eqs. (6.42) and (6.44) are compared with experimental data. Agreement with experimental data is seen to be good. Unfortunately, at present, a detailed comparison between experimental and calculated values for $d\gamma / dT$ is extremely difficult because experimentally derived values for $d\gamma / dT$, are, in general, highly uncertain, as shown in Table 6.5.

6.7 Adsorption of Solutes on Liquid Metallic Surfaces

The plane of separation between two phases is known as an interface. In particular, an interface between a condensed phase and its own vapour or an inert gas is called a surface. The surface is in a higher energy state than the bulk liquid phase, because coordination among the atoms at the surface is incomplete. In this sense the surface is unstable relative to the bulk liquid phase, so that it may be more favourable energetically for the liquid to be covered by a certain foreign element. The formation of a layer of the foreign element on the surface of a liquid metallic element generally causes a decrease in its surface tension. For example, oxygen and sulphur can selectively attach, and cause a particularly significant decrease in the surface tension of metallic liquids, as such, these elements are known as surface active components.

The formation of a layer of a foreign substance, or solute, on the impermeable surface of condensed matter is called 'adsorption'. The adsorption behaviour of solute in liquid metals has, and can be, investigated by measuring the surface tension of a liquid metal as a function of the concentration of solute. The results can then be interpreted on the basis of a thermodynamic treatment of interfaces.

Thus, the excess surface concentration of solute in a two-component (binary) system at constant temperature and pressure is given by

$$\Gamma_s = -\frac{d\gamma}{RT d(\ln a_s)} = -\frac{a_s}{RT}\frac{d\gamma}{da_s} \tag{6.56}$$

in which Γ_s is the excess quantity of solute s associated with unit area of surface (i.e. the excess surface concentration per unit area), and a_s is the activity of solute s in the system. Equation (6.56) is known as the Gibbs adsorption equation, or the Gibbs adsorption isotherm. In dilute solutions, where Henry's law is obeyed, the solute activity a_s can be replaced by the solute concentration in terms of mass (weight) per cent or mole (atomic) per cent. In other words, at low concentrations of solute, Γ_s can be taken to equal the surface concentration of solute per unit interfacial area. As is evident from Eq. (6.56), the excess surface concentration Γ_s can be evaluated from the slope of experimentally determined $d\gamma / d(\ln a_s)$ for $d\gamma / d(\ln x)$ values, where x is mole (atomic) fraction.

Halden and Kingery [37] carried out experimental determinations of the effect of carbon, nitrogen, oxygen, and sulphur additions on the surface tension of liquid iron. These surface tension measurements are plotted as a function of $\ln$(weight %), or $\ln$(mass %), in Figure 6.11. At these concentrations, the activity is essentially equal to concentration for all materials apart from carbon.

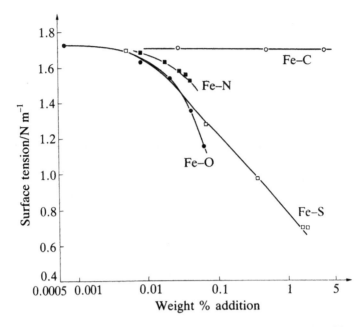

Figure 6.11 *Effect of C, N, S, and O on the surface tension of liquid iron (after Halden and Kingery [37]).*

Halden and Kingery [37] deduced excess surface concentrations of solutes using the slopes of the curves shown in Figure 6.11. Their calculated values are shown in Figure 6.12. As is obvious from this figure, they obtained the following results. The excess surface concentration of oxygen rapidly increases up to a value of 21.8×10^{-6} mol m^{-2} at about 0.04 weight (mass) per cent. The area per oxygen atom at the surface is 7.62×10^{-20} m^2, in reasonable agreement with the value of 8.12×10^{-20} m^2 per atom found in FeO for the plane of maximum packing, and with the value of 6.78×10^{-20} m^2 calculated from Pauling's radius of 1.40×10^{-10} m for O^{2-}.

Referring now to sulphur dissolved in iron, its excess surface concentration also rapidly increases up to a value of 11.6×10^{-6} mol m^{-2}. This concentration corresponds to an area of 14.4×10^{-20} m^2 per atom, which is somewhat larger than the value of 11.56×10^{-20} m^2 per atom in the plane of maximum packing for FeS as well as the value of 10.49×10^{-20} m^2 calculated from Pauling's radius of 1.84×10^{-10} m for S^{2-}. At low concentrations, sulphur is seen to be more highly surface active than oxygen. This is due to its large ionic size, which leads to the ion becoming highly polarized by the iron's ionic potential. Similarly, complete surface coverage occurs more rapidly with sulphur, and at a lower activity, than with oxygen.

Nitrogen has a smaller effect on surface tension, lowering the surface tension of iron by only 0.2 N m^{-1} at 1.01×10^5 Pa (1 atm) of nitrogen. At this concentration, the excess surface concentration was 8.3×10^{-6} mol m^{-2}, which is only a small fraction of a

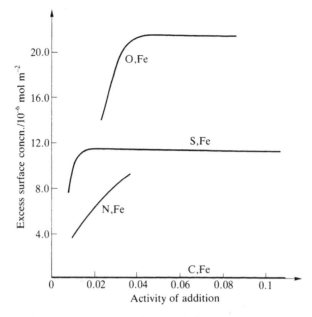

Figure 6.12 *Excess surface concentration of additions to iron at 1570 °C (after Halden and Kingery [37]).*

hexagonal close-packed monolayer (95×10^{-6} mol m^{-2}). Carbon has no effect on the surface tension of pure liquid iron at 1570 °C.

Subsequent to the work of Halden and Kingery [37], several investigators made experimental determinations of the effects of controlled solute additions on the surface tension of liquid iron. A part of their results is summarized in Figure 6.13 (by Allen [31]), in which surface tension is plotted versus solute concentration, in weight (mass) per cent. As seen, there are no large discrepancies among the data. Other examples for the effects of the periodic Group VIA element additions on the surface tension of liquid iron are given in Figure 6.14.

Olsen and Johnson [39] have studied the surface tension of mercury–thallium amalgams as a function of thallium content. Their experimental results are shown in Figures 6.15 and 6.16. They determined the excess surface concentration Γ_{Tl} from the slope of the plot of surface tension value for the amalgams, versus the logarithm of the mole fraction of thallium in mercury, i.e. ln x_{Tl} (Figure 6.16). As seen, a linear relation exists for this system at mole fractions below 0.084 (because of similar atomic weights,[3] the mole fraction values of mercury–thallium solutions are approximately equal to their mass, or weight, fractions). Analysis of the graph for this region yielded an excess surface concentration Γ_{Tl} of 1.78×10^{-6} mol m^{-2}. Olsen and Johnson's explanation for these results is as follows. The limiting concentration expected in a close-packed

---

[3] Former name for relative atomic mass.

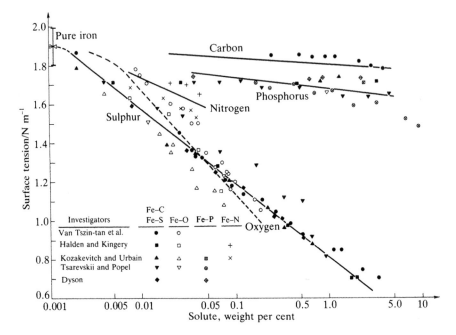

Figure 6.13 *Effect of various non-metals on the surface tension of liquid iron at 1550–1570 °C (after Allen [31]).*

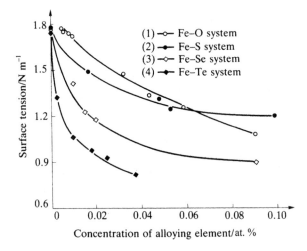

Figure 6.14 *Effect of alloying elements on the surface tension of liquid iron at 1600 °C (after Ogino et al. [38]).*

monolayer of thallium atoms with an atomic radius of 1.99×10^{-10} m is estimated to be 12×10^{-6} mol m^{-2}. A comparison of these values suggests that either an imperfect mono-layer is formed or that the assumption of close packing in the monolayer is incorrect. Thus, at concentrations less than 8.5 mass per cent, the thallium appears to concentrate in a surface layer on the mercury with an accompanying reduction in the surface tension of the amalgam. The increase in surface tension for amalgams with thallium content greater than that of the eutectic composition is more difficult to explain. (For a

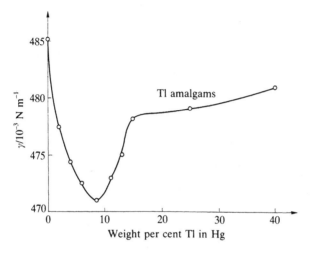

Figure 6.15 *Surface tension vs. weight (mass) per cent thallium in mercury (after Olsen and Johnson [39]).*

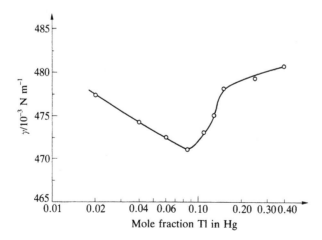

Figure 6.16 *Surface tension vs. logarithm of mole fraction thallium in mercury (after Olsen and Johnson [39]).*

number of systems, the slope remains constant over a considerable composition range, corresponding to adsorption of a monolayer at the surface [37,39].) However, if there are components from compounds which are less stable in the surface layer than in the bulk, the surface tension of the mixture may be higher than that of the pure components. It would appear that a compound (Hg_xTl_y) is formed which might be concentrated in the bulk of the amalgam. The formation of such a compound would remove thallium atoms from the surface layers and thereby raise surface tension values.

Thus, surface tension data for a solution as a function of concentration can offer potentially useful information. For example, in the rate of absorption of nitrogen into liquid iron, Belton [40] considered the adsorption of strongly surface active solutes in terms of the ideal Langmuir isotherm. The Langmuir isotherm is most simply expressed, at constant temperature, and for a single solute forming a monolayer as follows:[4]

$$\frac{\theta_s}{1 - \theta_s} = Ka_s \tag{6.57}$$

Here, θ_s is the fractional coverage by the single solute and K is called a coverage in-dependent adsorption coefficient. Belton has shown that combination of the Langmuir and Gibbs isotherms, assuming ideal isotherms hold at all compositions, leads to (cf. also March and Tosi [9])

$$\gamma_0 - \gamma = RT\Gamma_s^0 \ln(1 + Ka_s) \tag{6.58}$$

in which γ_0 is the surface tension of the pure solvent, $(\gamma_0 - \gamma)$ is the depression of sur-face tension of the pure liquid metal, and Γ_s^0 is the saturation coverage by the solute $(\theta_s = \Gamma_s/\Gamma_s^0)$. Belton reported that Eq. (6.58) can be used in conjunction with ex-perimental values of Γ_s^0 and K to give a very good description of the experimental results for surface tension–concentration curves, and presented the following isotherms (in units of mN/m).

Fe–S solutions at 1550 °C:

$$1788 - \gamma = 195 \ln(1 + 185a_S) \tag{6.59}$$

Fe–C(2.2 mass %)–S solutions at 1550 °C:

$$1765 - \gamma = 184 \ln(1 + 325a_S) \tag{6.60}$$

Fe–O solutions at 1550 °C:

$$1788 - \gamma = 240 \ln(1 + 220a_O) \tag{6.61}$$

---

[4] The Langmuir isotherm arises from the assumption that the energy of adsorption of the species is inde-pendent of surface coverage and atomic arrangement of adsorbing species on the surface. Incidentally, the same equation can be derived through statistical mechanics.

Fe–Se solutions at 1550 °C:

$$1788 - \gamma = 176 \ln(1 + 1200 a_{Se}) \tag{6.62}$$

Cu–S solutions at 1120 °C:

$$1276 - \gamma = 132 \ln(1 + 140 a_S) \tag{6.63}$$

Cu–S solutions at 1300 °C:

$$1247 - \gamma = 149 \ln(1 + 27 a_S) \tag{6.64}$$

Ag–O solutions at 980 °C:

$$923 - \gamma = 50 \ln(1 + 340 a_O) \tag{6.65}$$

Ag–O solutions at 1107 °C:

$$904 - \gamma = 55 \ln(1 + 57 a_O) \tag{6.66}$$

As examples, the values calculated by the above equations are shown together with experimental data in Figures 6.17 to 6.19. As indicated in these figures, the isotherms are found to represent the data closely.

The effect of alloying on the surface tension of liquid metals in dilute solutions has been compiled by Wilson [41] and reviewed by Allen [31]. Roughly speaking, as pointed out by Baes and Kellog [42], the elements most likely to be highly surface active in metallic liquids are those of limited solubility in the metallic liquid and possessing considerably weaker intermolecular bonding forces than metallic liquid itself. The elements of the

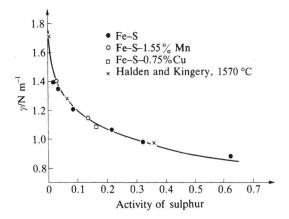

Figure 6.17 *The depression of the surface tension of iron by sulphur, and comparison with the ideal isotherm (Eq. (6.59)) for 1550 °C (after Belton [40]).*

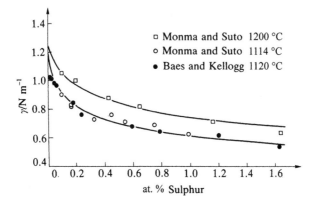

Figure 6.18 *Comparison of the isotherms (Eqs. (6.63) and (6.64)) for 1120 and 1300 °C with the data for the depression of the surface tension of copper by sulphur (after Belton [40]).*

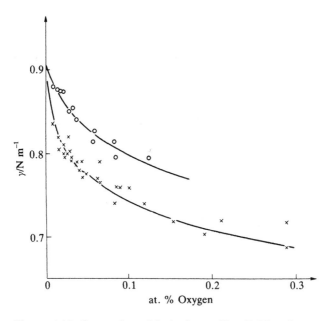

Figure 6.19 *Comparison of the isotherms (Eqs. (6.65) and (6.66)) with the measurements of Bernard and Lupis for the effect of oxygen on silver. Upper line and open circles, 1108 °C; lower line and crosses, 980 °C (after Belton [40]).*

periodic Group VIA are apt to be strongly surface active in metallic liquids (e.g. see Figure 6.14). On the other hand, in dilute solutions, a high surface tension solute (in general, a high melting point metal) is expected to have little effect on the surface tension of solutions. These considerations appear to be substantiated by the results of experimental investigations. However, common trace impurities may produce considerable effects on the surface tension of metallic liquids. Systematic investigations into the effects of solute additions on the surface tension of carefully purified metallic liquids are therefore highly desirable.

6.8 Equations for the Surface Tension of Binary Liquid Mixtures

The surface tensions of most binary liquid mixtures or alloys exhibit negative deviations from the proportional mathematical addition of pure components' surface tensions. This is often the result of the liquid surface becoming enriched with the component of lower surface tension.

Several theoretical equations describing the surface tension of binary liquid mixtures have been proposed. However, in view of liquid state physics, none is completely satisfactory, owing to a lack of information on the structures of (binary) liquid surfaces. Furthermore, no empirical relation has been presented for the surface tension of liquid mixtures or alloys. In the field of materials process science, thermodynamic approaches have been made to describe the surface tension of liquid alloys.

Let us now give brief outlines of two different approaches to the surface tension of liquid mixtures.

(1) By extending Gibbs treatment on the surface tension of mixtures, Butler [43] deduced a set of equations of the form

$$
\begin{aligned}
\gamma_M &= \gamma_1 + \frac{RT}{A_1} \ln \frac{a_1^S}{a_1} \\
&= \gamma_2 + \frac{RT}{A_2} \ln \frac{a_2^S}{a_2} \\
&= \ldots
\end{aligned}
\tag{6.67}
$$

in which γ_M and γ_1, γ_2, $\cdots$, represent the surface tension of the mixture and its components 1, 2, $\cdots$, respectively; A_1, A_2, $\cdots$, are the respective areas of components 1, 2, $\cdots$, occupied in monolayers, i.e. molar surface area; a_1, a_2, $\cdots$, are Raoultian or mole fraction activities of the components in the bulk of the mixture referred to the pure bulk components as the standard states; and a_1^S, a_2^S, $\cdots$, are the components' activities in the surface monolayer in which the surface monolayers of the pure components are taken as the standard states.

In deriving Eq. (6.67), the essential assumption made is that the difference in composition of the surface from that of the bulk is entirely restricted to the first layer of molecules. For ideal, or perfect, binary mixtures, Eq. (6.67) may be written:

$$\gamma_M = \gamma_1 + \frac{RT}{A_1} \ln \frac{x_1^s}{x_1}$$

$$= \gamma_2 + \frac{RT}{A_2} \ln \frac{x_2^s}{x_2} \qquad (6.68)$$

in which x_1, x_2 and x_1^s, x_2^s are the mole fractions in the bulk and in the surface monolayer, respectively.

Butler's model has been widely applied to calculate the surface tensions of binary liquid alloys (e.g. Lee et al. [44]).

(2) Guggenheim [45], using a statistical mechanical approach (i.e. the method of grand partition function), derived equations for the surface tension of binary solutions based on the assumption that the difference in composition of the surface from the bulk is confined to a unimolecular layer. Guggenheim's equations for binary systems can be expressed in the following form: for the surface tension of a perfect solution,

$$\exp\left(-\frac{\gamma_M A}{RT}\right) = x_1 \exp\left(-\frac{\gamma_1 A}{RT}\right) + x_2 \exp\left(-\frac{\gamma_2 A}{RT}\right) \qquad (6.69)$$

and for the surface tension of a regular solution,

$$\gamma_M = \gamma_1 + \frac{RT}{A} \ln \frac{x_1^s}{x_1} + \frac{W}{A} l \left\{ (x_2^s)^2 - x_2^2 \right\} - \frac{W}{A} m\, x_2^2$$

$$= \gamma_2 + \frac{RT}{A} \ln \frac{x_2^s}{x_2} + \frac{W}{A} l \left\{ (x_1^s)^2 - x_1^2 \right\} - \frac{W}{A} m\, x_1^2 \qquad (6.70)$$

where A is the molar surface area,[5] $W = H^E / x_1 x_2$, (H^E is the enthalpy of mixing), l and m are the fractions of the total of next-neighbour contacts made by a molecule (any molecule in the surface layer) within its own layer and with molecules in the next layer ($l + 2m = 1$).[6]

Bernard and Lupis [46] determined the surface tension of silver–gold alloys and demonstrated that the experimental results agree very well with those calculated from Eq. (6.69), as shown in Figure 6.20.

Guggenheim's treatment of the surface tension for a binary regular solution was modified by Hoar and Melford [47]; their modified expressions are

$$\gamma_M = \gamma_1 + \frac{RT}{A_1} \ln \frac{x_1^s}{x_1} + \frac{W}{A_1} \left\{ l'(x_2^s)^2 - x_2^2 \right\}$$

$$= \gamma_2 + \frac{RT}{A_2} \ln \frac{x_2^s}{x_2} + \frac{W}{A_2} \left\{ l'(x_1^s)^2 - x_1^2 \right\} \qquad (6.71)$$

---

[5] In Guggenheim's approach, the molar areas A_1 and A_2 are assumed to be equal, i.e. $A_1 = A_2 = A$.
[6] In a simple cubic lattice, $l = 2/3$ and $m = 1/6$, while in a close-packed lattice, $l = 1/2$ and $m = 1/4$.

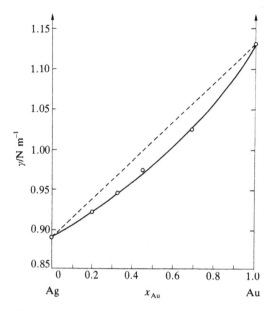

Figure 6.20 *Surface tension of silver–gold alloys at 1108 °C. The continuous line is predicted by the 'perfect solution' model (after Bernard and Lupis [46]).*

in which l' is the fractional factor. According to their investigations, the value of l' lies in the range of 0.5 to 0.75 for binary liquid alloys.

Equation (6.71) gives relatively good agreement with experimental data for the surface tensions of binary lead–tin and lead–indium alloys, while its performance has been rather poor for alloy systems which present large deviations from ideality, and for compound-forming alloy systems in the solid state.

Finally, we indicate some recent research papers on the surface tension of liquid alloys [48–52].

6.9 Methods of Surface Tension Measurement

The surface tensions of a large number of liquid metallic elements have been determined experimentally. Experimentally derived surface tension data are available for some 67 liquid metallic elements. However, the accurate measurement of a metallic liquids' surface tension is not easy, because of the difficulty of maintaining uncontaminated metallic liquids' surfaces, particularly from contamination by oxygen, at high temperatures. Similarly, some difficulties are involved in the adequate design of equipment for high temperatures. Figures 6.21 and 6.22, respectively, give values of the surface

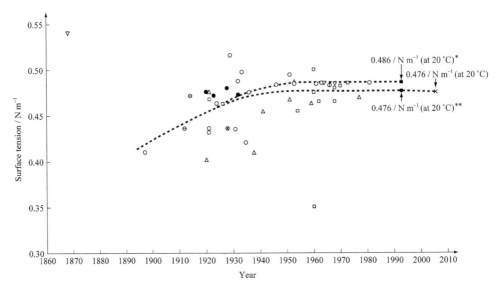

Figure 6.21 *Surface tension values for liquid mercury at room temperature as a function of year. Data from Lang [53] and Keene [21]. Methods of surface tension measurement:* ○, *sessile drop;* □, *maximum bubble pressure;* △, *maximum drop pressure;,* ●, *drop weight;* ▽, *pendant drop;* ⊖, *oscillating jet;* ⊗, *capillary depression,;* Φ, *contact angle* ∗, *mean of the five highest values for* γ, *and* ∗∗, *mean* γ *for all data (see the review by Keene [21]),* ×, *adopted by Mills and Su [36]:* γ (mN m⁻¹) = 489 − 0.23(T − 235 K).*

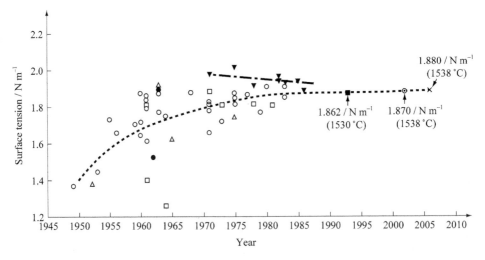

Figure 6.22 *Surface tension values for liquid iron at or near its melting point as a function of year. Data from Iida and Guthrie [54], Keene [21], and Mills and Su [36] Methods of surface tension measurement:* ○, *sessile drop;* □, *maximum bubble pressure;* △, *pendant drop;* ▼, *oscillating drop;* ●, *drop weight;* ■, *mean* γ *at 1530 °C (after Keene [21]).* ⊙, *recommended by Mills [55],* ×, *adopted by Mills and Su [36]:* γ (mN m⁻¹) = 1880 − 0.41(T − 1811 K).*

tensions of liquid mercury, and iron at, or near their melting point temperatures. Mercury appears to be the only liquid metallic element whose surface tension may be known accurately. In other words, the surface tensions of liquid metallic elements, particularly those with high melting points and/or those which are chemically reactive, have yet to be established.

In experimental determinations of the surface tension of metallic liquids, the surface tension of mercury at room temperature is employed as an important standard for calibration purposes.

As indicated in Figure 6.21, the surface tension of mercury has taken us many years to establish. As can be seen from Figures 6.21 and 6.22, the surface tension value of a liquid metallic element has a tendency to increase over a period of years, and to approach its presumably true value. This is supported by the fact that the extent of scatter in the values of surface tension measured has gradually diminished with time. This reduced scatter can be attributed mainly to decrease in impurities and surface contamination. However, as seen from Figure 6.22, the surface tension values of liquid iron using the oscillating drop technique seem to be decreasing with the passing of the years. Furthermore, as seen from Figures 6.21 and 6.22, the maximum bubble pressure technique has occasionally produced low surface tension values. Needless to say, precise metrological studies on methods to determine the surface tension of liquid metallic elements are indispensable.

Most methods for measuring surface tensions are based on some form of the Laplace equation for describing the pressure differences set up across curved interfaces. The surface tension pressure difference ΔP is given by

$$\Delta P = \gamma \left(\frac{1}{R_1} + \frac{1}{R_2} \right) \tag{6.72}$$

where γ is the surface tension or interfacial tension and R_1 and R_2 are the principal radii of curvature of the surface at the point considered. For a spherical interface, $R_1 = R_2 = R$, so that $\Delta P = 2\gamma / R$. These relations are relevant to the methods described below.

There are a number of methods whereby the surface tension of a liquid can be measured. When measuring the surface tension of metallic liquids, it is important to adopt a method which is appropriate to the experimental aim and metal-types. For most determinations of γ, one or more of the following methods have been employed:

(a) Sessile drop method
(b) Maximum bubble pressure method
(c) Pendant drop method
(d) Drop weight method
(e) Maximum drop pressure method
(f) Capillary rise method
(g) Oscillating (or Levitated) drop method

Of these methods, the sessile drop method, the maximum bubble pressure method, and the oscillating drop method are most suitably applied at elevated temperatures and have been most frequently used. For the high melting point refractory metals, the pendant drop method and the drop weight method have sometimes been employed.

The following are the brief outlines of these techniques.

6.9.1   Sessile Drop Method

As already mentioned in Subsection 3.6.6, the sessile drop method is based on the measurement of the dimensions of a stationary liquid drop (usually 3×10^{-7} to 5×10^{-7} m^3) resting on a horizontal substrate. This technique has been used extensively for measuring the surface tension of metallic liquids, because of the following advantages: (i) the sessile drop method allows for a more accurate determination of surface tensions over wide ranges of temperatures compared with the other methods used at high temperatures, (ii) it also allows simultaneous measurements of contact angle, spreading coefficient, work of adhesion, and density, (iii) lastly, both the mathematical treatment for calculating surface tensions from measured drop dimensions, as well as the experimental equipment needed for producing a sessile drop, are relatively simple. However, every precaution must be taken to ensure the absence of any source of contamination. An adequate cylindrical symmetry of the sample, as well as good optical equipment,[7] are necessary before accurate data can be obtained.

Several equations have been presented to calculate surface tension from the dimensions of a sessile drop. Figure 6.23 shows the dimensions to be measured.

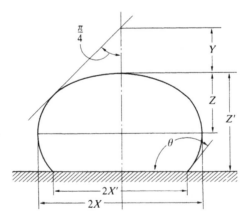

Figure 6.23 *Measurements of sessile drop for surface tension calculations.*

---

[7] The dimensions of a sessile drop are determined from an analysis of the profile of the drop (generally, its photograph).

6.9.1.1   *Bashforth and Adams Equation*

The surface tension of the liquid specimen can be obtained using the Bashforth and Adams [56] equation:

$$\gamma = \frac{g\rho b^2}{\beta} \tag{6.73}$$

where g is the gravitational acceleration and ρ is the liquid density. From the measured values of X and Z, parameters b and β can be determined using the Bashforth and Adams tables.

The difficulty in measuring Z entails an uncertainty of 2 to 3 per cent on γ. Incidentally, the drop volume V can be calculated from the relation

$$V = \frac{\pi b^2 X^2}{\beta}\left(\frac{2}{b} - \frac{2\sin\theta}{X} + \frac{\beta Z}{b^2}\right) \tag{6.74}$$

Density can be obtained from the drop volume and its weight.

6.9.1.2   *Dorsey Equation*

In this approach, the surface tension is calculated from the following relation (Dorsey's equation [57]):

$$\gamma = g\rho X^2\left(\frac{0.0520}{f} - 0.1227 + 0.0481f\right) \tag{6.75}$$

where the Dorsey factor f is given by

$$f = \frac{Y}{X} - 0.4142$$

This method yields better accuracy than the previous one, but the difficulty in the experimental measurement of Y (see Figure 6.23) remains an appreciable source of error, or uncertainty.

6.9.1.3   *Corrected Worthington Equation*

$$\gamma = \frac{1}{2}\rho g Z^2 \frac{1.641X}{1.641X + Z} \tag{6.76}$$

The mathematical treatment of this equation [58] appears to be satisfactory, but, once again, the difficulty in the determination of Z with high accuracy still remains.

6.9.2 Maximum Bubble Pressure Method

This technique involves measuring the maximum pressure attained in bubbles formed at the tip of a capillary tube immersed to different depths within a liquid, as described in Subsection 3.6.4.

The maximum bubble pressure method has frequently been employed in experimental determinations of the surface tensions of metallic liquids. Its popularity stems from the reason that each successive measurement is made on a freshly formed surface. As a result, surface contamination effects are reduced to a minimum. The method is thought to be particularly suitable for metallic liquids which are very sensitive to surface contamination, such as alkali metals, magnesium, aluminium, and calcium, which have a great affinity for oxygen. Needless to say, helium or argon (inert gas), used to form the bubbles must be carefully purified. The measurements are carried out remotely since no direct observations on the mechanics of bubble formation in metallic liquids are made. Methods have been established for obtaining absolute surface tension figures from the experimental data, which obviates any need for any experimental calibration procedure.

If the bubble detaching from the orifice tip is perfectly spherical, Eq. (3.17) holds. However, since the pressure on the liquid side of the bubble varies with the head of liquid, the bubble is not spherical even though it may approach a spherical shape when a capillary tube of very small radius is used. A correction must, therefore, be applied for any distortion of the bubble due to gravitational effects. Similarly, possible non-wetting effects may lead to low values of surface tension.

Several methods have been proposed for correcting experimental data to give absolute surface tension values. Of these, the Cantor relation corrected by Schrödinger [59] is probably the most frequently used [53]:

$$\gamma = \frac{rP_\gamma}{2}\left\{1 - \frac{2}{3}\left(\frac{r\rho g}{P_\gamma}\right) \times 10^{-3} - \frac{1}{6}\left(\frac{r\rho g}{P_\gamma}\right)^2 \times 10^{-6}\right\} \tag{6.77}$$

where $P_\gamma(= P_m - \rho g h)$ is in SI units, P_m is the maximum gas pressure (gauge) at an immersion depth h, and r is the capillary radius (mm).

The expression in the brackets is the correction factor for P_γ. Equation (6.77) holds well for small values of $r(2\gamma/\rho g)^{-1/2}$, i.e. $r < 0.2 \, (2\gamma/\rho g)^{1/2}$ (where $2\gamma/\rho g$ is called the 'capillary constant' or 'specific cohesion').

A method proposed by Sugden [60], which is mathematically more complicated, holds over a wide range of capillary diameters.

This method, which makes use of two capillary tubes of different radii, has been employed on occasion. One advantage of the method is that density data are not needed in calculating surface tensions.

6.9.3 Pendant Drop Method

Figure 6.24 indicates the shape of a drop hanging from the tip of a vertical capillary or rod. If the forces of gravity and surface tension are balanced in a static pendant drop, the surface tension is given by [61,62]

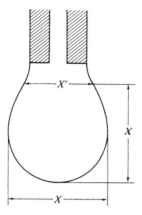

Figure 6.24 *Measurements of pendant drop for surface tension calculations.*

$$\gamma = \frac{\rho g X^2}{H} \tag{6.78}$$

in which X is the maximum drop diameter and $1/H$ is the shape factor of the drop. Values of $1/H$ have been calculated as a function of the experimentally determined parameters X'/X, and usually vary between 0.3 and 1.0 [61]. Providing equilibrium is attained and there are no errors in ρ, X', and X, Eq. (6.78) yields accurate surface tension values.

In several ways, the pendant drop method is similar to the sessile drop method.

6.9.4 Drop Weight Method

When a liquid drop formed at the tip of a vertical cylindrical rod or capillary grows large enough, the liquid drop falls as a result of its weight. The liquid surface tension can be computed from the mass of the drop which falls. If m is the mass of the drop separating from a vertical rod or capillary tube of radius r as a result of gravitational forces, the surface tension of the liquid is given by [62–64]

$$\gamma = \frac{mg}{2\pi r f_D} \tag{6.79}$$

where f_D is a function of $(r/V^{1/3})$, V being the volume of the falling drop. A drop whose weight is equal to $2\pi r\gamma$ (when $f_D = 1$, $2\pi r\gamma = mg$) is called the 'ideal drop'. Both observation and theory indicate that, while the major fraction of the drop falls, a part remains attached to the capillary. Equation (6.79) may be rewritten, as follows:

$$\gamma = \frac{mgF_D}{r} \tag{6.80}$$

in which F_D is a function of (V/r^3) or $(m/\rho r^3)$. The values of F_D corresponding to given values of (V/r^3) have been listed in tables [63].

The pendant drop method and the drop weight method have something in common with each other. Of the many techniques available, the pendant drop method and the drop weight method are applicable to most transition metals. In these methods, liquid drops are generally formed by electron bombardment heating in high vacuum, and their surfaces are free of any contamination. Contamination from the capillary tube can be eliminated by heating the end of a rod under study. In such procedures, since the liquid drops are suspended by a solid of the same composition, these methods only allow the measurement of melting point surface tension. The surface tension of highly reactive and refractory metals can be readily determined, provided their metals can be obtained in the form of smooth rods of a few millimetres diameter.

Neither the pendant drop technique nor the drop weight technique are suitable for determining variations of surface tension with temperature. Furthermore, they are limited to alloys having a narrow liquid-solid range.

Evidently, great care must be taken to avoid shaking the liquid drop during the course of experiment.

6.9.5 Maximum Drop Pressure Method

This method involves measuring the pressure needed to force a tiny liquid drop from the tip of an upwards-facing capillary. The technique involves similar principles to the maximum bubble pressure method.

Advantages claimed for the maximum drop pressure method are that it provides fresh and uncontaminated surfaces, avoids the introduction of contact angles, and is free of theoretical uncertainty.

This method has been employed for determining the surface tension of highly reactive, low melting point metals, but is limited to about 750 °C because of difficulties associated with the design of equipment for high temperatures [65].

6.9.6 Capillary Rise Method

The method represents an application of the well-known phenomenon of capillary action (capillarity). The liquid's surface tension can be calculated from the expression

$$\gamma = \frac{\rho g h r}{2 \cos \theta} \tag{6.81}$$

where *h* is the difference in height between the liquid surface within the capillary and the surface of the bulk liquid metal outside the capillary, *r* is the radius of the capillary, and θ is the contact angle between the liquid and the capillary wall.

The theory of the capillary rise method is simple. However, the technique requires an exact knowledge of contact angles. Since this information is lacking for most metallic liquids, this angle is often more difficult to measure than surface tension. As a result, the method is not commonly used for metallic liquids. However, measurements of γ for some low melting point metals, e.g. indium, tin, and lead, have been made up to 750 °C using this technique [63,66].

6.9.7 Oscillating (or Levitated) Drop Method

The above mentioned methods are static or quasi-static. Since the early 1970s, a dynamic method known as the oscillating (or levitated) drop method has been used for measuring the surface tension γ of liquid metals. This containerless method started with an application of the levitation technique in which the oscillation frequency of a drop (related to the surface tension), levitated in a high frequency magnetic field, is measured.

According to Ishikawa et al. [67] and Paradis et al. [68], a levitated sample is melted, and brought to a selected temperature. Subsequently, a $P_2(\cos\theta)$-mode of drop oscillation is induced in the sample by superimposing a small sinusoidal electric field on the levitation field. Using the characteristic oscillation frequency ω_c of the signal, after correcting for non-uniform surface charge distribution, the relationship between ω_c and γ for the liquid drop is given by

$$\omega_c^2 = \left(\frac{8\gamma}{\rho r_0^3}\right)Y \tag{6.82}$$

where r_0 is the radius of the liquid drop, ρ is its density, and *Y* is a correction factor that depends on the drop's charge, the permittivity of vacuum, and the applied electric field. Real-time values of the radius and density are used in Eq. (6.82) to prevent distortion due to sample evaporation. (The measurements of radius and density of a liquid metal sample, and the determination of the sample temperature, are described in Subsection 3.6.7.)

The uncertainty in the surface tension measurements was estimated to be better than 5 per cent from the response of the oscillation detector and from the density measurements [67,68].

This method has the advantage of eliminating persistent sources of contamination which arise through the use of substrates and/or capillary tubes associated with the sessile drop, the maximum bubble pressure, and the capillary rise methods. Thus, at present, the oscillating drop method is widely used for surface tension measurements of liquid metals, in particular, high melting point liquid metals over wide temperature ranges, including undercooled regions.

Methods for measuring surface tension of metallic liquids are described in several review papers [21,36] or books [55,69,70].

6.10   Experimental Data for the Surface Tension of Liquid Metallic Elements

As already mentioned, experimentally obtained surface tension data for a large number of liquid metallic elements have been reported; at present, the data are available for some 67 liquid metallic elements.

The surface tensions of liquid metallic elements usually decrease with increasing temperature (i.e. the temperature coefficient $d\gamma/dT$ is negative, and at the critical temperature, $\gamma = 0$); in general, their temperature variation observed experimentally can be represented by the following linear relation, namely

$$\gamma = \gamma_m + \frac{d\gamma}{dT}(T - T_m) \tag{6.83}$$

where $d\gamma/dT$ is a constant whose value depends on each metallic element, ranging from -0.05×10^{-3} to -0.7×10^{-3} N m^{-1} K^{-1}, T is the absolute temperature, and subscript m denotes the melting point.

It is extremely difficult to assign an overall uncertainty to the experimental data on the surface tensions of liquid metallic elements. Broadly speaking, however, references [21,36,70–72] indicate that the reported surface tension and temperature coefficient data for liquid metallic elements vary by about ±1 to ±10 per cent and ±10 to ±50 per cent around the mean, respectively.

For a few metallic elements, e.g. cadmium [21], there are sufficient data to indicate positive temperature coefficients over limited temperature ranges. Most of the positive coefficients reported can be attributed to impurity effects or to measurements made under non-equilibrium conditions.

A liquid metallic element's surface tension is an important thermophysical property, or quantity; more accurate and reliable data are needed for the surface tension of all, or almost all, liquid metallic elements.

Table 17.8 lists experimentally derived surface tension data for liquid metallic elements, together with temperature coefficient data.

....

REFERENCES

1. M. Inouye and T. Choh, *Trans. ISIJ*, 8 (1968), 134.
2. R.H. Fowler, *Proc. R. Soc. Lond.*, A, **159** (1937), 229.
3. J.G. Kirkwood and F.P. Buff, *J. Chem. Phys.*, **17** (1949), 338.
4. A. Harashima, *J. Phys. Soc. Japan*, 8 (1953), 343.
5. M.D. Johnson, P. Hutchinson, and N.H. March, *Proc. R. Soc. Lond.*, A, **282** (1964), 283.
6. Y. Waseda and K. Suzuki, *Phys. Stat. Sol. (b)*, **49** (1972), 643.
7. A.B. Bhatia and N.H. March, *J. Chem. Phys.*, **68** (1978), 1999; A.B. Bhatia and N.H. March, *J. Chem. Phys.*, **68** (1978), 4651.
8. P.A. Egelstaff and B. Widom, *J. Chem. Phys.*, **53** (1970), 2667.

9. N.H. March and M.P. Tosi, *Atomic Dynamics in Liquids*, MacMillan, London, 1976, Chap.10.

10. D.V. Gogate and D.S. Kothari, *Phil. Mag.*, **20** (1935), 1136.

11. R. Stratton, *Phil. Mag.*, **44** (1953), 1236.

12. I. Egry, *Scr. Metall. Mater.*, **26** (1992), 1349; I. Egry, *Scr. Metall. Mater.*, **28** (1993), 1273.

13. I. Egry, G. Lohöfer, and S. Sauerland, *J. Non-Cryst. Solids*, **156–158** (1993), 830.

14. I. Yokoyama, *Physica B*, **291** (2000), 145.

15. D.C. Wallace, *Proc. R. Soc. Lond.*, *A*, **433** (1991), 615.

16. A.S. Skapski, *J. Chem. Phys.*, **16** (1948), 386; A.S. Skapski, *J. Chem. Phys.*, **16** (1948), 389.

17. R.A. Oriani, *J. Chem. Phys.*, **18** (1950), 575.

18. T. Utigard, *Z. Metallkd.*, **84** (1993), 792.

19. G. Kaptay, *Mater. Sci. Forum*, **473–474** (2005), 1.

20. A.R. Miedema and R. Boom, *Z. Metallkd.*, **69** (1978), 183.

21. B.J. Keene, *Int. Mater. Rev.*, **38** (1993), 157.

22. F. Schytil, *Z. Naturforsch.*, **4** (1949), 191.

23. V.I. Kononenko, A.L. Sukhman, S.L. Gruverman, and V.V. Torokin, *Phys. Stat. Sol. (a)*, **84** (1984), 423.

24. N.H. March, *J. Non-Cryst. Solids*, **250–252** (1999), 1.

25. T. Iida, R.I.L. Guthrie, and M. Isac, in: *ICS Proceedings of the 3rd International Congress on Science and Technology of Steelmaking*, Association for Iron and Steel Technology, Charlotte, NC, 2005, p. 3.

26. T. Iida, R. Guthrie, M. Isac, and N. Tripathi, *Metall. Mater. Trans. B*, **37** (2006), 403.

27. I. Yokoyama and Y. Waseda, *High Temp. Mater. Process.*, **24** (2005), 213.

28. S. Blairs, *Int. Mater. Rev.*, **52** (2007), 321.

29. M. Hayashi, H. Yamada, N. Nabeshima, and K. Nagata, *Int. J. Thermophys.*, **28** (2007), 83.

30. L. Battezzati and A.L. Greer, *Acta Metall.*, **37** (1989), 1791.

31. B.C. Allen, in: *Liquid Metals – Chemistry and Physics*, edited by S.Z. Beer, Marcel Dekker, New York, 1972, p. 161.

32. N. Eustathopoulos, B. Drevet, and E. Ricci, *J. Cryst. Growth*, **191** (1998), 268.

33. B. Winet, L. Magnusson, H. Fredriksson, and P.J. Desré, *J. Colloid Interface Sci.*, **225** (2002), 363.

34. R.M. Digilov, *Int. J. Thermophys.*, **23** (2002), 1381.

35. H.M. Lu, T.H. Wang, and Q. Jiang, *J. Cryst. Growth*, **293** (2006), 294.

36. K.C. Mills and Y.C. Su, *Int. Mater. Rev.*, **51** (2006), 329.

37. F.A. Halden and W.D. Kingery, *J. Phys. Chem.*, **59** (1955), 557.

38. K. Ogino, K. Nogi, and O. Yamase, *Tetsu-to-Hagané*, **66** (1980), 179.

39. D.A. Olsen and D.C. Johson, *J. Phys. Chem.*, **67** (1963), 2529.

40. G.R. Belton, *Mater. Trans.*, **7B** (1976), 35.

41. J.R. Wilson, *Met. Rev.*, **10** (1965), 573.

42. C.F. Baes and H.H. Kellog, *Trans. AIME.*, **197** (1953), 643.

43. J.A.V. Butler, *Proc. R. Soc. Lond.*, *A*, **135** (1932), 348.

44. J. Lee, W. Shimoda, and T. Tanaka, *Meas. Sci. Technol.*, **16** (2005), 438.

45. E.A. Guggenheim, *Trans. Faraday Soc.*, **41** (1945), 150.

46. G. Bernard and C.H.P. Lupis, *Mater. Trans.*, **2** (1971), 555.

47. T.P. Hoar and D.A. Melford, *Trans. Faraday Soc.*, **53** (1957), 315.

48. L.C. Prasad and P.K. Jha, *Phys. Stat. Sol. (a)*, **202** (2005), 2709.

49. R. Novakovic, F. Ricci, D. Giuranno, and A. Passerone, *Surf. Sci.*, **576** (2005), 175.

50. I. Egry, J. Brillo, and T. Matsushita, *Mater. Sci. Eng. A*, **413–414** (2005), 460.

51. R. Novakovic and T. Tanaka, *Physica B*, **371** (2006), 223.

52. K. Mukai, T. Matsushita, K.C. Mills, S. Seetharaman, and T. Furuzono, *Metall.Mater. Trans. B*, **39** (2008), 561.
53. G. Lang, *J. Inst. Met.*, **101** (1973), 300.
54. T. Iida and R.I.L. Guthrie, *The Physical Properties of Liquid Metals*, Clarendon Press, Oxford, 1993, p. 113.
55. K.C. Mills, *Recommended Values of Thermophysical Properties for Selected Commercial Alloys*, Woodhead Publishing and ASM International, Cambridge, 2002, p. 112.
56. F. Bashforth and J.C. Adams, *An Attempt to Test the Theories of Capillary Action*, Cambridge University Press, 1883.
57. N.E. Dorsey, *J. Wash. Acad. Sci.*, **18** (1928), 505.
58. A.M. Worthington, *Phil. Mag.*, **20** (1885), 51.
59. E. Schrödinger, *Ann. Phys.*, **46** (1915), 413.
60. S. Sugden, *J. Chem. Soc.*, **121** (1922), 858; S. Sugden, *J. Chem. Soc.*, **125** (1924), 27.
61. S. Fordham, *Proc. R. Soc. Lond.*, *A*, **194** (1948), 1.
62. B.C. Allen, *Trans. Mater. Soc. AIME*, **227** (1963), 1175.
63. W.D. Harkins and F.E. Brown, *J. Am. Chem. Soc.*, **41** (1919), 499.
64. A. Calverley, *Proc. Phys. Soc.*, **70B** (1957), 1040.
65. T.R. Hogness, *J. Am. Chem. Soc.*, **43** (1921), 1621.
66. D.A. Melford and T.P. Hoar, *J. Inst. Met.*, **85** (1956–7), 197.
67. T. Ishikawa, P.-F. Paradis, and S. Yoda, *Appl. Phys. Lett.*, **85** (2004), 5866.
68. P.-F. Paradis, T. Ishikawa, R. Fujii, and S. Yoda, *Appl. Phys. Lett.*, **86** (2005), 41901.
69. K.C. Mills, in *Fundamentals of Metallurgy*, edited by S. Seetharaman, Woodhead Publishing, Cambridge, 2005, p. 109.
70. H. Fukuyama and Y. Waseda (eds.), *High-Temperature Measurements of Materials*, Springer, Berlin, 2009.
71. T. Matsumoto, H. Fujii, T. Ueda, M. Kamai, and K. Nogi, *Meas. Sci. Technol.*, **16** (2005), 432.
72. T. Ishikawa, P.-F. Paradis, T. Itami, and S. Yoda, *Meas. Sci. Technol.*, **16** (2005), 443.

7

Viscosity

7.1 Introduction

When the gradient of a property such as temperature exists in a liquid, it is well known that a transport process occurs in that liquid. The transport process is a non-equilibrium, or irreversible, process in which the property (e.g. temperature) can change with time if spatial variations of the property exist within the liquid. The well-known transport processes of momentum, mass, and energy involve viscosity, diffusion, and thermal conduction, respectively. This chapter is devoted to a discussion of the viscosity of metallic liquids. Chapters 8 and 9 cover diffusion and electrical and thermal conductivities (i.e. electronic transport processes), respectively.

Let us consider, therefore, momentum transport processes that occur when an incompressible liquid is subjected to a uniform shear stress. A velocity gradient is set up perpendicular to the direction of the applied stress, as a result of the fluid's resistance to the applied motion. This resistance is known as a viscous force. Thus, when adjacent parts of a liquid move at different velocities (i.e. a velocity gradient is present), viscous forces act so as to cause the slower-moving regions to move more rapidly, and the faster-moving ones to move more slowly. Thus, viscosity is a physical property which only manifests itself when a relative motion between different layers of fluid is set up.

We now consider the problem of viscosity from a microscopic viewpoint. Although the nearest-neighbour distances and coordination numbers in the liquid state at or near the melting temperature are closely similar to those in the solid state, the dynamic behaviour of atoms in the two states is entirely different. From the microscopic point of view, the most characteristic feature of a liquid is the high mobility of its individual atoms. However, the motions of atoms through a liquid are impeded by frictional forces set up by their nearest neighbours. Viscosity is, therefore, also a measure of the friction among atoms. Consequently, a liquid's viscosity is of great interest in both the technology and theory of metallic liquids' behaviour.

From a practical standpoint, viscosity plays an important role as a key to solve quantitatively problems in fluid behaviour, as well as those related to the kinetics of reactions in materials processes. For example, a metallic liquid's viscosity is one of the main factors dominating the rise of small gas bubbles and non-metallic inclusions through it. Besides, useful information on the rates of slag (or flux)/metal reactions or the rates

The Thermophysical Properties of Metallic Liquids: Volume 1 – Fundamentals. First Edition.
Takamichi Iida and Roderick I. L. Guthrie. © Takamichi Iida and Roderick I. L. Guthrie 2015.
Published in 2015 by Oxford University Press.

of transfer of impurity elements from metal to slag and attendant composition changes can be obtained by continuous monitoring of a slag's viscosity during the reactions. Similarly, metallic glass formation in alloy systems is promoted by a rapid increase in viscosity during undercooling of the liquid.

From the standpoint of theory, other dynamic properties of liquids, e.g. diffusion, involve viscosity as an essential quantity. Several theoreticians have proposed equations for the viscosity of liquids based on statistical kinetic theories, or non-equilibrium statistical mechanics. Consequently, scientists also have a keen interest in the viscosity of liquid metals, since they represent the simplest forms of monatomic substances.

Numerous experimental measurements of the viscosities of liquid metallic elements have been made over the last 100 years or more. Even so, accurate and reliable data are still not in abundance. Large discrepancies exist between experimental viscosities obtained for some liquid metals, particularly iron, aluminium, zinc, zirconium, and titanium. The reason for these discrepancies has been attributed to the high reactivity of metallic liquids, and the technical difficulty of taking precise measurements at elevated temperatures. In addition to these, investigations by Iida et al. (e.g. see Figure 7.19) have indicated that a part of these large discrepancies is due to the lack of a rigorous working formula for calculating viscosities. More importantly, however, in the early part of this century, experimental studies carried out on the basis of a metrological theory have made steady progress in the experimental determination of the viscosities of metallic liquids [1].

7.2   Theoretical Equations for Viscosity

Exact expressions for the equilibrium properties of liquids have been formulated on the basis of statistical mechanical theory. If two basic quantities, the pair distribution function $g(r)$ and the pair potential $\phi(r)$, are available, the thermodynamic properties of liquids can be readily calculated (i.e. on the basis of the pair theory of liquid). However, the derivation of calculable, rigorous expressions for the dynamic properties of liquids is extremely difficult since atomic motion in liquids cannot be precisely described as a function of time.[1] Consequently, various approximate expressions for transport coefficients have been proposed, which rest on different concepts.

7.2.1   Equations Based on the Statistical Kinetic Theories of Liquids

(1) Born and Green [2], using their kinetic theory, derived an expression for the viscosity of liquids in terms of the pair distribution function and the pair intermolecular potential:

---

[1] In the theory of liquids, the space- and time-dependent correlation functions are the most fundamental properties. A complete knowledge of these functions at different densities and temperatures is sufficient to calculate all thermodynamic and transport properties of classical liquids. Unfortunately, there is, at the present time, no practical theory, other than the machine solution of Newton's equation of motion for each atom, for computing time-dependent correlation functions for liquids from first principles.

$$\mu = \frac{2\pi}{15}\left(\frac{m}{kT}\right)^{1/2} n_0^2 \int_0^\infty g(r)\frac{\partial\phi(r)}{\partial r} r^4 \mathrm{d}r \tag{7.1}$$

in which m is the mass of a molecule (or an atom).

Values for viscosity of liquid metals have been calculated through the use of Eq. (7.1) by Johnson et al. [3,4]. In calculating viscosity values, available experimental data for pair distribution functions, and pair potentials (i.e. ion–ion oscillatory potentials) calculated from the Born–Green (B-G) and the Percus–Yevick (P-Y) integral equations connecting $g(r)$ and $\phi(r)$[5] were used. As shown in Table 7.1, values calculated from the Born–Green integral equation appear to coincide reasonably with experimental data for the metals. The temperature dependence of viscosity is also semi-quantitatively correct. From a theoretical point of view, however, a thorough investigation is still needed for the validity of Eq. (7.1) and accuracy of the calculations for the pair potentials.

Table 7.1 *Comparison of calculated values (B-G, P-Y) for viscosity of liquid metals with experimental data.*

Metal	Temperature		Viscosity mPa s		
	K	(°C)	B-G	P-Y	expt.
Li	453	(180)	0.79	1.20	0.59
	473	(200)	0.43	1.08	0.55
Na	387	(114)	0.70	1.74	0.68
	476	(203)	0.59	1.53	~0.40
K	343	(70)	0.68	0.94	~0.51
	618	(345)	0.44	0.88	0.25
Rb	313	(40)	0.78	1.43	0.67
	513	(240)	0.53	1.26	~0.32
Cs	303	(30)	0.84	1.42	0.63
	573	(300)	0.52	1.28	~0.34
Hg	273	(0)	1.78	1.24	1.68
	423	(150)	1.57	1.09	~1.1
Al	973	(700)	0.95	1.35	2.9
	1123	(850)	0.88	1.22	~1.3
Pb	623	(350)	1.84	1.29	~2.2
	823	(550)	1.60	1.18	1.7

After Johnson et al. [3].

As already mentioned, we see that the integrands of Eqs. (6.1) and (7.1) are exactly equal. Hence, combination of these two formulae leads to the following simple expression:

$$\mu = \frac{16}{15}\left(\frac{m}{kT}\right)^{1/2}\gamma \tag{7.2}$$

At the melting point temperature, we have

$$\mu_m = \frac{16}{15}\left(\frac{m}{kT_m}\right)^{1/2}\gamma_m \tag{7.3}$$

or

$$\mu_m = 0.370\left(\frac{M}{T_m}\right)^{1/2}\gamma_m \tag{7.4}$$

where M is the molar mass and subscript m denotes the melting point. Equation (7.2) is called the Fowler–Born–Green relation.

(2) Rice and co-workers [6,7] have also proposed a viscosity equation for dense fluids, in terms of interatomic pair potentials (presented as a rigid core plus an attraction) and pair distribution functions, using their statistical kinetic theory. According to their considerations, the viscosity μ is divided into three parts: (a) the kinetic contribution μ_k, (b) the part of momentum transfer due to a hard-sphere collision $\mu_\phi(\sigma)$, and (c) a contribution arising from interactions between the soft attractions $\mu_\phi(r > \sigma)$.

The values for the viscosity of five liquid metals (sodium, potassium, zinc, indium, and tin) were calculated using Rice's formulae and compared with experimental data by Kitajima et al. [8]. The comparison between calculated and measured values indicates that the kinetic contribution to shear viscosity is negligibly small ($\mu_k \ll \mu_\phi(\sigma) + \mu_\phi(r > \sigma)$), and that the contribution due to the pair interactions is dominant for the liquid state (thus, the magnitude of $\mu_\phi(r>\sigma)$ represents 70 to 80 per cent of the total shear viscosity).

Rice and Kirkwood [7] derived an approximate equation for viscosity arising from interactions between soft attractions.

$$\mu_\phi(r > \sigma) = \frac{2\pi mn_0^2}{15\varsigma_f}\int_0^\infty r^4\left\{\frac{\partial^2\phi(r)}{\partial r^2} + \frac{4}{r}\frac{\partial\phi(r)}{\partial r}\right\}g(r)dr \tag{7.5}$$

in which ς_f is the friction coefficient.

Nevertheless, the ratio of calculated to measured values is of the order of 0.5, while the agreement between calculation and experiment is still not satisfactory from the viewpoint of materials process science and engineering.

7.2.2　An Equation Based on a Hard-Sphere Model

An expression for the (shear) viscosity of a dense fluid of non-attracting hard spheres due to Longuet-Higgins and Pople has been discussed by Faber [4] and is expressed in terms of the packing fraction η.

$$\mu = 3.8 \times 10^{-8} \frac{(MT)^{1/2}}{V^{2/3}} \frac{\eta^{4/3}(1 - \eta/2)}{(1 - \eta)^3}, \quad \text{(in Pa.s)} \tag{7.6}$$

where M is the molar mass, V is the molar volume, and T is the absolute temperature. Substitution of the value of 0.46 for the packing fraction of liquid metals at their melting point temperatures into Eq. (7.6) yields an equation for the melting point viscosity of a liquid metal:

$$\mu_m = 0.66 \times 10^{-7} \frac{(MT_m)^{1/2}}{V_m^{2/3}} \tag{7.7}$$

in which subscript m represents the melting point. This formula underestimates μ_m by a factor of about 0.37.

7.2.3　The Principle of Corresponding States

Several workers have developed correlations for the transport coefficients of liquid metallic elements as a function of temperature, using the theory of corresponding states.

Helfand and Rice [9] have shown that by assuming a pair potential, $\phi(r) = \varepsilon\phi^*(r/\sigma_a)$, where ε and σ_a are characteristic energy and distance constants and ϕ^* is a universal function of r/σ_a, it is possible to apply the theorem of corresponding states for transport properties, i.e. viscosity and diffusion. The viscosity equation based on the corresponding states theory is given by [10]

$$\mu = \mu^*(V^*)^{2/3} \frac{(MR\varepsilon/k)^{1/2}}{N_A^{1/3} V^{2/3}} \tag{7.8}$$

where the energy parameter $\varepsilon/k = 5.2T_m$

In Figure 7.1, values of $\mu^*(V^*)^{2/3}$ for various liquid metals are plotted as a function of the reciprocal of the reduced temperature T^*, where $T^* = Tk/\varepsilon$. Pasternak [10] calculated viscosity values for pure liquid iron using Eq. (7.8) and Figure 7.1. Calculated values are in good agreement with Cavalier's experimental data (see Figures 7.21 and 7.22); the calculated melting point viscosity is approximately 15 per cent lower than that of the recommended data.

The corresponding-states correlations for the transport coefficients have spreads of ± 20 per cent or more, as a result of the simplicity of the assumptions made for the pair potential.

We should note that at the melting temperature, Eq.(7.8) becomes very similar to the Andrade formula for melting point viscosity.

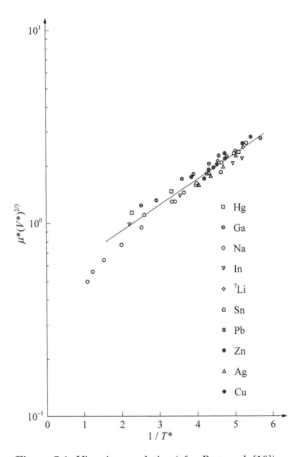

Figure 7.1 *Viscosity correlation (after Pasternak [10]).*

7.2.4   Eyring's Model Theory (the Theory of Significant Structure)

Breitling and Eyring [11] proposed the theory of significant structure by extending Eyring's early hole theory. According to this model theory, the viscosity for a simple liquid can be expressed by

$$\mu = \frac{V_s}{V}\mu_s + \frac{V - V_s}{V}\mu_g \tag{7.9}$$

where V, V_s, μ_s and μ_g are the molar volumes of liquid and solid, and the viscosities of solid-like and gas-like structures, respectively. In Eq. (7.9), a fraction V_s/V of the molecules manifests solid-like behaviour, while the remaining fraction $(V - V_s)/V$ is gas like (where $V - V_s$, is the excess volume of the liquid; the excess volume can be considered as dynamic vacancies spread throughout the liquid). The expressions for μ_s and μ_g are given by

$$\mu_s = \frac{6N_A h}{\sqrt{2}Z\kappa} \frac{V}{V_s(V - V_s)} \frac{1}{1 - e^{-\theta/T}} \exp\left\{\frac{d' \Delta_s^g H V_s}{(V - V_s)RT}\right\} \tag{7.10}$$

$$d' = \frac{(n-1)}{2Z} \frac{(V_m - V_s)^2}{V_m V_s}$$

$$n = \frac{Z V_s}{V_m} \tag{7.11}$$

$$\mu_g = \frac{2}{3\sigma^2}\left(\frac{mkT}{\pi^3}\right)^{1/2}$$

where h is Planck constant, Z is the number of nearest neighbours, κ is the transmission coefficient, θ is the Einstein characteristic frequency, $\Delta_s^g H$ is the molar sublimation energy, and σ is the diameter of the molecule.

Viscosity values for various liquid metals have been computed by Breitling and Eyring [11] using the above equations. The results show that agreement with experimental data is poor. Their conclusion was that, since all the parameters are fixed, little can be done to calculate better values for liquid metal viscosities until an improved partition function or improved parameters are used.

7.2.5 An Application of the Stokes–Einstein (or the Sutherland–Einstein) Relation

Combining the Stokes–Einstein, or the Sutherland–Einstein, relation represented by Eq. (8.24), with a hard-sphere model for self-diffusivity, Yokoyama [12,13] proposed the following viscosity equation.

$$\mu = \frac{kT}{2\pi\sigma D_{HS}} \tag{7.12}$$

where D_{HS} is the self-diffusivity in the hard-sphere fluid, with σ being a hard-sphere diameter. The values of D_{HS} can be calculated from Eq. (8.16).

7.3 Semi-Empirical or Semi-Theoretical Equations for Viscosity

Numerous expressions for transport coefficients, i.e. viscosity and diffusivity, have been proposed on the basis of a variety of models for liquids. Well-known examples are the hole theory of Frenkel [14], the reaction-rate theory of Eyring [15], the quasi-crystalline theory of Andrade [16], and the free-volume theory of Cohen and Turnbull [17]. All these model theories are based on phenomenological parameters. In many cases, the parameters must be determined experimentally.

The weakness inherent with all the model theories of a liquid is that one feature is overemphasized at the expense of another. This weakness, or problem, may be overcome only by an adequate basic theoretical treatment. Nevertheless, in view of materials process science and engineering, progress in understanding the transport coefficients depends upon phenomenological analysis based on the theory of liquids.

7.3.1   The Andrade Formula

According to Andrade [16], the atoms in the liquid state at the melting point may be regarded as executing vibrations about equilibrium positions with random directions and periods, just as if they were in the solid state (i.e. Einstein oscillators). Assuming that on the basis of such a quasi-crystalline model, the viscosity of a liquid is produced by the transfer of momentum of atomic vibrations from one layer to a neighbouring one. Andrade derived the following equation for the melting point viscosity μ_m of simple (or monatomic) liquids

$$\mu_m = \frac{4}{3}\frac{vm}{a}\bigg|_{T=T_m} \tag{7.13}$$

where v is the characteristic frequency of vibration and a is the average distance between the atoms. The numerical factor of 4/3 was a rough estimate.

In calculating the viscosity, Andrade used Lindemann's formula for v (i.e. Eq. (1.16)) and $(V_m/N_A)^{1/3}$ for a. Thus Eq. (7.13) was rewritten as:

$$\mu_m = C_A\frac{(MT_m)^{1/2}}{V_m^{2/3}} \tag{7.14}$$

where C_A is called the Andrade coefficient and V_m is the molar volume at the melting point T_m. As Born and Green [2] stated, the Andrade derivation was an ingenious dimensional consideration, in which the value for the proportionality factor C_A requires experimental or theoretical determination. The average value for the proportionality factor C_A, found by dividing experimental viscosity values μ_m by the values of $(MT_m)^{1/2}V_m^{-2/3}$ for the various liquid metals for which data are available, is 1.80×10^{-7} $kg^{1/2}$ m s^{-1} $K^{-1/2}$ $mol^{-1/6}$. As such, Eq. (7.14) can be written as:

$$\mu_m = 1.80 \times 10^{-7}\frac{(MT_m)^{1/2}}{V_m^{2/3}} \tag{7.15}$$

Equation (7.15) is called the Andrade formula (or the Andrade equation).

Early in this century, Iida et al. [18,19] found that the values of the Andrade coefficient C_A vary periodically with atomic number. In Figure 7.2, the Andrade coefficient of liquid metallic elements is shown as a function of their atomic number. It can be seen that there is a periodic variation in C_A values, in which the periodic Group IIB metals, i.e. zinc group metals, occupy the peaks, and semiconductors (silicon, germanium) and semimetals (antimony, bismuth) occupy the valleys of the curve. Moreover, Figure 7.2 indicates that (i) the Group IA metals, i.e. alkali metals, have approximately the same values of C_A ($\approx 1.94 \times 10^{-7}$), (ii) the d-block transition metals in Period

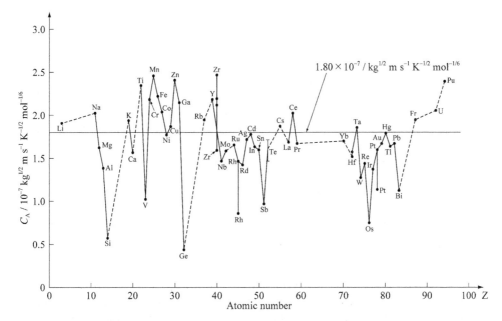

Figure 7.2 *The Andrade coefficient, C_A, appearing in Eq. (7.14), for 52 liquid metallic elements plotted against atomic number Z. The selenium (Z = 34) point falls outside the range of the figure; its C_A value is 29×10^{-7} $kg^{1/2}$ ms^{-1} $K^{-1/2}$ $mol^{-1/16}$.*

4 have C_A values of roughly 2.1×10^{-7}, while (iii) the d-block transition metals in Periods 5 and 6 appear to have roughly 1.5×10^{-7}, where the values of C_A are in SI units, i.e. $kg^{1/2}$ m s^{-1} $K^{-1/2}$ $mol^{-1/6}$. It seems that vanadium and osmium display anomalous behaviour, although experimental re-examinations are also needed to confirm their viscosities. Unfortunately, the experimental viscosity data for several transition metals are highly uncertain. Accurate and reliable data are needed before devoting time to subsequent studies in the field of materials process science.

The Andrade formula for viscosity was presented in 1934. It now stands in need of an overhaul in the light of the modern day theory of liquids, although Eq. (7.15) reproduces the experimental data for liquid metals at their melting point temperatures with comparatively good agreement: the periodic variation in C_A values, shown in Figure 7.2, suggests that some unidentified parameters are probably involved in the Andrade coefficient C_A.

From the standpoint of material process science, the Andrade formula for the viscosity of monatomic liquids is worthy of further comment for the following reasons. The Andrade formula is expressed in terms of well-known parameters (i.e. M, T_m, and V_m, or ρ_m) and gives relatively good agreement with experimental data. At melting temperatures, other expressions for viscosity, e.g. the Born–Green equation based on statistical kinetic theory [2], the equation based on a hard-sphere model, i.e. Eq. (7.7), the equation based on the corresponding-states principle, i.e. Eq. (7.8), closely resemble the Andrade formula; only the numerical factor of each theory or model is different from that of the Andrade formula. In addition, Osida [20] applied Andrade's considerations of viscosity to the conduction of heat in dielectric liquids and obtained satisfactory results.

From the viewpoint of materials process science, the present authors [21] reconsidered the model theory of Andrade and have proposed an alternative expression for the viscosity of liquid metallic elements, using more fundamental physical parameters, including the pair distribution function and the average interatomic frequency.

In reviewing Andrade's derivation of the viscosity equation, the following aspects were re-examined by the authors: (a) his assumption of a quasi-crystalline structure for liquids (which is equivalent to a close-packed crystal structure); (b) his neglect of atomic migration, or self-diffusion; (c) his employment of Lindemann's formula for the atomic frequency (of vibration) in liquids; (d) his derivation of the numerical factor of 4/3.

Our treatment for these controversial problems is as follows: (a) the structure of liquids, i.e. the distribution of atoms in a liquid metallic element, may be represented by the pair distribution function $g(r)$ obtained experimentally; (b) the distribution of atoms represented by $g(r)$ provides only the time-averaged distribution of atoms. On a microscopic time scale, atoms in the liquid state can easily diffuse due to a fluctuation in temperature or in kinetic energy. In the authors' approach, however, the phenomenon of free movement, or diffusion, of atoms is not directly considered, but an assumption is introduced that the frequency of atomic vibrations decreases with increasing temperature (for simplicity and convenience, a time averaged, apparent frequency is introduced). (c) a modified Lindemann's formula represented by Eq. (6.36) is used for the atomic frequency in liquid metallic elements. (d) Osida's treatment for the thermal conductivity of dielectric liquids is applied in the deduction of a numerical factor.

The phenomenon of viscosity in monatomic liquids, in the authors' work, as in Andrade's, is based on the transfer of momentum of atomic vibrations from one layer to a neighbouring one. In other words, momentum transfer is produced by collisions, or temporary unions, with neighbours at every displacement of atomic vibrational motion. Momentum is therefore transported in the presence of a velocity gradient in the liquid. As such, our treatment of the mechanism of viscosity is similar to that of Andrade. We can also introduce this modelling approach to readily formulate the viscosity of monatomic liquids. Further, to obtain a tractable solution to an evaluation of these momentum interchanges, it is assumed that all atoms in a state of vibration are identical harmonic oscillators. While this assumption is only approximate, it is supported by the fact that both solid and liquid metals on either side of their melting points exhibit molar heat capacity values similar to those of harmonic oscillators (i.e. $C_V \approx 3R$), and further, metals also have similar phonon dispersion curves (i.e. the relation of frequency versus wave number) based on inelastic scattering data of neutrons. [5,22].

Now, imagine that a monatomic liquid flows in the x-axis direction with a velocity v and a velocity gradient dv/dz, i.e. the flow rate increases by an amount dv for each increment of distance dz in the z-axis direction perpendicular to the x-y plane, as shown in Figure 7.3. When an atom (A) at a distance (at a given position) l away from the surface $z = z_0$ (or the x–y plane) of equal velocity vibrates at an inclination θ to the z-axis, the atom (A) makes contact in the case of a liquid of close-packed structure, at every vibration with two atoms in neighbouring planes which are at distances r away from the atom (A), where $l \leq r$. Referring to Figure 7.3, the atom (A) conveying momentum crosses the plane, $z = z_0$, twice for each complete vibration. Thus, in the case of close-packed liquids, the transfer of momentum per vibration per atom is given by

$$4m\left(\frac{dv}{dz}\right)r\cos\theta \qquad (7.16)$$

where m is the mass of the atom. Since there are 12 nearest neighbours in a close-packed structure of liquid metals, the solid angle occupied by one atom (refer to the atom (A)) can be taken to be equal to $\pi/3 (= 4\pi/12)$. The number of atoms dN within a spherical shell of solid angle $\pi/3$ between the distances r and $r+dr$ from the reference atom (A) becomes

$$\frac{\pi}{3}n_0 g(r)r^2 dr \qquad (7.17)$$

where n_0 represents the average number density. Consequently, the transfer of momentum per vibration per atom is given by the product of Eqs. (7.16) and (7.17)

$$\frac{4\pi}{3}m\left(\frac{dv}{dz}\right)n_0 g(r)r^3 dr\cos\theta \qquad (7.18)$$

Since all directions of interatomic vibration may be regarded as equally probable, the fraction of atoms vibrating at angles between θ and $\theta + d\theta$ is $\sin\theta d\theta$ (i.e.

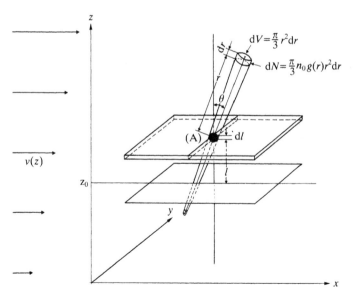

Figure 7.3 *Explanation of viscosity as communication of momentum at every displacement of an atom (A) (after Iida et al. [21]).*

$2 \times 2\pi r \sin\theta \, rd\theta / 4\pi r^2$), where $\theta \le \cos^{-1}(l/r)$. The number of atoms in a tube of unit cross section between $z = l$ and $z = l + dl$ planes is $n_0 dl$, so that the number of atoms vibrating at angles between θ and $\theta + d\theta$ within the reference tubular volume element is

$$n_0 dl \sin\theta \, d\theta \qquad (7.19)$$

Since viscous forces arise as a result of the rate of momentum interchange between atoms of adjacent layers, the frequency of atomic vibrations is needed. In our approach, we do not consider directly the phenomenon of movement of atoms by diffusion. We therefore compensate by taking a time-averaged, or apparent, frequency v of atomic vibrations. On melting, an increase in temperature fluctuations, or kinetic energy fluctuations, is to be expected in view of the increase in disorder and thermal motion within the liquid as compared to the solid. Assuming that an atom vibrates around its mean position until a kinetic energy fluctuation causes it to diffuse in a linear manner to its next place of oscillation, the probability $W(\phi)$ of a thermal fluctuation ϕ occurring may be expressed by the Gaussian approximation.

$$W(\phi)d\phi = (2\pi\bar{\phi}^2)^{-1/2} \exp\left(-\frac{\phi^2}{2\bar{\phi}^2}\right)d\phi \qquad (7.20)$$

The fluctuation in kinetic energy may be replaced by

$$\phi = \frac{E - \bar{E}}{\bar{E}} \qquad (7.21)$$

where $\bar{E}$ is the average kinetic energy, i.e. $\bar{E} = 3kT/2$, and E is the kinetic energy required for the thermal fluctuation. Supposing that E is the kinetic energy equivalent to the boiling point T_b, E is given by

$$E = \frac{3}{2}kT_b \qquad (7.22)$$

Similarly, $\bar{\phi}^2$ can be represented by

$$\bar{\phi}^2 = \frac{\overline{E^2} - \bar{E}^2}{\bar{E}^2} \approx \frac{4}{3} \qquad (7.23)^2$$

Using these relations, Eq. (7.20) can be written as

$$W(\phi)d\phi = \left(\frac{3}{8\pi}\right)^{1/2} \exp\left\{-\frac{3}{8}\left(\frac{T_b - T}{T}\right)^2\right\} d\left(\frac{T_b - T}{T}\right) \qquad (7.24)$$

---

[2] $\overline{E^2} - \bar{E}^2 = kT^2 c_v \approx 3k^2 T^2$, where c_v is the heat capacity per atom at constant volume.

Consequently, the probability, denoted by $P(T)$, that the atom will stay in a state of oscillation around a fixed coordinate position, is given by

$$P(T) = 1 - \int \left(\frac{3}{8\pi}\right)^{1/2} \exp\left\{-\frac{3}{8}\left(\frac{T_b - T}{T}\right)^2\right\} d\left(\frac{T_b - T}{T}\right) \qquad (7.25)$$

or

$$P(T) = 1 - \int_{\psi}^{\infty} (2\pi)^{-1/2} \exp\left(-\frac{\psi^2}{2}\right) d\psi \qquad (7.26)$$

where

$$\psi = \frac{\sqrt{3}}{2}\left(\frac{T_b - T}{T}\right)$$

We see that the probability function, $P(T)$, becomes less than unity with increasing temperature. Assuming that the apparent atomic frequency at equilibrium bulk temperature T may be expressed by

$$\nu = \nu_0 P(T) \qquad (7.27)$$

where ν_0 is a constant (ν_0 corresponds to the frequency of oscillation when no net displacement, i.e. diffusion, of the oscillating atom takes place), we are finally able to write an expression for viscous force, f, per unit area. Thus, by multiplying Eqs. (7.18), (7.19), and (7.27) and integrating with respect to r, l and θ, the following expression is obtained for viscous force f per unit area:

$$\begin{aligned}
f &= \frac{4\pi}{3}\nu_0 P(T) m n_0^2 \left(\frac{dv}{dz}\right) \int_0^a g(r) r^3 dr \int_{-r}^r dl \int_0^{\cos^{-1}\frac{1}{r}} \sin\theta \cos\theta\, d\theta \\
&= \frac{8\pi}{9}\nu_0 P(T) m n_0^2 \left(\frac{dv}{dz}\right) \int_0^a g(r) r^4 dr
\end{aligned} \qquad (7.28)$$

where $\underline{a}$ is the distance over which the transfer of momentum take place. Comparing Eq. (7.28) with Newton's law of viscosity given by Eq. (7.52), we obtain

$$\mu = \frac{8\pi}{9}\nu_0 P(T) m n_0^2 \int_0^a g(r) r^4 dr \qquad (7.29)$$

It can be assumed that momentum interactions with atoms in monatomic liquids occur mostly between nearest-neighbour atoms. We can estimate their maximum distance of separation as being represented by the minimum between the first and second peaks in the $g(r)$ curve of Figure 7.4, and use this as the upper limit $\underline{a}$ of the integral in Eq. (7.29). The integral itself can be estimated, to a good approximation, using the empirical relation

$$\int_0^a g(r)r^4 \mathrm{d}r \approx 1.57g(r_\mathrm{m})r_\mathrm{m}^4(r_\mathrm{m}-r_0) \tag{7.30}$$

where r_m and r_0 are the positions of the first peak, or the main peak, and its left-hand edge in the $g(r)$ curve, respectively, of the liquid metallic element under consideration (see Figure 7.4). Substituting Eq. (7.30) into Eq. (7.29), we have the desired expression for the viscosity of liquid metallic elements (or monatomic liquids).

$$\mu \approx 4.38v_0P(T)mn_0^2g(r_\mathrm{m})r_\mathrm{m}^5(1-r_0/r_\mathrm{m}), \qquad \text{(in mPa.s)} \tag{7.31}$$

Using the relationship described by Eq. (7.27), we obtain

$$\mu \approx 4.38vmn_0^2g(r_\mathrm{m})r_\mathrm{m}^5(1-r_0/r_\mathrm{m}) \tag{7.32}$$

Since r_m and n_0 are approximately given by the relations $r_\mathrm{m} \approx a$ and $n_0 \approx a^{-3}$ (where a is the average interatomic distance), Eq. (7.32) can be simplified to

$$\mu \approx 4.38g(r_\mathrm{m})(1-r_0/r_\mathrm{m})\frac{vm}{a} \tag{7.33}$$

The average frequency of atoms in liquid metallic elements at their melting temperatures may be given by the modified Lindemann's formula described by Eq. (6.21) (i.e. $v = \beta v_\mathrm{L}$).[3] Thus, using the relation given by Eq. (6.21), we have an expression for the melting point viscosity μ_m

$$\mu_\mathrm{m} \approx 4.38\beta g(r_\mathrm{m})(1-r_0/r_\mathrm{m})\frac{v_\mathrm{L}m}{a}\bigg|_{T=T_\mathrm{m}} \tag{7.34}$$

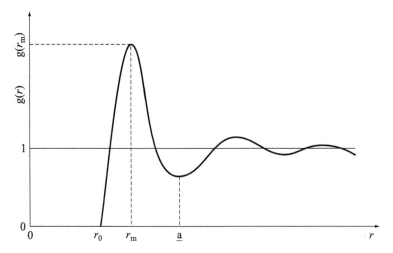

Figure 7.4 *Positions of the parameters r_0, r_m, and a in a pair distribution function curve.*

[3] At the melting temperature, $v = v_0P(T) = \beta v_\mathrm{L}$.

Since the average values of β, $g(r_m)$, and $(1 - r_0 / r_m)$ at the melting point are 0.55, 2.64, and 0.18, respectively, Eq. (7.34) can be roughly approximated by

$$\mu_m \approx 1.1 \frac{v_L m}{a} \bigg|_{T=T_m} \tag{7.35}$$

This relation is then very similar to Andrade formula given by Eq. (7.13) for the melting point viscosity of monatomic liquids.

Viscosity values calculated from Eq. (7.29) depend upon the integral term appearing there; since its evaluation is rather complex and arbitrary, liquid metal viscosities will be calculated using Eq. (7.31) or Eq. (7.32).

Substituting Eq. (1.16) into Eq. (7.34), and taking $m = M/N_A$ and $a \approx (V_m / N_A)^{1/3}$, Eq. (7.34) can be rewritten as

$$\mu_m \approx 4.38 \beta C_L N_A^{-2/3} g(r_m)(1 - r_0 / r_m) \frac{(M T_m)^{1/2}}{V_m^{2/3}} \bigg|_{T=T_m} \tag{7.36}$$

Comparison of Eq. (7.36) with Eq. (7.14) gives

$$C_A \approx 4.38 \beta C_L N_A^{-2/3} g(r_m)(1 - r_0/r_m)\big|_{T=T_m} \tag{7.37}$$

Equation (7.37) indicates that the Andrade coefficient C_A depends on the correction factor β for Lindemann's equation and the pair distribution function $g(r)$, both of which are closely related to the structure of liquid metallic elements. Conceivably, βC_L could be related to the structure of a solid at the melting point.

Djemili et al. [23] have proposed an extension of this viscosity model to binary liquid alloys in the form

$$\mu = \frac{8\pi}{9} v_0 n_0^2 \left[x_A^2 P(A) m_A \int_0^a g_{AA}(r) r^4 dr + x_B^2 P(B) m_B \int_0^b g_{BB}(r) r^4 dr \right.$$
$$\left. + x_A x_B \{ m_A P(A) + m_B P(B) \} \int_0^c g_{AB}(r) r^4 dr \right. \tag{7.38}$$

where $P(A)$ and $P(B)$ are the proportions of vibrators A and B in the liquid, x_A and x_B the concentrations of components A and B, and a, b, and c the values of r corresponding, respectively, to the first minimum of the pair distribution functions $g_{AA}(r)$, $g_{BB}(r)$, and $g_{AB}(r)$.

In evaluating values for $P(A)$ and $P(B)$, Djemili et al. have taken the statistical thermodynamic representation of a simple liquid, proposed by Hichter et al. [24], instead of choosing the Gaussian function of our approach. They applied the above expression to the cadmium–indium system. The agreement of the viscosity values calculated from Eq. (7.38) with experimental data is fairly good.

7.3.2   The Hybrid Equation of Macedo and Litovitz

Macedo and Litovitz [25] reconsidered the reaction-rate (or rate process) theory of Eyring and the free-volume theory of Cohen and Turnbull for liquid viscosities. According to Macedo and Litovitz, expressions for the viscosity of liquid can be formulated by assuming that the following two events must simultaneously occur before a molecule can undergo a diffusive jump: (a) the molecule must attain sufficient energy to break away from its neighbours, and (b) it must have an empty site large enough in which to jump. They have, thus, proposed a hybrid expression for the viscosity of liquids using Eyring's rate theory to calculate (a), and the Cohen and Turnbull approach for (b). This hybrid equation contains both the activation energy and the free volume, and is given by:

$$\mu = A_0 \exp\left(\frac{E_V^*}{RT} + \frac{\gamma_\mu v_0}{v_f}\right) \tag{7.39}$$

$$A_0 = \left(\frac{RT}{E_V^*}\right)^{1/2} \frac{(2mkT)^{1/2}}{v^{2/3}}$$

where E_V^* is the height of the potential barrier between equilibrium positions, γ_μ the constant between 0.5 and 1 (a numerical factor introduced to correct the overlap of free volume), v_0 the close-packed molecular volume, v_f the average free volume per molecule, and v a quantity roughly equal to the volume of a molecule.

In Eq. (7.39), the value of A_0 varies with temperature but is usually far less than the exponential term.

In calculating the viscosity, values of E_V^*, v_0, v_f, and v must generally be determined experimentally.

7.3.3   Chhabra and Tripathi Model

Chhabra and Sheth [26] reviewed and compiled available literature data on the viscosity of 27 liquid metals; three models, i.e. the Arrhenius, the Andrade, and the Hildebrand fluidity equations, were investigated to describe their temperature dependence. Subsequently, Chhabra and Tripathi [27] derived the following equation, based on the Mehrotra model [28]:

$$\log(\mu + 1) = 10^{b_1} T^{b_2} \tag{7.40}$$

where b_1 and b_2 are fitting parameters. Thus, they determined values of the parameters b_1 and b_2, so as to fit experimental data for 28 liquid metallic elements. Chhabra and Tripathi claim that Eq. (7.40) gives a somewhat better description to the viscosity–temperature data than the Arrhenius equation. Chhabra [29] extended his viscosity model to liquid alloys.

7.3.4   Morioka et al.'s Model

Morioka et al. [30,31] presented a 'gas-like' model for the viscosity of metallic liquids. The parameters used in their model are not arbitrarily adjustable parameters to fit experimental viscosity data, but can be related to fundamental physical quantities such as the collision cross-section of atoms, and the coordination number of nearest-neighbour atoms. Furthermore, the characteristic feature of this model is that it is capable of describing viscosities consistently, from pure liquid metals, or monatomic liquid metals, through to multicomponent liquid alloys, using parameters related to the known fundamental physical quantities. In calculating the viscosities, however, the values of a few adjustable parameters are determined from experimental viscosity data. The model was applied to 11 cases of monatomic liquid metals, and to binary and ternary alloy systems of silver, gold, and copper, over wide temperature ranges. The results of calculations indicate that the model can successfully describe the experimental viscosity data.

7.3.5   Kaptay's Unified Equation

Kaptay [32] presented a unified equation for describing the viscosity of liquid metallic elements as a function of the absolute temperature. His model is based on a combination of the Andrade formula with the model theories of the activation concept and the free volume concept. Incidentally, in Kaptay's paper, it is shown that the activation energy and the free volume concepts have identical roots and lead to identical results; in addition, it contains a history of the model theories for the viscosity of liquid metallic elements that is well outlined.

Finally, Budai et al. [33] and Živković [34] have proposed an extension of Kaptay's model for liquid alloys.

7.4   Viscosity Equation in Terms of a New Dimensionless Parameter $\xi_T^{1/2}$

In the early part of the twenty-first century, Iida et al. [18,19] proposed the following equation for the melting point viscosity of liquid metallic elements, in which they introduced a new dimensionless parameter, or dimensionless number, $\xi_T^{1/2}$. Combining Eqs. (5.32) and (6.37), they obtained

$$\mu_m = C_0 \left(\frac{M}{\xi_T T_m} \right)^{1/2} \gamma_m \tag{7.41}$$

$$C_0 \equiv \frac{1}{9.197 R^{1/2} k_0} = 3.771 \times 10^{-2} k_0^{-1} / \text{kg}^{-1/2} \text{m}^{-1} \text{sK}^{1/2} \text{mol}^{1/2}$$

where C_0 (or $0.03771/k_0$) is a constant which is approximately the same for all liquid metallic elements, apart from germanium, silicon, and selenium. In their work, the constant of proportionality C_0 was determined, so as to give a minimum S value for the

37 liquid metallic elements (as listed in Table 7.7). The value of 0.379 kg$^{-1/2}$ m^{-1} s K$^{1/2}$ mol$^{1/2}$ was obtained for C_0. Consequently, Eq. (7.41) can be written as

$$\mu_{\mathrm{m}} = 0.379 \left(\frac{M}{\xi_{\mathrm{T}} T_{\mathrm{m}}} \right)^{1/2} \gamma_{\mathrm{m}} \tag{7.42}$$

For 35 liquid metallic elements, excluding germanium and silicon (see Table 7.7), the C_0 value of 0.384 kg$^{-1/2}$ m^{-1} s K$^{1/2}$ mol$^{1/2}$ was obtained.

Using the relationship between the numerical factor k_0 and the atomic number shown in Figure 6.6, C_0 ($= 0.03771 / k_0^{-1}$) values for all metallic elements can be determined; in other words, their accurate viscosity values may be predicted. Volume 2 of the present book deals with this subject in more detail.

7.5   Temperature Dependence of Viscosity

The temperature dependence of the viscosity of liquid metallic elements can be described, to a good approximation, by an Arrhenius type relationship, namely

$$\mu = A \exp\left(\frac{H_{\mu}}{RT} \right) \tag{7.43}$$

where A and H_{μ} are constants. H_{μ} is sometimes called the apparent activation energy (for viscous flow). As is evident from Eq. (7.43), only the parameter H_{μ} provides the temperature dependence of respective liquid metallic elements. This relationship is shown in Figure 7.5 for several liquid metallic elements, over a wide temperature range. Until now, this simple expression has still not received a rigorously theoretical explanation.

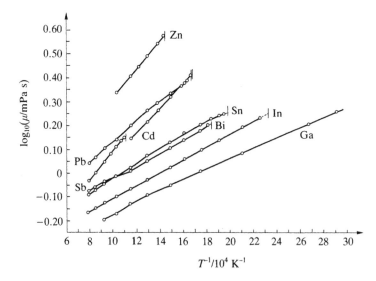

Figure 7.5 *Common logarithm of viscosity for liquid metallic elements vs. reciprocal of the absolute temperature (after Iida et al. [35]).*

Eyring and co-workers [15] proposed a viscosity equation for a very wide range of liquids, based on the theory of rate processes. The liquid viscosity in the Eyring model is

$$\mu = \frac{hN_A}{V} \exp\left(\frac{E}{RT}\right) \tag{7.44}$$

where h is Planck constant and E is the molar activation energy for surmounting the energy barrier. For a very large range of liquids, $E \approx (0.3 \text{ to } 0.4)\Delta_l^g H_b$, but for liquid metals, E is very much less than $0.4\Delta_l^g H_b$. Figure 7.6 shows a plot of E, or H_μ, against $\Delta_l^g H_b$ for 29 liquid metallic elements. As can be seen from Figure 7.6, the trend for E, or H_μ, to increase with increase in $\Delta_l^g H_b$ is confirmed; however, the overall scatter is rather high.

In order to estimate viscosity values for liquid metallic elements, Grosse [36] attempted to represent the constants A and H_μ appearing in Eq. (7.43), using well-known parameters. According to Grosse's approach, the constant A can be obtained by using the Andrade formula for the melting point viscosity. This constant then becomes

$$A = \frac{5.7 \times 10^{-2}(MT_m)^{1/2}}{V_m^{2/3} \exp(H_\mu/RT_m)}, \text{ (in cP)} \tag{7.45}$$

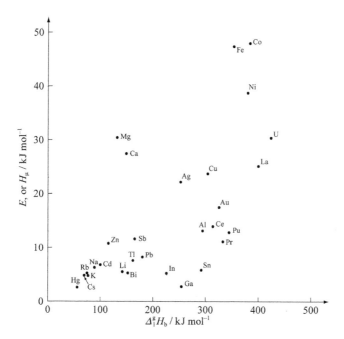

Figure 7.6 *The parameter E, appearing in Eq. (7.44), or H_μ, plotted against evaporation enthalpy, $\Delta_l^g H_b$.*

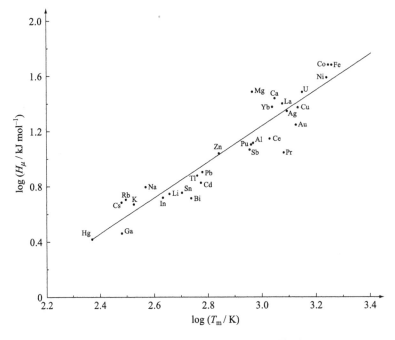

Figure 7.7 *Plot of log H_μ vs. log T_m for 30 liquid metallic elements.*

Further, Grosse, and Hirai [37], indicated that a simple empirical relationship exists between H_μ for liquid metallic elements and their melting point temperature T_m.

The present authors have recently carried out a similar analysis and provided a plot of log H_μ versus log T_m for various liquid metallic elements, as illustrated in Figure 7.7. The data for H_μ were taken from Table 17.9(a). The linear relationship between the two variables, shown in Figure 7.7, is

$$H_\mu = 2.32\,T_m^{1.29} \quad \left(\text{in J mol}^{-1}\right) \tag{7.46}$$

The numerical factors in Eq. (7.46) were determined so as to give the minimum S value (see Appendix 3) for the 30 liquid metallic elements plotted in Figure 7.7. This minimization approach is also used in Eq. (8.33).

Using Eqs. (7.14) or (7.41), (7.43), and (7.46), and applying the conditions $T = T_m$, $\mu = \mu_m$, the authors have obtained the following equation for the viscosity of liquid metallic elements.

$$\mu = \mu_0 \exp\left(\frac{2.32\,T_m^{1.29}}{RT}\right) \tag{7.47}$$

$$\mu_0 = \frac{\mu_m}{\exp\left(\dfrac{2.32\,T_m^{0.29}}{R}\right)} = \mu_m\left(0.279\,T_m^{0.29}\right)^{-1}$$

Incidentally, the temperature dependence of liquid metallic elements can also be expressed using Eq. (7.40).

7.6 Assessment of Viscosity Models

The performance of several viscosity models described in this chapter are now evaluated using the parameters δ_i, Δ, and S.

In Table 7.2, calculated values of viscosity, using the statistical kinetic theory of liquids (shown in Table 7.1) are compared with experimental data, together with δ_i, Δ, and S

Table 7.2 *Comparison of calculated values for viscosity of liquid metals with experimental data. Calculated values are taken from Johnson et al. [3] (see Table 7.1).*

Metal		Temperature °C	Viscosity / mPa S			δ_i %	
			Measured	Calculated			
				B-G	P-Y	B-G	P-Y
Aluminium	Al	700	1.29	0.95	1.35	36	−4.4
		850	1.04	0.88	1.22	18	−14.8
Caesium	Cs	30	0.679	0.84	1.42	−19	−52.2
		300	0.277	0.52	1.28	−47	−78.4
Lead	Pb	350	2.45	1.84	1.29	33.2	89.9
		550	1.67	1.60	1.18	4.4	41.5
Lithium	Li	180	0.600	0.79	1.20	−24	−50.0
		200	0.564	0.43	1.08	31	−47.8
Mercury	Hg	0	1.69	1.78	1.24	−5.1	36.3
		150	1.13	1.57	1.09	−28.0	3.7
Potassium	K	70	0.517	0.68	0.94	−24	−45
		345	0.249	0.44	0.88	−43	−72
Rubidium	Rb	40	0.665	0.78	1.43	−15	−53.5
		240	0.309	0.53	1.26	−42	−75.5
Sodium	Na	114	0.639	0.70	1.74	−8.7	−63.3
		203	0.444	0.59	1.53	−25	−71.0
					$\Delta(16)$ %	25.2	50.0
					$S(16)$	0.283	0.558

values. In calculating δ_i values, experimentally derived viscosity data listed in Table 17.9 were used: Table 7.1 was given in 1964, i.e. about a half century ago, so that recent and more reliable experimental data are adopted in Table 7.2. The calculated values based on the theory of liquids show reasonable results, particularly using the Born–Green integral equation, which give $\Delta(16)$ and $S(16)$ values of 25.2 per cent and 0.283, respectively; however, the Andrade model represented by Eq. (7.15), being restricted to melting point viscosity, provides $\Delta(8)$ and $S(8)$ values of 8.7 per cent and 0.107 respectively for the same eight liquid metals.

Figure 7.8 shows the Andrade correlation for 52 liquid metallic elements for which experimental data are available. Table 7.3 gives a comparison of viscosity values calculated using the Andrade formula (i.e. Eq.(7.15) with experimental values, together with the corresponding δ_i, Δ, and S values. We note that germanium, osmium, selenium, silicon, and vanadium are excluded, because these elements exhibit anomalous viscosity behaviour, although experimental re-examinations are also needed for the high melting point metals (i.e. osmium and vanadium). The data used for calculating the viscosities

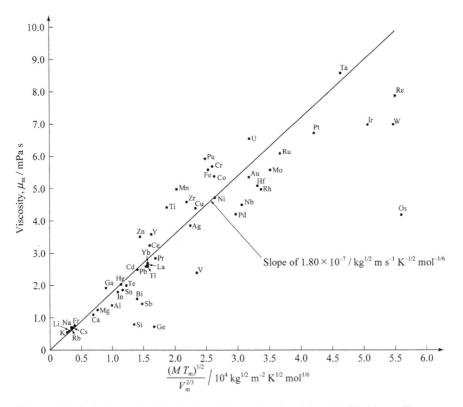

Figure 7.8 *Andrade relationship for the melting point viscosities of 52 liquid metallic elements. The selenium point falls outside the range of the figure; its ordinate, i.e. μ_m, is 24.8 mPa s and its abscissa, i.e. $(MT_m)^{1/2} V_m^{-2/3}$, 0.856×10^4 kg$^{1/2}$ m^{-2} K$^{1/2}$ mol$^{1/6}$.*

Table 7.3 *Comparison of experimental values for the melting point viscosity of liquid metallic elements with those calculated from the Andrade formula, together with δ_i, Δ, and S values.*

Element		Viscosity μ_m/mPa s		δ_i %
		Experimental	Calculated	
Aluminium	Al	1.38	1.79	−22.9
Antimony	Sb	1.44	2.67	−46.1
Bismuth	Bi	1.59	2.54	−37.4
Cadmium	Cd	2.50	2.53	−1.2
Caesium	Cs	0.686	0.657	4.4
Calcium	Ca	1.10	1.26	−12.7
Cerium	Ce	3.25	2.90	12.1
Chromium	Cr	5.7	4.69	22
Cobalt	Co	5.4	4.76	13
Copper	Cu	4.38	4.20	4.3
Francium	Fr	0.765	0.707	8.2
Gallium	Ga	1.95	1.63	19.6
Gold	Au	5.37	5.79	−7.3
Hafnium	Hf	5.1[†]	5.97	−15
Indium	In	1.79	1.97	−9.1
Iridium	Ir	7.0	9.10	−23
Iron	Fe	5.6	4.55	23
Lanthanum	La	2.66	2.84	−6.3
Lead	Pb	2.59	2.78	−6.8
Lithium	Li	0.602	0.567	6.2
Magnesium	Mg	1.25	1.38	−9.4
Manganese	Mn	5	3.66	37
Mercury	Hg	2.048	2.06	−0.6
Molybdenum	Mo	5.6	6.34	−12
Nickel	Ni	4.7	4.76	−1.3

continued

Table 7.3 *(continued)*

Element		Viscosity μ_m / mPa s		δ_i %
		Experimental	Calculated	
Niobium	Nb	4.5	5.52	−18
Palladium	Pd	4.22	5.36	−21.3
Platinum	Pt	6.74	7.58	−11.1
Plutonium	Pu	5.95	4.49	32.5
Potassium	K	0.537	0.499	7.6
Praseodymium	Pr	2.85	3.05	−6.6
Rhenium	Re	7.9	9.91	−20
Rhodium	Rh	5	6.08	−18
Rubidium	Rb	0.674	0.623	8.2
Ruthenium	Ru	6.1	6.62	−7.9
Silver	Ag	3.88	4.05	−4.2
Sodium	Na	0.695	0.617	12.6
Tantalum	Ta	8.6	8.33	3.2
Tellurium	Te	2.0[†]	2.20	−9.1
Thallium	Tl	2.59	2.85	−9.1
Tin	Sn	1.87	2.11	−11.4
Titanium	Ti	4.42[‡]	3.39	30.4
Tungsten	W	7.0	9.85	−29
Uranium	U	6.57	5.73	14.7
Ytterbium	Yb	2.67	2.83	−5.7
Yttrium	Y	3.6	2.96	22
Zinc	Zn	3.50	2.62	33.6
Zirconium	Zr	4.6	3.95	16
			$\Delta(48)$ %	14.9
			$S(48)$	0.182

[†] Mean values (see Tables 17.9(a) and (b)).
[‡] From Iida et al. [19] (see also Paradis et al. [38]).

and the experimental data on viscosity are given in Tables 17.1, 17.2, 17.4, and 17.9, respectively. The Andrade formula performs well with melting point viscosities of liquid metallic elements, giving global $\Delta(48)$ and relative standard deviation $S(48)$ values of 14.9 per cent and 0.182, respectively. It has been recognized for a long time that the viscosity values calculated from the Andrade formula most closely fit experimental data for liquid metals.

Table 7.4 *Experimental and calculated viscosities for liquid metallic elements at their melting point temperatures, together with δ_i, Δ, and S values.*

Element		$(\mu_m)_{exp}$ mPa s	$(\mu_m)_{cal}$/mPa s		δ_i/pct	
			Eq. (7.15)[†]	Eq. (7.32)	Eq. (7.15)[†]	Eq. (7.32)
Aluminium	Al	1.38	1.79	1.84	−22.9	−25.0
Antimony	Sb	1.44	2.67	1.96	−46.1	−26.5
Bismuth	Bi	1.59	2.54	2.10	−37.4	−24.3
Cobalt	Co	5.4	4.76	5.89	13	−8.3
Copper	Cu	4.38	4.20	4.02	4.3	9.0
Gallium	Ga	1.95	1.63	1.95	19.6	0
Gold	Au	5.37	5.79	5.51	−7.3	−2.5
Indium	In	1.79	1.97	1.94	−9.1	−7.7
Iron	Fe	5.6	4.55	6.05	23	−7.4
Lead	Pb	2.59	2.78	2.56	−6.8	1.2
Magnesium	Mg	1.25	1.38	1.21	−9.4	3.3
Mercury	Hg	2.048	2.06	2.20	−0.6	−6.9
Nickel	Ni	4.7	4.76	5.44	−1.3	−14
Potassium	K	0.537	0.499	0.516	7.6	4.1
Silver	Ag	3.88	4.05	3.40	−4.2	14.1
Sodium	Na	0.695	0.617	0.675	12.6	3.0
Thallium	Tl	2.59	2.85	2.47	−9.1	4.9
Tin	Sn	1.87	2.11	1.86	−11.4	0.5
Zinc	Zn	3.50	2.62	2.58	33.6	35.7
			$\Delta(19)$ %		14.7	10.4
			$S(19)$		0.192	0.144

[†] Andrade formula.

As described earlier, Iida et al. reconsidered the Andrade model and proposed Eq. (7.32) for the viscosity of liquid metallic elements. Table 7.4 lists the results of calculations for the melting point viscosities using Eq. (7.32), together with values calculated from the Andrade formula (Eq. (7.15)) and δ_i, Δ, and S values. The data used for viscosity calculations using Eq. (7.32) are given in Table 7.5. As is clear from Table 7.4, Eq. (7.32) is a definite improvement over the Andrade formula, giving $\Delta(19)$ and

Table 7.5 *Values of the parameters used for calculating the viscosities of liquid metallic elements at their melting points.*

Element		T_m °C	ν 10^{12} s^{-1}	m 10^{-26} kg	n_0^2 10^{57} m^{-6}	$g(r_m)^\dagger$	r_m^5 10^{-48} m^5	$\left(1 - \dfrac{r_0}{r_m}\right)$
Aluminium	Al	660.3	3.86	4.48	2.84	2.87	1.66	0.180
Antimony	Sb	630.6	1.19	20.22	1.03	2.35	3.68	0.209
Bismuth	Bi	271.4	0.919	34.70	0.839	2.57	4.16	0.168
Cobalt	Co	1495	3.84	9.79	6.30	2.50	0.938	0.242
Copper	Cu	1084.6	3.08	10.55	5.75	2.86	0.977	0.176
Gallium	Ga	29.8	2.17	11.58	2.79	2.65	1.66	0.144
Gold	Au	1064.2	1.64	32.71	2.82	2.95	1.72	0.164
Indium	In	156.6	1.50	19.07	1.36	2.67	3.05	0.140
Iron	Fe	1538	3.93	9.27	5.75	2.64	1.10	0.227
Lead	Pb	327.5	1.01	34.41	0.963	3.10	3.68	0.153
Magnesium	Mg	650	3.31	4.04	1.55	2.50	2.86	0.187
Mercury	Hg	−38.8	1.06	33.31	1.69	2.73	2.43	0.127
Nickel	Ni	1455	3.75	9.75	6.58	2.43	0.901	0.236
Potassium	K	63.4	1.14	6.49	0.162	2.36	19.72	0.211
Silver	Ag	961.8	1.99	17.91	2.70	2.67	1.78	0.170
Sodium	Na	97.7	1.99	3.82	0.590	2.46	6.75	0.207
Thallium	Tl	304	1.02	33.94	1.12	2.82	3.46	0.149
Tin	Sn	231.9	1.48	19.71	1.25	2.62	3.05	0.146
Zinc	Zn	419.5	2.36	10.86	3.67	2.50	1.33	0.188

$\dagger$ Extrapolated values (see Figures 2.20(a–c) and Table 2.1). Data for $g(r)$ are taken from Waseda (cf. Chapter 2).

Table 7.6 *Comparison of experimental and Yokoyama's calculated values for the viscosity of liquid metallic elements near their melting point temperatures, together with δ_i, Δ, and S values.*

Element		Temperature		Viscosity / mPa s		δ_i %
		K	(°C)	Experimental	Calculated	
Aluminium	Al	943	(670)	1.36	1.62	−16.0
Antimony	Sb	933	(660)	1.37	2.44	−43.9
Bismuth	Bi	573	(300)	1.50	2.34	−35.9
Cadmium	Cd	623	(350)	2.35	2.33	0.9
Caesium	Cs	303	(30)	0.679	0.59	15
Cobalt	Co	1823	(1550)	4.9	4.33	13
Copper	Cu	1423	(1150)	3.97	3.86	2.8
Gallium	Ga	323	(50)	1.81	1.51	19.9
Gold	Au	1423	(1150)	4.87	5.34	−8.8
Indium	In	433	(160)	1.77	1.77	0
Iron	Fe	1833	(1560)	5.4	4.11	31
Lead	Pb	613	(340)	2.51	2.51	0
Lithium	Li	463	(190)	0.581	0.51	14
Magnesium	Mg	953	(680)	1.17	1.26	−7.1
Mercury	Hg	238	(−35)	2.01	1.87	7.5
Nickel	Ni	1773	(1500)	4.4	4.28	2.8
Potassium	K	343	(70)	0.517	0.46	12
Rubidium	Rb	313	(40)	0.665	0.56	19
Silver	Ag	1273	(1000)	3.69	3.71	−0.5
Sodium	Na	378	(105)	0.669	0.56	19
Thallium	Tl	588	(315)	2.50	2.57	−2.7
Tin	Sn	523	(250)	1.79	1.93	−7.3
Zinc	Zn	723	(450)	3.23	2.41	34.0
					$\Delta(23)$ %	13.6
					S(23)	0.183

$S(19)$ values of 10.4 per cent and 0.144, respectively. According to Wallace [39] and Yokoyama [13], antimony, bismuth, gallium, indium, mercury, and tin can be classified as anomalous metals. Nevertheless, Eq. (7.32) also provides much better results for the anomalous metals, as shown in Table 7.4. These results suggest that the Andrade coefficient C_A is dependent on the structure of liquid metallic elements and on the correction factor for Lindemann's melting formula (i.e. $g(r_m)(1 - r_0/r_m)$ and β. These are dimensionless quantities). Unfortunately, Eq. (7.32) is not well-suited as a predictive model for the viscosities of metallic liquids, because the available structural data are sparse.

Table 7.7 *Comparison of experimental and calculated values for the melting point viscosity of pure metallic liquids using relationships between viscosity and surface tension, together with δ_i, Δ, and S values.*

Element		$(\mu_m)_{exp}$ mPa s	$(\mu_m)_{cal}$ /mPa s		δ_i /%	
			Eq. (7.4)[†]	Eq. (7.42)	Eq. (7.4)[†]	Eq. (7.42)
Aluminium	Al	1.38	2.09	2.26	−34.0	−38.9
Antimony	Sb	1.44	1.59	1.95	−9.4	−26.2
Bismuth	Bi	1.59	2.77	2.34	−42.6	−32.1
Cadmium	Cd	2.50	3.24	2.86	−22.8	−12.6
Caesium	Cs	0.686	0.536	0.705	28.0	−2.7
Calcium	Ca	1.10	0.805	1.22	36.6	−9.8
Cerium	Ce	3.25	3.36	4.71	−3.3	−31.0
Cobalt	Co	5.4	4.06	4.73	33	14
Copper	Cu	4.38	3.34	3.86	31.1	13.5
Gallium	Ga	1.95	4.06	2.53	−52.0	−22.9
Germanium	Ge	0.73	1.74	2.26	−58	−68
Gold	Au	5.37	5.16	4.50	4.1	19.3
Indium	In	1.79	3.39	2.43	−47.2	−26.3
Iron	Fe	5.6	3.86	4.50	45	24
Lanthanum	La	2.66	2.91	3.62	−8.6	−26.5
Lead	Pb	2.59	3.14	2.52	−17.5	2.8
Lithium	Li	0.602	0.577	0.879	4.3	−31.5

Table 7.7 *(continued)*

Element		$(\mu_m)_{exp}$ mPa s	$(\mu_m)_{cal}$/mPa s		δ_i/%	
			Eq. (7.4)[†]	Eq. (7.42)	Eq. (7.4)[†]	Eq. (7.42)
Magnesium	Mg	1.25	1.10	1.43	13.6	−12.6
Manganese	Mn	5	2.56	3.42	95	46
Mercury	Hg	2.048	5.29	3.25	−61.3	−37.0
Molybdenum	Mo	5.6	4.43	4.64	26	21
Nickel	Ni	4.7	3.87	4.46	21	5.4
Platinum	Pt	6.74	6.31	5.75	6.8	17.2
Plutonium	Pu	5.95	3.33	4.62	78.7	28.8
Potassium	K	0.537	0.439	0.587	22.3	−8.5
Praseodymium	Pr	2.85	2.87	3.74	−0.7	−23.8
Rubidium	Rb	0.674	0.548	0.720	23.0	−6.4
Silicon	Si	0.8	1.28	2.18	−38	−63
Silver	Ag	3.88	3.20	3.34	21.3	16.2
Sodium	Na	0.695	0.574	0.785	21.1	−11.5
Tantalum	Ta	8.6	5.63	6.23	53	38
Tellurium	Te	2.0	1.17	2.70	71	−26
Thallium	Tl	2.59	3.20	2.80	−19.1	−7.5
Tin	Sn	1.87	3.15	2.27	−40.6	−17.6
Tungsten	W	7.0	6.03	7.08	16	−1.1
Ytterbium	Yb	2.67	1.49	2.52	79.2	6.0
Zinc	Zn	3.50	2.84	2.78	23.2	25.9
			$\Delta(37)$ %		32.7	22.2
			S(37)		0.400	0.269
			$\Delta(35)$[‡] %		31.8	19.7
			S(35)[‡]		0.394	0.228

[†] The Fowler–Born–Green relation.
[‡] Excluding germanium and silicon.

Table 7.8 *Comparison of the measured and calculated energies of apparent activation for viscous flow of liquid metallic elements, together with δ_i, Δ, and S values.*

Element		H_μ / kJ mol^{-1}		δ_i %
		Measured	Calculated	
Aluminium	Al	13.08	15.76	−17.0
Antimony	Sb	11.67	15.11	−22.8
Bismuth	Bi	5.192	7.862	−34.0
Cadmium	Cd	6.698	8.798	−23.9
Caesium	Cs	4.79	3.67	30.5
Calcium	Ca	27.51	19.81	38.9
Cerium	Ce	13.97	18.81	−25.7
Cobalt	Co	48.13	35.91	34.0
Copper	Cu	23.85	25.55	−6.7
Gallium	Ga	2.880	3.689	−21.9
Gold	Au	17.62	25.05	−29.7
Indium	In	5.244	5.791	−9.4
Iron	Fe	47.44	37.04	28.1
Lanthanum	La	25.22	21.56	16.9
Lead	Pb	8.142	8.920	−8.7
Lithium	Li	5.52	6.21	−11.1
Magnesium	Mg	30.5	15.5	96.8
Mercury	Hg	2.614	2.649	−1.3
Nickel	Ni	38.85	34.87	11.4
Plutonium	Pu	12.88	15.31	−15.9
Potassium	K	4.69	4.22	11.1
Praseodymium	Pr	11.18	21.88	−48.9
Rubidium	Rb	5.13	3.84	33.6
Silver	Ag	22.2	22.6	−1.8
Sodium	Na	6.25	4.79	30.5

Table 7.8 *(continued)*

Element		$H_\mu / \text{kJ mol}^{-1}$		δ_i
		Measured	Calculated	%
Thallium	Tl	7.638	8.471	−9.8
Tin	Sn	5.833	7.134	−18.2
Uranium	U	30.45	26.77	13.7
Ytterbium	Yb	23.77	19.29	23.2
Zinc	Zn	10.91	10.72	1.8
			Δ (30) %	22.6
			$S(30)$	0.289

On the basis of the Stokes–Einstein formula represented by Eq. (7.12), Yokoyama calculated the viscosities of 23 liquid metallic elements near their melting point temperatures. In viscosity calculations using Eq. (7.12), Yokoyama employed Eq. (8.16) for the self-diffusivity D_{HS} of the hard-sphere fluid, based on Speedy's analysis. Table 7.6 shows experimental and Yokoyama's calculated viscosity values, together with δ_i, Δ, and S values. (The experimental values are calculated, at given temperatures, from experimental viscosity data listed in Table 17.9.) With the exception of semimetals, iron, and zinc, the Stokes–Einstein formula performs well for the pure liquid metals listed in Table 7.6.

Let us now evaluate the viscosity equations expressed in terms of surface tension, or the relationship between viscosity and surface tension, of liquid metallic elements. Table 7.7 shows the viscosity values calculated from Eq. (7.3), or (7.4), and Eq. (7.42), and experimentally derived viscosity data, together with δ_i, Δ, and S values for assessing these two models. The data for calculating viscosities are given in Tables 5.5, 17.1, 17.2, and 17.8. As can be seen from Table 7.7, the relationship between viscosity and surface tension in terms of the new dimensionless parameter $\xi_T^{1/2}$, i.e. Eq. (7.42), performs well with $\Delta(35)$ and $S(35)$ values of 19.7 per cent and 0.228, respectively. It proves to be much better than the Fowler–Born–Green relation, based on statistical kinetic theory of liquids, which gives $\Delta(35)$ and $S(35)$ values of 31.8 per cent and 0.394, respectively. When the Fowler–Born–Green relation, i.e. Eq. (7.3) or (7.4), is corrected by the parameter $\xi_T^{1/2}$, statistically, the results of calculations for the viscosity of liquid metallic elements become closer to experimental values; in other words, using the parameter $\xi_T^{1/2}$, the accuracy of calculations for the viscosities of liquid metallic elements, including the anomalous metals, is considerably improved. In addition, the relationship between k_0, or C_0, and the atomic number, illustrated in Fig.6.6, allows us to estimate (through interpolation) values of the numerical factor C_0, appearing in Eq. (7.41), for respective metallic elements for those not yet available. If both values of $\xi_T^{1/2}$ and C_0 are known, more accurate viscosity values can readily be calculated using Eq. (7.41). As already noted, some examples of this approach will be given in Volume 2. Thus, Eq. (7.41) can be regarded as a modified Fowler–Born–Green relation.

Values for the parameter H_μ appearing in Eq. (7.43) have been calculated from Eq. (7.46), and its performance has been assessed by determining δ_i, Δ, and S values for the 30 liquid metallic elements plotted in Figure 7.7. Table 7.8 shows the experimental and calculated values for H_μ, together with δ_i, Δ, and S values. Equation (7.46) performs reasonably well with the parameter H_μ, giving global delta $\Delta(30)$ and relative standard deviation $S(30)$ values of 22.6 per cent and 0.289, respectively. Incidentally, accurate experimental determination of H_μ for liquid metallic elements is a non-trivial task. Similarly, the values of H_μ for several elements (e.g. magnesium, praseodymium) are highly uncertain.

7.7   Viscosity of Liquid Alloys

7.7.1   Viscosity of Dilute Liquid Alloys

In materials processing operations, the thermophysical properties of dilute liquid alloys are important. A number of experimental studies have been carried out on the viscosities of dilute alloys, particularly on dilute, iron-based, alloys. Unfortunately, experimental results are frequently inconsistent. The experimental uncertainties in the viscosity measurements are estimated to be about ± 1 to ± 20 per cent; the great majority of data for dilute liquid alloys lacks reliability.

Iida et al. [40] have carefully measured the viscosities of mercury-based dilute binary alloys (i.e. amalgams) containing 1 to 3 atomic per cent solute, using the capillary method. (The solute elements are bismuth, cadmium, gallium, gold, indium, lead, thallium, tin, and zinc.) The experimental uncertainties with the viscosity measurements were about ± 0.5 per cent. According to the experimental results, the changes in the viscosity of pure liquid mercury are about 1 to 5 per cent with the addition of 1 atomic per cent solute. The differences in the viscosities between the pure mercury and each dilute alloy are comparatively small. In general, the liquid viscosity of a dilute alloy which contains a small amount of solute element (~ 1 to ~ 3 per cent) is not likely to be greatly different from that of the base metal (or the pure metal).

7.7.2   Viscosity of Binary Liquid Alloys

Numerous attempts have been made to describe the viscosities of binary liquid mixtures or alloys. For example, Moelwyn-Hughes [41] proposed that

$$\mu_A = (x_1\mu_1 + x_2\mu_2)\left(1 - 2x_1x_2\frac{\Delta u}{kT}\right) \tag{7.48}$$

where μ_A is the viscosity of binary liquid mixtures (or alloys), x is the mole (or atomic) fraction, Δu is the interchange energy,[4] and subscripts 1 and 2 refer to the components. Equation (7.48) may be written in the form of the excess viscosity,[5] μ^E:

---

[4] $\Delta u = H^E / x_1 x_2 N_A$, where H^E is the (integral) enthalpy of mixing.
[5] $\mu^E = \mu_A - (x_1\mu_1 + x_2\mu_2)$.

$$\mu^{E} = -2(x_1\mu_1 + x_2\mu_2)\frac{H^{E}}{RT} \tag{7.49}$$

This simple expression derived by Moelwyn-Hughes indicates that the sign (i.e. + or −) of the excess viscosity depends only upon that of the enthalpy of mixing, H^{E}. Table 7.9 shows the signs of the excess viscosity and the enthalpy of mixing. The excess viscosity μ^{E} appears to be influenced by the enthalpy of mixing, but the correlation of μ^{E} with H^{E} is not particularly satisfactory. Evidently, for transport coefficients such as viscosity and diffusivity, they cannot be formulated in terms of thermodynamic properties alone. Differences in atomic size and atomic mass of the components must also be important factors. As indicated in Table 7.9, μ^{E} tends to become negative as the difference in atomic size increases.

Iida et al. [42,43] modified the Moelwyn-Hughes model to include the effect of both the size and mass of atoms (or ions), and proposed an expression in the form of the excess viscosity of binary liquid alloys, as follows:

$$\mu^{E} = (x_1\mu_1 + x_2\mu_2)\left[-\frac{5x_1x_2(d_1 - d_2)^2}{x_1 d_1^2 + x_2 d_2^2} \right.$$
$$\left. + 2\left\{ \left(1 + \frac{x_1 x_2\left(m_1^{1/2} - m_2^{1/2}\right)^2}{\left(x_1 m_1^{1/2} + x_2 m_2^{1/2}\right)^2}\right)^{1/2} - 1 \right\} - \frac{0.12 x_1 x_2 \Delta u}{kT} \right], \text{ (in cP)} \tag{7.50}$$

or

$$\mu^{E} = (x_1\mu_1 + x_2\mu_2)\left[-\frac{5x_1x_2(d_1 - d_2)^2}{x_1 d_1^2 + x_2 d_2^2} \right.$$
$$\left. + \left\{ \left(1 + \frac{x_1 x_2\left(m_1^{1/2} - m_2^{1/2}\right)^2}{\left(x_1 m_1^{1/2} + x_2 m_2^{1/2}\right)^2}\right)^{1/2} - 1 \right\} - 0.12(x_1 \ln f_1 + x_2 \ln f_2) \right], \text{ (in cP)} \tag{7.51}$$

where d is the diameter of an atom (in calculating viscosities, they used Pauling ionic radii for d), m is the mass of an atom, and f is the activity coefficient. In the square brackets of the above equations, the first ($\mu_{h,d}^{E}$) and the second ($\mu_{h,m}^{E}$) terms represent the hard parts, and the third (μ_{s}^{E}) term the soft part of the friction constant for viscous movements of atoms (or ions). As is obvious from these equations, $\mu_{h,d}^{E}$ tends to be become less negative as the difference between d_1 and d_2 increases. On the other hand, $\mu_{h,m}^{E}$ tends to become more positive as the difference between m_1 and m_2 increases.

Values calculated from Eqs. (7.50) and (7.51) are shown in Figures 7.9 to 7.12, together with experimental data. Calculated results coincide qualitatively with experimental data and, in particular, for regular or nearly regular solutions, excellent agreement between calculation and experiment is obtained.

Early this century, Prasad and Jha [44] extended Iida et al.'s model on the basis of the idea that viscosity, which involves movement of species, must depend not only upon thermodynamic properties but also upon hard sphere diameters, masses,

Table 7.9 *Relationship between excess viscosity of various binary liquid alloys and difference in the ionic radii (Pauling ionic radii) / the integral enthalpy of mixing.*

Alloy	$\|d_1 - d_2\|$ 10^{-10} m	μ^E	H^E
Au–Sn	0.66	–	–
Ag–Sb	0.64	–	–
Ag–Sn	0.55	–	–, +[a]
Al–Cu	0.46	–	–
Au–Cu	0.46	0, –	–
K–Na	0.38	–	+
Cd–Sb	0.35	–	+, –[a]
Cu–Sb	0.34	–	–
Ag–Cu	0.30	–	+
Hg–In	0.29	–, +[a]	–
Cu–Sn	0.25	–	–
Al–Zn	0.24	–	+
Cd–Bi	0.23	–	+
K–Hg	0.23	+	–
Pb–Sb	0.22	–	+, –[a]
Mg–Pb	0.19	+, –[a]	–
Na–Hg	0.15	+	–
Al–Mg	0.15	+	–
Cd–Pb	0.13	–	+
Pb–Sn	0.13	0	+
Sb–Bi	0.12	–	+
Ag–Au	0.11	+	–
Fe–Ni	0.10	–, +[a]	–
In–Bi	0.07	+	–
Mg–Sn	0.06	+	–
In–Pb	0.03	–	+
Sn–Zn	0.03	–	+
Sn–Bi	0.03	+	+

[a] Change in the signs of the excess viscosity and the integral enthalpy of mixing with composition (after Iida et al. [42]).

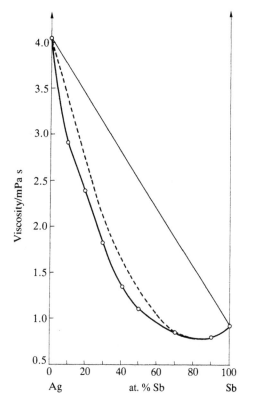

Figure 7.9 *Viscosities of silver–antimony alloys at 1000 °C.* o, *experimental values,* – – –, *values calculated from Eq. (7.50) (after Iida et al. [42]).*

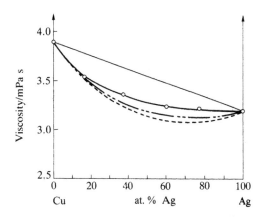

Figure 7.10 *Viscosities of copper–silver alloys at 1100 °C.* o, *experimental values (Gebhardt and Worwag);* – – –, *values calculated from Eq. (7.50);* – · · –, *values calculated from Eq. (7.51). (after Iida et al. [42] and Morita et al. [43]).*

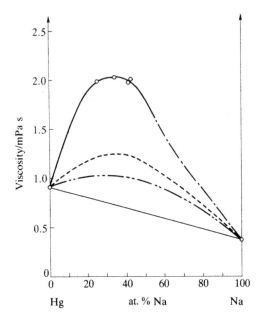

Figure 7.11 *Viscosities of mercury–sodium alloys at 370 °C. ○, experimental values (Degenholde and Sauerwald); – · –, values interpolated on the experimental data; – – –, values calculated from Eq. (7.50); – · · –, values calculated from Eq. (7.51) (after Iida et al. [42]).*

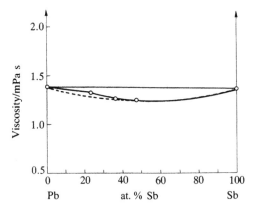

Figure 7.12 *Viscosities of lead–antimony alloys at 700 °C. ○, experimental values (Gebhardt and Köstlin); – – –, values calculated from Eq. (7.50) (after Iida et al. [42] and Morita et al. [43]).*

and some microscopic functions. In their approach, the hard sphere diameter of the constituent atoms in the binary alloy systems was calculated by minimizing the pair potential obtained from pseudopotential formalism. This theoretical investigation reproduces experimental data for the viscosity of tin-based binary liquid alloys. Moreover, it gives a rational explanation for the behaviour of viscosity in these alloys. Incidentally, in Prasad and Jha's paper [44], the concentration-dependent variation of surface tension of these liquid alloys was also investigated theoretically, using the grand partition function approach.

Various models have been proposed to describe the concentration dependence of the viscosity of liquid alloys. They are able to successfully reproduce the viscosity of some binary and ternary (Cu–Au–Ag alloys) liquid alloys, but none of them are universally applicable. As already mentioned, in the area of materials process science, both accuracy and universality are required of any model for predicting the thermophysical properties of metallic liquids. Nevertheless, it is extremely difficult to present such a model (i.e. satisfying both accuracy and universality) for the viscosity of liquid alloys.

Some models for the viscosity of liquid metallic elements and multicomponent alloy systems have been succinctly reviewed by Brooks et al. [1].

7.8   Methods of Viscosity Measurement

The definition of a coefficient of viscosity, dynamic viscosity (or shear viscosity), or simply 'viscosity' is based on the following mathematical expression derived by Sir Isaac Newton:

$$f = \mu \frac{dv}{dz} \qquad (7.52)[6]$$

where f is the force exerted by the fluid per unit area of a plane parallel to the x direction of motion when the velocity v is increasing with distance z measured normal to the plane, at the rate dv/dz. Thus, viscosity represents the constant of proportionality. Equation (7.52) is known as Newton's law of viscosity. If a liquid obeys Eq. (7.52), i.e. the shearing force f is proportional to the velocity gradient dv/dz, it is known as a Newtonian liquid. All pure liquid metals are believed to be Newtonian liquids.[7]

Incidentally, it should be kept in mind that the definition of viscosity by Eq. (7.52) only holds for laminar, or streamline, flow. Turbulent viscosities are often defined using

---

[6] Newton's law of viscosity is sometimes written as

$$f_{zx} = -\mu \frac{dv_x}{dz}$$

This states that the shear force, acting in the x direction, per unit area, f_{zx} (due to a gradient of x velocity in the z direction), is proportional to the negative of the local velocity gradient dv_x/dz.

[7] In general, liquids with comparatively small molecules (e.g. monatomic liquids) are considered to be Newtonian liquids.

a similar equation and apply when the flow is turbulent. The latter turbulent viscosities are typically 10^4 to 10^6 times greater than their laminar counterparts.

The reciprocal of viscosity is known as fluidity. The ratio of viscosity to density, i.e. $v = \mu / \rho$, is known as the fluid's kinematic viscosity (denoted by v), This is an important physical quantity in fluid mechanics. The dimensions of kinematic viscosity ($L^2\,T^{-1}$) are equivalent to those of diffusivity, so that v can also be considered to represent the transverse diffusion of momentum down a perpendicular velocity gradient.

A variety of methods exist for measuring viscosities of liquids. However, suitable techniques for determination of the viscosity of metallic liquids are restricted to the following on account of their low viscosities,[8] their chemical reactivity, and their generally high melting points. They are:

(a)　Capillary method
(b)　Oscillating vessel method
(c)　Oscillating (or levitated) drop method
(d)　Rotational method
(e)　Oscillating plate method.

7.8.1　Capillary Method

When a liquid is made to flow through a capillary tube under a given pressure, the time required for a definite volume of the liquid to be discharged depends on the viscosity of the liquid. In the capillary method, viscosities can be determined by measuring efflux times of liquid samples through the capillary tube. In practice, as can be easily understood from Figure 7.13, a definite volume of liquid is introduced into the reservoir bulb, and then sucked into the measuring bulb. The efflux time t for the meniscus to pass between the fiducial marks m_1 and m_2 is then observed. The meniscus can be observed directly by eye, or indirectly using an electrical method. The relation between viscosity and efflux time is then given by the modified Poiseuille equation, or Hagen–Poiseuille equation, as follows:

$$\mu = \frac{\pi r^4 \rho \bar{g} h t}{8V(l + nr)} - \frac{m\rho V}{8\pi(l + nr)t} \qquad (7.53)$$

where r and l are the radius and length of the capillary, respectively, $\bar{h}$ is the effective height of column of liquid, ρ is the liquid density, V is the volume discharged in time t (i.e. the volume of the measuring bulb), and m and n are constants (in general, $m = 1.1$ to 1.2, $n = 0$ to 0.6). In Eq. (7.53), the second term is called the kinetic-energy correction which is important in determining the viscosities of metallic liquids because of their low kinematic viscosities, and nr is called the end-correction.

This technique has usually been used as a relative, rather than absolute, method. This is especially true since the experimental procedures are simple, and any errors incidental

---

[8] Metallic liquids are particularly notable for their low kinematic viscosities.

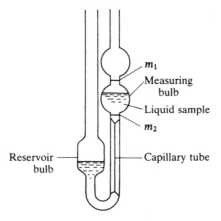

m_1

Measuring bulb

Liquid sample

m_2

Reservoir bulb

Capillary tube

Figure 7.13 *Schematic illustration of a capillary viscometer.*

to the measurement of dimensions are thereby avoided. For a viscometer in which r, l, $\bar{h}$, and V are fixed, Eq. (7.53) reduces to

$$\frac{\mu}{\rho} = v = C_1 t - \frac{C_2}{t} \tag{7.54}$$

$$C_1 = \frac{\pi r^4 g \bar{h}}{8V(l+nr)}, \quad C_2 = \frac{mV}{8\pi(l+nr)}$$

where C_1 and C_2 are constants. The values of C_1 and C_2 are easily evaluated using viscosity standard reference samples.

In determining the viscosities of metallic liquids by the capillary method, an especially fine and long-bore tube (in general, r <0.15 to 0.2 mm, l > 70 to 80 mm) is needed, so as to satisfy the condition of a low Reynolds number for ensuring laminar flow. Material limitations for fine, long-bore tubes, and the resulting need for particularly clean liquid metal specimens, represent two of the disadvantages of this technique. Only capillary tubes of heat-resisting glass and quartz glass have been used to date; the method has, therefore, been applied to determinations of viscosity of metallic liquids with melting points below 1200 °C or so.

In measuring viscosities, the problem of possible contamination of liquid metal samples is of critical importance. Blockage of the capillary tube by minute oxide inclusions and/or bubbles can have a pronounced effect upon the efflux time. Therefore, clean metallic liquid perfectly free from any contamination is essential. Metrologically, however, the capillary method was already established and all relevant experimental corrections were known. Careful measurements using this technique can provide reliable viscosity values for low melting metallic liquids and give an accuracy better than ± 0.5 per cent. Viscosities of some 15 liquid metallic elements have been measured successfully using the capillary method (e.g. antimony, bismuth, copper, indium, lead, tin). [35,45].

7.8.2   Oscillating Vessel Method

When a liquid is placed in a vessel hung by a torsional suspension, and the vessel is set in oscillation about a vessel axis, the resulting motion is gradually damped on account of frictional energy absorption and dissipation within the liquid. The viscosity of the liquid sample can be calculated by observing the decrement and the time period of the oscillations. This is the principle of the oscillating vessel methods. Diagrams for the oscillation system of measurement are shown in Figures 7.14 and 7.15.

For high temperature viscosity measurements of metallic liquids, this method has been most frequently used [1,46]. Its practical advantages are that the apparatus, and in particular the shape of vessel or crucible, are simple and a closed vessel can be used. Similarly, the decrement (i.e. the amplitude) and the periodic time of the oscillations can be measured with great accuracy. The main disadvantage of this technique is that a rigorous analytical formula for calculating viscosity from the observed decrement and period of the oscillations is lacking, owing to the mathematical difficulty of solving the differential equation of motion for this oscillating system. A number of theoretical and experimental investigations have been carried out and several analytical equations proposed for calculating viscosity from experimental data. Individual workers have assessed the uncertainties, or errors, in measuring liquid metal viscosities, and have concluded that they range between 1 to 5 per cent. However, discrepancies as large as 30 to 50 per cent or more exist between the experimental results of different workers (e.g.

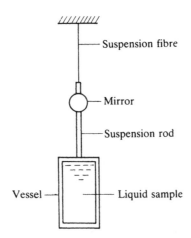

Figure 7.14 *Schematic illustration of an oscillating vessel viscometer.*

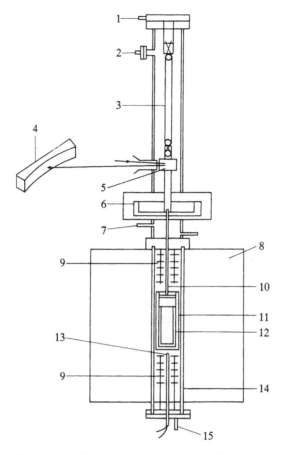

Figure 7.15 *Schematic diagram of an oscillating vessel (or crucible) viscometer (used to measure the viscosity of metallic liquids at high temperatures).*

see Figures 7.21 to 7.23). Thus, experimental and metrological (theoretical) investigations on this problem have been made by several workers; the essential points of their research work are described in next section.

7.8.3   Oscillating (or Levitated) Drop Method

The oscillating (or levitated) drop method is widely used today for viscosity measurements of high melting point liquids. This containerless method is based on the fact that the oscillations of levitated liquid drop are damped owing to the viscosity of the liquid. Thus, the liquid viscosity can be determined from the damping; the relationship between viscosity μ and the damping constant Γ (or decay time $\tau = \Gamma^{-1}$) is expressed in the form

$$\Gamma = \frac{5\mu}{\rho r_0^2} = \frac{20\pi r_0 \mu}{3m} \tag{7.55}$$

where ρ, r_0, and m are the density, the radius, and the mass of the levitated liquid drop, respectively. According to Paradis et al. [47], in Eq. (7.55), real-time values of ρ and r_0, obtained in earlier experiments, are used to prevent any distortion due to sample evaporation.

The viscosities of some 15 high melting point metals (transition metals) have been measured in the last ten years or so, using the oscillating drop method.

The essentials for successful viscosity measurements for refractory liquid metals using the oscillating drop technique have been described in several papers [1,46].

7.8.4   Rotational Method

Let us now consider a liquid filling the space between two coaxial cylinders (or sphere), as shown in Figure 7.16. When the outer cylinder rotates with a constant angular velocity, and the inner cylinder remains fixed, the viscous liquid exerts a revolving force on the inner cylinder. If the inner cylinder is suspended by a fibre, the revolving force, i.e. the torque, can be evaluated by measuring the angular displacement of the fibre; the viscosity can be calculated from observation of the torque. Incidentally, various types of viscometers based on the principle of the rotational technique exist: rotation of sphere, rotation of disk, and rotation of cylinder (outer cylinder rotated, inner cylinder suspended; outer cylinder fixed, inner cylinder rotated).

The application of the rotational method to metallic liquids is technically difficult, because, in the case of low viscosity liquids such as metallic liquids, the requirement for a low Reynolds number means that the clearance allowed between the two cylinders must be very small, and the rotating and stationary parts, therefore, must be truly coaxial. Only a few investigators have made use of this technique [48,49].

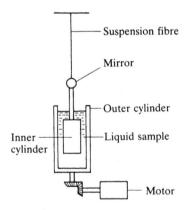

Figure 7.16 *Schematic illustration of a rotational viscometer.*

7.8.5 Oscillating Plate Method

When a flat plate executing linear oscillations is immersed in a liquid, as shown in Figure 7.17, its motions are impeded by the retarding force which the viscous liquid exerts on the oscillating plate. If the plate is now vibrating in the liquid with a constant driving force, the amplitude of motion of the plate is reduced to a degree dependent on the viscosity of the liquid. The oscillating plate method is based on measurements of the amplitudes of plate oscillations in air and in a liquid sample. The relation between viscosity and amplitudes is expressed in the following form:

$$\rho\mu = K_0 \left(\frac{f_a E_a}{fE} - 1 \right)^2 \tag{7.56}$$

$$K_0 \equiv \frac{R_M^2}{\pi f A^2}$$

where f_a and f are the resonant frequencies for the plate in air and in the liquid, respectively, E_a and E are the resonant amplitudes of plate oscillation in air and in the liquid, respectively. K_0 is a constant of the apparatus, R_M is the real component of the mechanical impedance, and A is the (effective) area of the plate.

Except for liquids having high values of $\rho\mu$, f may, to a good approximation, be assumed to be equal to f_a. Equation (7.56) then becomes

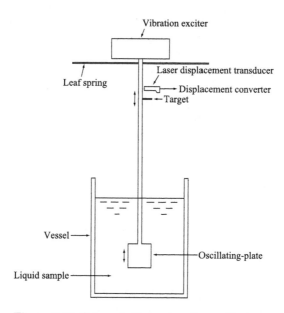

Figure 7.17 *Schematic illustration of an oscillating plate viscometer.*

$$\rho\mu = K_0 \left(\frac{E_a}{E} - 1 \right)^2 \qquad (7.57)$$

The apparatus' constant K_0 is determined experimentally using viscosity standard reference samples.

The advantages of this method are that it allows instantaneous and continuous indications of the product $\rho\mu$ over a wide range, and further, its construction and operation are comparatively simple. This technique was applied to determine the viscosity of liquid iron by Arsentiev et al. [50]. However, the oscillating plate method is unsuitable for measurements with low viscosity liquids such as metals, because a thin oscillating plate of large area must be vibrated slowly within the liquid, so as to ensure laminar flow. This technique has been used for viscosity measurements of slag systems and mould fluxes [1].

Methods of viscosity measurements for metallic liquids are adequately described in several papers [1,46,51].

7.9 Determination of Viscosity Using the Oscillating Vessel Method

7.9.1 Working Formulae for Viscosity Determination by the Oscillating Vessel Method

Although there are various techniques for viscosity measurement, the oscillating vessel (cylindrical vessel) method has been used most extensively for metallic liquids. Unfortunately, the calculation of the viscosity of a liquid from the observed logarithmic decrement and period of the oscillations are extremely complicated. One of the major reasons for large discrepancies among experimental viscosity data is the result of approximations in working formulae used to connect the observed damping of oscillations and dimensions of the apparatus with the viscosity of the liquid.

In the field of liquid metals, the following working formulae have frequently been employed for determinations of the viscosity of metallic liquids using the oscillating cylindrical vessel method.

7.9.1.1 *Knappwost's Equation*

Knappwost [52,53] proposed a semi-empirical equation for the calculation of viscosity from the measured logarithmic decrement δ and time period T of a system filled with a liquid sample for a cylindrical vessel of small aspect ratio (i.e. height to radius).

$$\delta T^{3/2} = K(\rho\mu)^{1/2} \qquad (7.58)$$

in which K is an apparatus constant. The value of K is determined, previous to viscosity measurements, by using viscosity standard reference samples (e.g. mercury, tin, lead). Knappwost's equation has often been used for relative viscosity determinations of metallic liquids because of its simplicity.

7.9.1.2 Roscoe's Equation

Roscoe [54,55] derived working formulae for absolute viscosity determinations. For a cylindrical vessel, Roscoe's equation is given by[9]

$$\mu = \left(\frac{I\delta}{\pi R^3 HZ}\right)^2 \frac{1}{\pi\rho T} \tag{7.59}$$

where

$$Z = \left(1 + \frac{R}{4H}\right)a_0 - \left(\frac{3}{2} + \frac{4R}{\pi H}\right)\frac{1}{p} + \left(\frac{3}{8} + \frac{9R}{\pi H}\right)\frac{a_2}{2p^2}$$

$$p = \left(\frac{\pi\rho}{\mu T}\right)^{1/2} R$$

$$a_0 = 1 - \frac{3}{2}\Delta - \frac{3}{8}\Delta^2 \cdots$$

$$a_2 = 1 + \frac{1}{2}\Delta + \frac{1}{8}\Delta^2 \cdots$$

$$\Delta = \frac{\delta}{2\pi}$$

I is the moment of inertia of the suspended system, R is the radius of the cylinder, and H is the height of the liquid sample in the cylinder.

In general, the value of Δ is small compared to 1, since δ is of the order of 10^{-2} to 10^{-3}, and so Eq.(7.59) may be assumed, to a good approximation, to be

$$\frac{\delta}{\rho} = A\left(\frac{\mu}{\rho}\right)^{1/2} - B\left(\frac{\mu}{\rho}\right) + C\left(\frac{\mu}{\rho}\right)^{3/2} \tag{7.60}$$

where

$$A = \frac{\pi^{3/2}}{I}\left(1 + \frac{R}{4H}\right)HR^3 T^{1/2}$$

$$B = \frac{\pi}{I}\left(\frac{3}{2} + \frac{4R}{\pi H}\right)HR^2 T$$

$$C = \frac{\pi^{1/2}}{2I}\left(\frac{3}{8} + \frac{9R}{4H}\right)HRT^{3/2}$$

---

[9] Ferris and Quested found that the coefficient of the second term of the right-hand side in equation $a_0 = 1 - \frac{3}{2}\Delta - \frac{3}{8}\Delta^2\cdots$had been printed incorrectly in the original Roscoe's paper, and should have been $(-3/2)$, as given above and not $(-1/2)$. The present authors would like to thank Drs P. Quested and K.C. Mills for informing us about this misprint.

In the field of liquid metals, Roscoe's absolute formula has been considered to provide remarkably accurate viscosity values.

7.9.1.3 *Shvidkovskii's Equation*

According to Shvidkovskii, if the parameter $R(2\pi/\nu T)^{1/2}$ is greater than 10 and H is greater than or equal to $1.85R$, the working formula connecting the observed data and the dimensions of the apparatus with the kinematic viscosity of the liquid sample can be expressed by [56]

$$\nu = \frac{I^2(\delta - T\delta_0/T_0)^2}{\pi(MR)^2 TW^2} \tag{7.61}$$

where

$$W = 1 - \frac{3}{2}\Delta - \frac{3}{8}\Delta^2 - a + (b - c\Delta)\frac{2nR}{H}$$

M represents the mass of the liquid sample; a, b, and c are constants tabulated as a function of $(2\pi R^2/\nu^* T)$ (ν^* is the value of ν when $W = 1$); n is the number of planes contacted horizontally with the liquid sample (i.e. in the case of a vessel having its lower end closed and its upper surface free, $n = 1$, if the vessel encloses the fluid top and bottom, $n = 2$); while the subscript 0 refers to an empty vessel.

Shvidkovskii's equation was used exclusively for absolute or relative viscosity determinations in the USSR.

Hopkins and Toye [57] and Toye and Jones [58] have also presented working formulae for the calculation of viscosity from measured damping data and dimensions of apparatus.

Andrade and Chiong [59] proposed a formula for absolute viscosity determinations in the case of the oscillating sphere method, and made careful measurements of the viscosities of the liquid alkali metals [60,61]. Although their working formula for a spherical vessel appears to provide accurate and reliable viscosity data, their technique is not popular because of the limitation of materials for the construction of spherical vessels.

7.9.2 Experimental Investigations of Working Formulae

Iida et al. [62,63] examined, experimentally, the validity of the above mentioned formulae. The samples used in these investigations were mercury, indium, tin, bismuth, and lead (vacuum melted, 99.99 to 99.999% purity) whose viscosities had been determined precisely using capillary viscometers.[10] The logarithmic decrements (i.e. the amplitudes) and time periods of the oscillations for these samples were measured carefully using an oscillating cylindrical viscometer. The outlines and experimental conditions of the viscometer employed in this work are shown in Table 7.10.

---

[10] In experimental determinations of the viscosity of metallic liquids, these metals are frequently employed as viscometer-calibrating liquids.

Table 7.10 *Outlines and experimental conditions of the oscillating vessel viscometer.*

Suspension system	Unifiler, ϕ 0.3 mm Mo (inverse suspension type)
Vessel	• graphite
	• ϕ 20 mm $\times$ H 30 mm (H is the height of liquid sample)
	• liquid top and bottom are closed
	• possible to use the same vessel repeatedly
Temperature range	194–640 °C
Atmosphere	He (0.1 MPa)

The results obtained are as follows:

(1) In Knappwost's formula, the value of K is assumed to be a constant. However, as pointed out by Kleinschmit and Grothe [64], the apparatus constant K can be expressed in terms of kinematic viscosity: by comparing Eqs. (7.58) and (7.60), we have

$$K = T^{3/2}\left\{A - B\left(\frac{\mu}{\rho}\right)^{1/2} + C\left(\frac{\mu}{\rho}\right)\right\} \qquad (7.62)^{11}$$

Figure 7.18 shows the apparatus constants plotted as a function of kinematic viscosity. The values of K for each plot were calculated from Knappwost formula represented by Eq. (7.58). As shown in Figure 7.18, the values of the apparatus constant seem to depend not only on kinematic viscosities but also on the kinds of metals used in the experiments. In conclusion, Knappwost's working formula, in general, gives roughly approximate values for viscosity.

(2) Values for viscosity of liquid bismuth were calculated from Eqs. (7.58), (7.60), and (7.61), using identical experimental data on the logarithmic decrements and period of oscillations. Relative viscosities were calculated from Knappwost's equation in which a value for the apparatus constant was determined using mercury at room temperature. Absolute viscosities were calculated from Roscoe's and from Shvidkovskii's formulae. Figure 7.19 shows calculated values for the viscosity of liquid bismuth, together with independent reference data determined using a capillary viscometer. Although identical experimental data were used in the calculations of viscosities, we see that considerable discrepancies exist among not only viscosity values, but also temperature dependence (i.e. apparent activation

---

[11] The variations of time periods T with temperature are negligibly small. In addition, if the moment of inertia of the whole suspended system is large, the differences in time periods between liquid metals are also negligibly small.

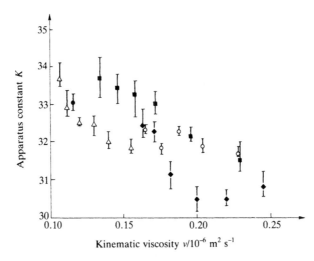

Figure 7.18 *Relations between the apparatus constant K in Eq. (7.58) and kinematic viscosity of viscometer-calibrating liquids:* ○, *lead;* ●, *mercury;* △, *bismuth;* ■, *indium;* ◆, *tin. The bars represent the scatter in measured values (after Iida et al. [62]).*

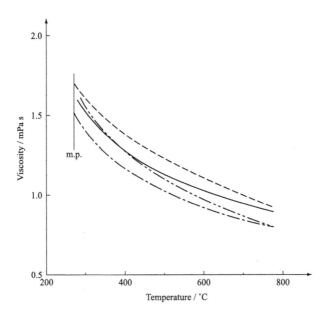

Figure 7.19 *Calculated values for viscosity of liquid bismuth. Working formulae used are:* ---, *Knappwost's formula;* -·-, *Roscoe's formula;* -··-, *Shvidkovskii's formula;* ——, *measured by a capillary viscometer (after Iida and Morita [65]).*

energies). For example, the calculated value for the viscosity of liquid bismuth at 430 °C, using Roscoe's absolute formula is about 20 per cent lower than the result using Knappwost's relative formula.

(3) Roscoe's equation has been considered to give accurate viscosity values; however, the agreement with the results by the capillary method is not good, as shown in Figure 7.19. For this reason, further experimental studies were made on the accuracy of Roscoe's equation. From the standpoint of metrology, two factors, i.e. end and slipping (or wetting) effects, should be investigated for the oscillating vessel method. Figure 7.20 gives a result of the determination of the end effects. As is obvious from this figure, the values calculated from Roscoe's absolute formula do not coincide with experimental data. Since the logarithmic decrements $(\delta - \delta_0)$ at $H = 0$ comes from the end effect, Roscoe'e formula provides insufficient weighting of the end correction. If we now assume that the end effect corresponds to the increase in the height of the liquid sample then, in the case of liquid lead contained in a cylindrical vessel of graphite, the value of the end correction ΔH to be added to the Roscoe's absolute formula is about 3 mm, according to the experimental result in Figure 7.20. The value of 3 mm for the end correction (in this investigation, $(H + \Delta H)/H = 1.1$) should not be neglected.

However, the major disagreement in the slopes $((\delta-\delta_0)$ vs. H, in Figure 7.20) between the calculated and experimental values may be due to a slipping phenomenon. A possible

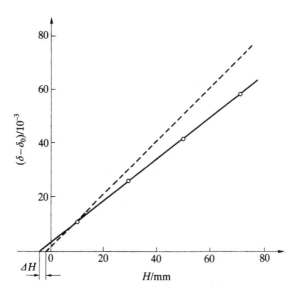

Figure 7.20 *Comparison of the calculated and measured values of end correction for liquid lead in a cylindrical vessel of graphite at 400° C: −−−, calculated; o, measured (after Iida et al. [63]).*

explanation for the smaller slope exhibited by the experimental data is that the liquid lead did not completely wet the sidewalls of the graphite vessel. The resulting slippage between the liquid interface and adjacent graphite walls would provide smaller damping than anticipated. The slipping, or the wetting, and end effects are extremely difficult problems. Progress in understanding the complicating factors such as slipping and end effects depend upon carefully constructed experimental investigations.

We have repeatedly mentioned that, in the field of liquid metals, Roscoe's formula has been considered to provide accurate viscosity values. Nevertheless, in the early part of this century, Ferris and Quested [66] pointed out that a more complete model of oscillating vessel (or cup) viscometers has been described by Beckwith, Kestin and Newell at around the same time. Furthermore, Ferris and Quested have investigated the Roscoe and the Beckwith–Newell working formulae for the viscosity determination of liquid metals by the oscillating cylinder method. According to their investigation, the Roscoe and the Beckwith–Newell working equations for the oscillating vessel viscometer appear to give almost identical results. However, Quested and co-worker's [1,66] research into the working equations for the determination of viscosity of liquid metals using the oscillating cylinder method suggest that the Beckwith–Newell model is more accurate and comprehensive than that of Roscoe and should therefore be adopted for future analysis. Thus, their (theoretical) metrological and experimental studies have made steady progress on the determination of metallic liquids' viscosities, using the oscillating cylinder method. Nevertheless, further work on the working equations for the viscosity measurements of metallic liquids by the oscillating vessel method is required for the further development of materials process science (e.g. in order to elucidate slipping and end effects).

7.10 Experimental Data for the Viscosity of Liquid Metallic Elements

At present, experimentally obtained viscosity data are available for some 53 liquid metallic elements. Although the evaluations of their reliability are of great importance, it is very difficult to state definitely their accuracies or uncertainties. Nevertheless, it appears reasonable to estimate that for low melting point liquid metallic elements, the uncertainties of viscosity values lie in the range 1–10 per cent, and for high melting point (or refractory) liquid metals, in the range 5–20 per cent, with exception of several liquid metallic elements. Unfortunately, the viscosity values for chromium, manganese, osmium, silicon, vanadium, and zirconium remain very uncertain. Experimental re-examinations for these elements must be performed in the near future.

The experimental data determined by a number of workers for the viscosity of liquid iron are shown in Figure 7.21. These viscosity values were measured using oscillating vessel viscometers, except for the data of Arsentiev et al. [50] (using an oscillating plate viscometer) and Krieger and Trenkler [49] (using a rotational viscometer). In calculating the viscosities, the following equations were used: the investigators in

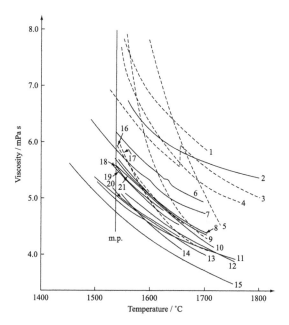

Figure 7.21 *Viscosity of liquid iron, as determined by a number of workers. −−−, values obtained by workers in the USSR. 1. Arsentiev et al., 2. Barfield and Kitchener, 3. Nobohatskii et al., 4. Romanov and Kochegarov, 5. Samarin, 6. Ogino et al., 7. Ogino et al., 8. Nakanishi et al., 9. Vatolin et al., 10. Frohberg and Cakici, 11. Cavalier, 12. Saito and Watanabe, 13. Lucas, 14. Kawai et al., 15. Thiele, 16. Avaliani et al., 17. Wen and Arsentiev, 18. Schenck et al., 19. Frohberg and Weber, 20. Narita and Onoe, 21. Krieger and Trenkler (after Iida and Morita [65]).*

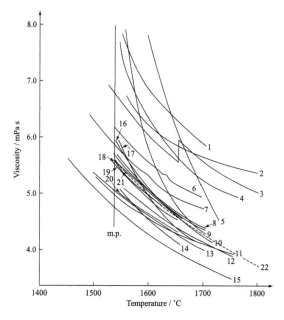

Figure 7.22 *Viscosity of liquid iron as a function of temperature. Identification numbers are the same as those in Figure 7.21. −−−, values measured by Andon et al. (from Mills [67]): values recommended by Mills [67] and Iida et al. [19].*

USSR, Shvidkovskii's equation; Ogino et al. and Frohberg and Cakici, Roscoe's equation; the other investigators, Kappwost's equation. As is obvious from Figure 7.21, workers in the USSR reported viscosity values which are higher near the melting point and exhibit greater temperature dependence in comparison with the data of others. This discrepancy may mainly be attributed to the approximate nature of the working formulae used. Figure 7.22 gives the present recommended data for the viscosity of liquid iron, together with the data reported by a number of workers.

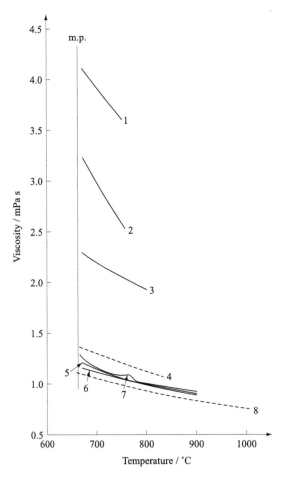

Figure 7.23 *Viscosity of liquid aluminium as a function of temperature, as determined by several workers. 1. Jones and Bartlett, 2. Yao and Kondic, 3. Yao, 4. Rothwell: values recommended by Iida et al. [19], 5. Gebhardt and Detering, 6. Gebhardt et al., 7. Lihl et al., 8. Andon et al. (from Mills [67]): values recommended by Mills [67].*

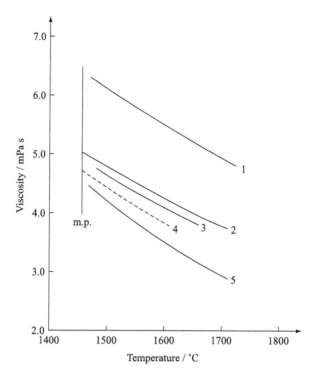

Figure 7.24 *Viscosity of liquid nickel as a function of temperature, as determined by several workers. 1. Vertman and Samarin, 2. Cavalier, 3. Schenck et al., 4. Andon et al. (from Mills [66]): values recommended by Mills [67] and Iida et al. [19], 5. Samarin.*

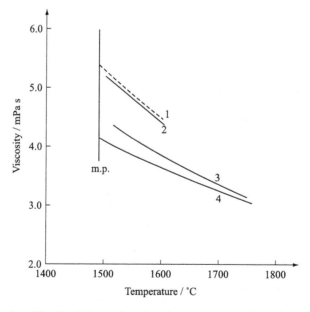

Figure 7.25 *Viscosity of liquid cobalt as a function of temperature, as determined by several workers. 1. Sato and Yamamura (from Mills [67]): values recommended by Mills [67] and Iida et al. [19], 2. Frohberg and Weber, 3. Watanabe and Saito, 4. Cavalier.*

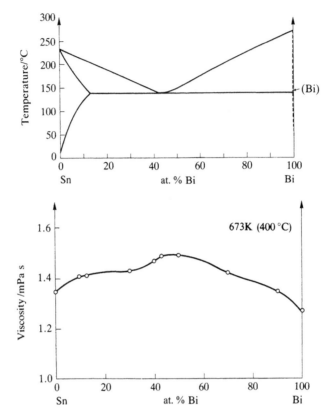

Figure 7.26 *Phase diagram and isothermal viscosities of the binary tin–bismuth system (after Iida and Guthrie [70]).*

As shown in Figure 7.23, experimental viscosity values for liquid aluminium exhibit even larger discrepancies. The authors' and Mills' recommended values are also indicated; the value for the melting point viscosity of liquid aluminium may probably lie in the range 1.0–1.5 mPa s.

Experimental viscosity values of liquid nickel and cobalt are given in Figures 7.24 and 7.25, respectively, in which recommended values are also shown.

Brooks et al. [68] have carried out viscosity measurements for liquid mercury, tin, aluminium, copper, nickel, and iron using an oscillating crucible (cylindrical vessel) viscometer; the viscosity values obtained are in good agreement with the recommended values of Iida and Shiraishi except for the case of iron.

Experimentally derived viscosity data for liquid metallic elements are listed in Tables 17.9(a) and (b).

Experimental data for the viscosity of binary liquid alloys are also in some confusion. For example, even in an amenable, low melting alloy system such as lead–tin, the composition dependence of isothermal viscosity is inconsistent. In this system, some researchers reported anomalous viscosity changes at compositions corresponding to the

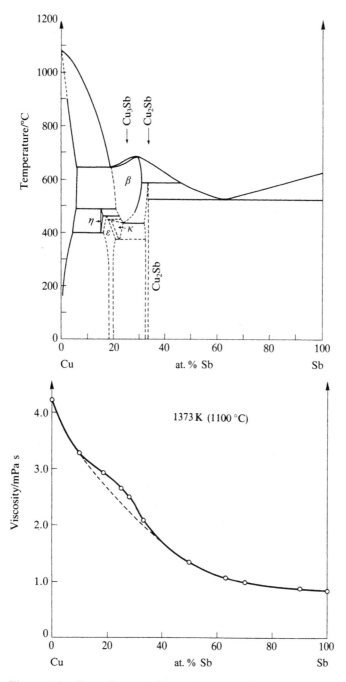

Figure 7.27 *Phase diagram and isothermal viscosities of the binary copper–antimony system (after Iida and Guthrie [70]).*

eutectic and at limits of solid solubility. On the other hand, other researchers obtained a linear or a slightly curved dependence on composition [69].

It has been pointed out that eutectic systems show negative deviations in isothermal viscosities. However, this relationship between excess viscosity and phase diagram has some exceptions. An example is shown in Figure 7.26. Compound-forming alloy systems are apt to show maxima or local maxima in their isothermal viscosities in those composition ranges where compounds are formed in the solid state. This is indicated in Figure 7.11. In addition, another example is shown in Figure 7.27, together with its phase diagram. In the copper–antimony system, a maximum in the isothermal viscosity-composition curve is not found, although the curve has a hump in the composition range where compounds (Cu_3Sb and Cu_2Sb) are formed in the solid state.

In general, discontinuous or sharp variations in the viscosities of metallic liquids with respect to temperature or composition should, at first, be doubted. Careful repetition of measurements is therefore necessary for any metallic liquids showing sharp changes in viscosities.

REFERENCES

1. R.F. Brooks, A.T. Dinsdale, and P.N. Quested, *Meas. Sci. Technol.*, **16** (2005), 354.
2. M. Born and H.S. Green, *Proc. R. Soc. Lond.*, A, **190** (1947), 455.
3. M.D. Johnson, P. Hutchinson, and N.H. March, *Proc. R. Soc. Lond.*, A, **282** (1964), 283.
4. T.E. Faber, *Introduction to the Theory of Liquid Metals*, Cambridge University Press, 1972, pp.144 and 165.
5. P.A. Egelstaff, *An Introduction to the Liquid State*, Academic Press, London, 1967, Chaps. 5 and 14.
6. S.A. Rice and A.R. Allnat, *J. Chem. Phys.*, **34** (1961), 2144; S.A. Rice and A.R. Allnat, *J. Chem. Phys.*, **34** (1961), 2156.
7. S.A. Rice and J.G. Kirkwood, *J. Chem. Phys.*, **31** (1959), 901.
8. M. Kitajima, K. Saito, and M. Shimoji, *Trans. JIM.*, **17** (1976), 582.
9. E. Helfand and S.A. Rice, *J. Chem. Phys.*, **32** (1960), 1642.
10. A.D. Pasternak, *Phys. Chem. Liq.*, **3** (1972), 41.
11. S.M. Breitling and H. Eyring, in *Liquid Metals, Chemistry and Physics*, edited by S.Z. Beer, Marcel Dekker, New York, 1972, Chap. 5.
12. I. Yokoyama, *Physica B*, **271** (1999), 230.
13. I. Yokoyama, *Physica B*, **291** (2000), 145.
14. J. Frenkel, *Kinetic Theory of Liquids*, Oxford University Press, 1946.
15. S. Glasstone, K.J. Laidler, and H. Eyring, *The Theory of Rate Processes*, McGraw-Hill, New York, 1941, Chap. IX.
16. E.N. da C. Andrade, *Phil. Mag.*, **17** (1934), 497; E.N. da C. Andrade, *Phil. Mag.*, **17** (1934), 698.
17. M.H. Cohen and D. Turnbull, *J. Chem. Phys.*, **31** (1959), 1164.
18. T. Iida, R.I.L. Guthrie, and M. Isac, in *ICS Proceedings of the 3rd International Congress on Science and Technology of Steelmaking*, Association for Iron and Steel Technology, Charlotte, NC, 2005, p.57.

19. T. Iida, R. Guthrie, M. Isac, and N. Tripathi, *Metall. Mater. Trans. B*, 37 (2006), 403.
20. I. Osida, *Proc. Phys.-Math. Soc. Japan*, 21 (1939), 353.
21. T. Iida, R.I.L. Guthrie, and Z. Morita, *Can. Mater. Q.*, 27 (1988), 1.
22. S.J. Cocking, *J. Phys. C* 2 (1969), 2047.
23. B. Djemili, L. Martin-Garin, R. Martin-Garin, and P. Desré, *J. Less-Common Met.*, 79 (1981), 29.
24. P. Hichter, F. Durand, and E. Bonnier, *J. Chim. Phys.*, 68 (1971), 804.
25. P.B. Macedo and T.A. Litovitz, *J. Chem. Phys.*, 42 (1965), 245.
26. R.P. Chhabra and D.K. Sheth, *Z. Metallkd.*, 81 (1990), 264.
27. R.P. Chhabra and A. Tripathi, *High Temp.-High Press.*, 25 (1993), 713.
28. A.K. Mehrotra, *Ind. Eng. Chem. Res.*, 30 (1991), 1367.
29. R.P. Chhabra, *Alloys Comp.*, 221 (1995), L1-3.
30. S. Morioka, X. Bian, and M. Sun, *Z. Metallkd.*, 93 (2002), 288.
31. S. Morioka, *Mater. Sci. Eng., A*, 362 (2003), 223.
32. G. Kaptay, *Z. Metallkd.*, 96 (2005), 24.
33. I. Budai, M.Z. Benkö, and G. Kaptay, *Mater. Sci. Forum*, 537–538 (2007), 489.
34. D. Živković, *Metall. Mater. Trans. B*, 39 (2008), 395.
35. T. Iida, Z. Morita, and S. Takeuchi, *J. Japan Inst. Metals*, 39 (1975), 1169.
36. A.V. Grosse, *J. Inorg. Nucl. Chem.*, 23 (1961), 333.
37. M. Hirai, *ISIJ Int.*, 33 (1993), 251.
38. P.-F. Paradis, T. Ishikawa, and S. Yoda, *Int. J. Thermophys.*, 23 (2002), 825.
39. D.C. Wallace, *Proc. R. Soc. Lond., A*, 433 (1991), 615.
40. T. Iida, A. Kasama, Z. Morita, I. Okamoto, and S. Tokumoto, *J. Japan Inst. Met.*, 37 (1973), 841.
41. E.A. Moelwyn-Hughs, *Physical Chemistry*, 2nd ed., Pergamon Press, Oxford, 1961, Chap. XVII.
42. T. Iida, M. Ueda, and Z. Morita, *Tetsu-to-Hagané*, 62 (1976), 1169.
43. Z. Morita, T. Iida, and M. Ueda, in *Liquid Metals 1976*, Institute of Physics Conference Series No.30, 1977, p.600.
44. L.C. Prasad and R.K. Jha, *Phys. Stat. Sol. (a)*, 202 (2005), 2709.
45. W. Menz and F. Sauerwald, *Acta Metall.*, 14 (1966), 1617.
46. H. Fukuyama and Y. Waseda (eds.), in *High-Temperature Measurements of Materials*, Springer, Berlin, 2009, Chap.2.
47. P.-F. Paradis, T. Ishikawa, and S. Yoda, *J. Appl. Phys.*, 97 (2005), 106101; see also T. Ishikawa, P.-F. Paradis, T. Itami, and S. Yoda, *Meas. Sci. Technol.*, 16 (2005), 443.
48. W.R.D. Jones and W.L. Bartlett, *J. Inst. Met.*, 81 (1952–3), 145.
49. W. Krieger and H. Trenkler, *Arch. Eisenhüttenw*, 42 (1971), 175; W. Krieger and H. Trenkler, *Arch. Eisenhüttenw*, 42 (1971), 685.
50. P.P. Arsentiev, B.G. Vinogradov, and B.S. Lisichkii, *Izu. Vuzov. Chern. Met.*, (1974), 181.
51. H.R. Thresh, *Trans. AMS.*, 55 (1962), 790.
52. A. Knappwost, *Z. Metallkd.*, 39 (1948), 314.
53. A. Knappwost, *Z. Phys. Chem.*, 200 (1952), 81.
54. R. Roscoe, *Proc. Phys. Soc.*, 72 (1958), 576.
55. R. Roscoe and W. Bainbridge, *Proc. Phys. Soc.*, 72 (1958), 585.
56. A.A. Vertman and A.M. Samarin, in *Metody Issledovaniya Svoisty Metallicheskikh Rasplavov*, Nauka, Moscow, 1969, p.28.
57. M.R. Hopkins and T.C. Toye, *Proc. Phys. Soc., B*, 63 (1950), 773.
58. T.C. Toye and E.R. Jones, *Proc. Phys. Soc.*, 71 (1958), 88.

59. E.N. da C. Andrade and Y.S. Chiong, *Proc. Phys. Soc.*, **48** (1936), 247.

60. Y.S. Chiong, *Proc. R. Soc. Lond.*, *A*, **157** (1936), 264.

61. E.N. da C. Andrade and E.R. Dobbs, *Proc. R. Soc. Lond.*, *A*, **211** (1952), 12.

62. T. Iida, A. Satoh, S. Ishiura, S. Ishiguro, and Z. Morita, *J. Japan Inst.Met.*, **44** (1980), 443.

63. T. Iida, T. Kumada, M. Washio, and Z. Morita, *J. Japan Inst. Met.*, **44** (1980), 1392.

64. P. Kleinschmit and K.H. Grothe, *Z. Metallkd.*, **61** (1970), 378.

65. T. Iida and Z. Morita, *Bull. Japan Inst. Met.*, **19** (1980), 655.

66. D.H. Ferris and P. Quested, NPL Report CMMT (A) 306, December 2000.

67. K.C. Mills, *Recommended Values of Thermophysical Properties for Selected Commercial Alloys*, Woodhead Publishing and ASM International, Cambridge, 2002.

68. R.F. Brooks, A.P. Day, R.J.L. Andon, L.A. Chapman, K.C. Mills, and P.N. Quested, *High Temp.-High Press.*, **33** (2001), 73.

69. H.R. Thresh and A.F. Crawley, *Met. Trans.*, **1** (1970), 1531.

70. T. Iida and R.I.L. Guthrie, *The Physical Properties of Liquid Metals*, Clarendon Press, Oxford, 1988, Chap. 6.

8

Diffusion

8.1 Introduction

Diffusion is the transport of mass from one region to another on an atomic scale. Diffusivities in the liquid state are much higher than those in the solid state. In the case of metallic elements, diffusivities in the two states can differ by a factor of 100 to 1000. The high atomic mobility of most metallic elements just above their melting temperatures, with diffusivities in the order of $10^{-9} \, \mathrm{m^2 \, s^{-1}}$, is one of the most characteristic properties of liquids.

Knowledge of diffusivities is needed for many fields of engineering. For example, for most materials processes, heterogeneous chemical reactions play an important role. The rates of heterogeneous reactions, e.g. between two liquid phases such as a slag and a metal, are limited by the diffusion of the reactant species to reaction sites at interfaces. Similarly, many other processes, such as crystal growth, or the distribution of solute elements during solidification, or corrosion, and so forth, also depend upon the rate of atomic diffusion. Although many other good examples are not hard to find, we can conclude, in general, that mass transfer through normally very thin boundary layers is typical of heterogeneous reactions, and that the mass transport is controlled by atomic diffusion phenomena.

Unfortunately, there are problems associated with experimental work relating to diffusion measurements in liquid metallic elements. First, experimental data for the self-diffusivities of liquid metallic elements are extremely scarce, mainly because of a lack of specific radioisotopes (i.e. tracers). At present, it would appear that experimentally derived self-diffusivity data are available only for some 20 metallic elements. For a clear understanding of those phenomena related to diffusion, a study of self-diffusion in liquid metallic elements is of critical importance. More reliable and accurate experimental data for self-diffusivities of various liquid metallic elements are therefore needed, so as to test and develop theories, and to predict solute or impurity diffusivities.

Remarkable progress has been made, over the past 50 years or so, in the development of diffusion theories for metallic liquids. The hard-sphere model in particular seems to be able to provide an accurate means for calculating self-diffusivities in metallic liquids. However, it may be premature to conclude its total success. A comparison between theory and experiment for various liquid metallic elements, i.e. liquid metals, semimetals,

The Thermophysical Properties of Metallic Liquids: Volume 1 – Fundamentals. First Edition.
Takamichi Iida and Roderick I. L. Guthrie. © Takamichi Iida and Roderick I. L. Guthrie 2015.
Published in 2015 by Oxford University Press.

and semiconductors (e.g. silicon, germanium, tellurium), is needed. Corresponding-sates methods appear to have been deemed more useful for estimating transport coefficients in the field of materials process science, although this does not coincide with the authors' views. For example, experimental values for liquid lead self-diffusivities (or liquid sodium viscosities) exhibit considerably larger deviations from self-diffusivity correlations (or from viscosity correlations) expected from the corresponding-states theory (see Figures 7.1 and 8.2). The hard-sphere theory, as well as the corresponding-states theory, give simple explanations for empirical models or equations and act as guides to experimental studies. In these respects, these theories are important for researchers in the area of materials process science. To the authors' knowledge, experimental data for pure liquid magnesium, aluminium, titanium, iron, cobalt, nickel, etc. have not yet been reported. Self-diffusivity data for these liquid metals are badly needed.

A second difficulty with diffusion data is that while many experimental investigations have been carried out on solute diffusivities in liquid metallic elements, the data are scarcely of sufficient accuracy. Similarly, while a variety of expressions describing solute diffusivities have been proposed, most of these equations are, unfortunately, unreliable in predicting solute diffusivities in many important liquid metallic systems. Also, for some equations, the judgment of their validity is difficult because of extremely large uncertainties in the corresponding experimental data. This is especially true of diffusivity data for gases in liquid metallic elements. The failure of theoretical calculations may well be due to a lack of knowledge of the physical state of solutes in liquid metallic elements, e.g. their atomic or ionic size, the viscous forces or friction constants between solute/solvent metallic atoms (or ions), and their atomic (or ionic) motions related to diffusion.

8.2   Theoretical Equations for Self-Diffusivity

In atomic (or molecular) dynamic theories, atoms within a liquid are assumed to interact in accordance with a certain type of interatomic pair potential. Equations for dynamical properties are solved numerically, using a computer of large storage capacity. These calculations provide useful information on structures and atomic motions in liquids.

8.2.1   Linear Trajectory Theory

In this theoretical analysis, the pair potential and friction coefficient are conveniently divided into two and into three parts, respectively. The pair potential $\phi(r)$ is written in the form [1]:

$$\phi(r) = \phi_H(r) + \phi_S(r) \tag{8.1}$$

$$\left.\begin{array}{ll} \phi_H(r) = \infty, & r < \sigma \\ \phi_H(r) = 0, & r \geq \sigma \\ \phi_S(r) = 0, & r < \sigma \\ \phi_S(r) = f(r), & r \geq \sigma \end{array}\right\} \tag{8.2}$$

where $\phi_H(r)$ is the hard-sphere potential, $\phi_S(r)$ is the soft potential, and σ is the effective core diameter.

Similarly, the friction coefficient is separated into three parts;

$$\varsigma_f = \varsigma_H + \varsigma_S + \varsigma_{SH} \tag{8.3}$$

where ς_H, ς_S, and ς_{SH} are friction coefficients due to the hard-core collision, to soft interaction between neighbouring atoms, and to cross effect between hard and soft forces in the pair potential, respectively. According to Helfand [2] and Davis and Palyvos [3], the friction coefficients are given by

$$\varsigma_H = \frac{8}{3} n_0 \sigma^2 g(\sigma)(\pi m k T)^{1/2} \tag{8.4}$$

$$\varsigma_S = -\frac{n_0}{12\pi^2}\left(\frac{\pi m}{kT}\right)^{1/2} \int_0^\infty Q^3 \tilde{\phi}_S(Q) \tilde{G}(Q) dQ \tag{8.5}$$

$$\varsigma_{SH} = -\frac{1}{3} n_0 g(\sigma)\left(\frac{m}{\pi kT}\right)^{1/2} \int_0^\infty \{Q\sigma \cos(Q\sigma) - \sin(Q\sigma)\}\tilde{\phi}_S(Q) dQ \tag{8.6}$$

where $g(\sigma)$ is the value of the pair distribution function at contact (i.e. $r = \sigma$), $\tilde{\phi}_S(Q)$ is the Fourier transform of the long-range part of the potential, and $\tilde{G}(Q)$ is the Fourier transform of $\{g(r) - 1\}$, with $g(r)$ the pair distribution function. Thus, from the Einstein relation (see Eq. (8.42) and these three expressions, the self-diffusivities are obtained:

$$D = \frac{kT}{\varsigma_H + \varsigma_S + \varsigma_{SH}} \tag{8.7}$$

8.2.2 Small Step Diffusion Theory

According to the small step diffusion theory of Rice and Kirkwood [4], the soft part of the friction coefficient can be expressed by

$$\varsigma_S = \left\{\frac{4\pi m n_0}{3} \int_0^\infty \nabla_r^2 \phi_S(r) g(r) r^2 dr\right\}^{1/2} \tag{8.8}$$

In this approach, the self-diffusivity is expressed in terms of ς_H and ς_S as follows:

$$D = \frac{kT}{\varsigma_H + \varsigma_S} \tag{8.9}$$

Several investigators have calculated self-diffusivities of pure liquid metals, using the linear trajectory and the small step diffusion theories. The results of their theoretical analyses provide good qualitative agreement with experiment. Calculated results indicate that the soft part of the effective interatomic potentials play a dominant role in determining the magnitude of self-diffusivities in liquid metals. However, the accuracy of the calculations, or rather, the uncertainty of the pair potentials used, needs more thorough examination.

8.2.3 Equations Based on the Hard-Sphere Theories (or Models)

Metallic atoms exhibit symmetrical, surrounding force fileds. These are approximately equivalent to the presence of hard spheres. During the last 40 or 50 years, considerable progress has been made in calculating self-diffusivities in liquid metals through development of the hard-sphere model (or theory).

Several expressions (in units of cgs; $cm^2 s^{-1} = 10^{-4} m^2 s^{-1}$) for self-diffusivity in liquid metals, which are based on hard-sphere models, have been proposed:

(1) The expression derived by Vadovic and Colver [5]:

$$D = 0.365r \left(\frac{\pi k T}{m} \right)^{1/2} \frac{\eta_m / \eta}{9.385 (T_m \rho / T \rho_m) - 1} \tag{8.10}$$

$$r = \left\{ \frac{3}{4} \left(\frac{\eta_m M}{\pi \rho_m N_A} \right) \right\}^{1/3}$$

where r is the atomic (or hard sphere) radius, i.e. $r = \sigma / 2$, and η is the packing fraction.

(2) The expression derived by Faber [6]:

$$D = 4.9 \times 10^{-6} \left(\frac{T}{M} \right)^{1/2} V^{1/3} \frac{(1 - \eta)^3}{\eta^{5/3} (1 - \eta / 2)} \tag{8.11}$$

with $\eta_m = 0.46$ at the melting temperature, Eq. (8.11) becomes

$$D_m = 3.6 \times 10^{-6} \left(\frac{T_m}{M} \right)^{1/2} V_m^{1/3} \tag{8.12}$$

(3) The expression derived by Protopapas et al. [7]:

$$D = \sigma C_{AW}(\eta) \left(\frac{\pi R T}{M} \right)^{1/2} \frac{(1 - \eta)^3}{8\eta(2 - \eta)} \tag{8.13}$$

In the approach of Protopapas et al., values of σ, η and $C_{AW}(\eta)$ can be calculated using the following relations:

$$\sigma = 1.126\sigma_m \left\{ 1 - 0.112(T / T_m)^{1/2} \right\} \tag{8.14}$$

in which σ_m is the values of σ at the melting point. With η_m taken as 0.472:

$$\sigma_m = \{6(0.472)M/\pi\rho_m N_A\}^{1/3}$$
$$= 1.41(M/\pi\rho_m N_A)^{1/3}$$
$$\eta = \frac{0.472\rho\sigma^3}{\rho_m\sigma_m^3} \tag{8.15}$$

The correction factor $C_{AW}(\eta)$ is obtained from Figure 8.1, once the value of η is known.

(4) An expression based on molecular dynamics simulations. The values for the self-diffusivity of a fluid of hard spheres have been re-assessed using molecular dynamics simulations by Speedy [8], and can be calculated from the expression

$$D_{HS} = \left(\frac{\pi D_0}{6\eta}\right)\left(1 - \frac{6\eta}{1.09\pi}\right)\left[1 + \left(\frac{6\eta}{\pi}\right)^2\left\{0.4 - 0.83\left(\frac{6\eta}{\pi}\right)^2\right\}\right] \tag{8.16}$$

$$D_0 = \frac{3}{8}\sigma\left(\frac{kT}{\pi m}\right)^{1/2}$$

where m is the mass of a sphere.

Equations (8.10), (8.11), (8.13), and (8.16), in general, give good results for the magnitudes and temperature dependence of the self-diffusivity for pure liquid metals.

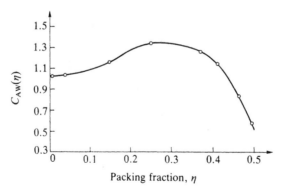

Figure 8.1 *The Alder–Wainwright correction C_{AW} as a function of the packing fraction η. The correction factor is defined as the ratio of the self-diffusion coefficient of a hard-sphere fluid to the value predicted by the Enskog theory (after Protopapas et al. [7]).*

8.2.4   The Principle of Corresponding States

Self-diffusivity data can also be correlated by the use of the corresponding-states principle. According to Pasternak [9], self-diffusivity can be expressed by

$$D = \frac{D^*}{(V^*)^{1/3}} \frac{(R\varepsilon/k)^{1/2}}{N_A^{1/3} M^{1/2}} V^{1/3} \tag{8.17}$$

where $\varepsilon/k = 5.2 T_m$. The $D^*/(V^*)^{1/3}$ versus $1/T^*$ correlation is illustrated in Figure 8.2 (where $T^* = T/(\varepsilon/k)$) for the self-diffusivities of ten liquid metals. As can be seen,

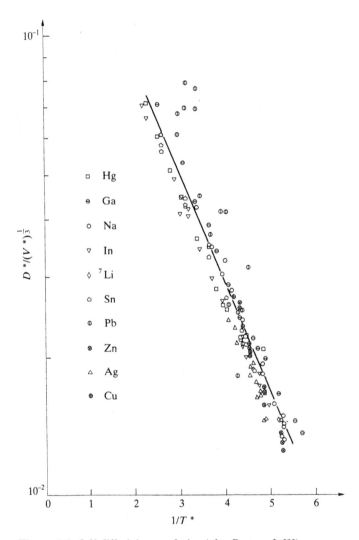

Figure 8.2 *Self-diffusivity correlation (after Pasternak [9]).*

this correlation has a spread of ± 20 per cent for self-diffusivities, provided the high temperature values for lead are neglected. The large deviations of lead from the correlation are, as yet, unexplained by the theory of liquids.

8.2.5 Model Theories

8.2.5.1 The Fluctuation Model

Swalin [10] proposed a model for diffusion in liquids from the standpoint of a fluctuation model theory. According to Swalin, diffusion in liquids occurs by the movement of atoms through small, variable distances rather than by jumps of atoms into discrete-sized 'holes' of the order of the atomic diameter. These small, variable distances are caused by a local density fluctuation about the diffusing atoms. The energy required for atomic motion is the energy required for a density fluctuation and may be expressed in terms of a Morse function. The equation for self-diffusivity in liquid metallic elements derived on this basis is represented either by

$$D = \frac{Zk^2 T^2}{8hK_f} \tag{8.18}$$

or

$$D = 1.29 \times 10^{-8} \frac{T^2}{\Delta_l^g H \alpha_p^2}, \text{ (in units of cgs)} \tag{8.19}$$

where Z is the coordination number, h is the Planck constant, K_f is the force constant obtained from the data of Waser and Pauling [11], $\Delta_l^g H$ is the enthalpy of evaporation, and α_p is related to the curvature of the potential versus distance curve. Equations (8.18) and (8.19) suggest that the temperature dependence of the self-diffusivity is approximately proportional to the square of the absolute temperature, i.e. $D \propto T^2$.

Swalin warns that the fluctuation model may be more accurate in calculating relative values of the diffusivity than absolute values since, in calculating relative values, many of the quantities which are difficult to evaluate quantitatively, cancel out [12].

8.2.5.2 The Significant Structure Model

In the significant model theory, the self-diffusivity is expressed as [13]:

$$D = \frac{kT}{\varsigma_A (V_s / N_A)^{1/3} \mu} \tag{8.20}$$

in which V_s is the molar volume in the solid and ς_A is a parameter expected to have values near six.

Breitling and Eyring [13] calculated self-diffusivities for six liquid metals and compared them with experimental data. Calculated values rarely differ from experimental values by more than an order of magnitude. Since the viscosities used were poor, their self-diffusivities would, as a result, also seem to be poor.

A number of other model theories for diffusion have been proposed. Other well-known examples are the activation-state theory (or model) or the jump diffusion model of Eyring and co-workers [14], the quasi-crystalline model theory of Careri and Paoletti [15], the free volume model of Cohen and Turnbull [16], the vibrational random walk model theory of Nachtrieb [17], and so forth.

In order to understand the mechanism of liquid diffusion, Cahoon [18] investigated the hole model. In the hole model, diffusivity in a liquid can be described by the Arrhenius equation:

$$D = D_0 \exp\left(-\frac{Q_E}{RT}\right) \tag{8.21}$$

$$D_0 = \gamma_g f \alpha_j v e^{\Delta S / R}$$

where D_0 is the frequency factor, Q_E is the activation energy, γ_g is a geometrical factor, f is a correlation coefficient, α_j is the atom 'jump' distance, v is the vibrational frequency, and ΔS is the entropy change associated with the thermally activated process. Cahoon made an investigation into the entropy change ΔS in the frequency factor; as a result, he concluded that while liquid diffusion via the hole mechanism is improbable, diffusion via a fluctuation mechanism remains feasible.

These model theories have been discussed in some review articles (e.g. Nachtrieb [17]).

In the case of self-diffusion, remarkable progress has been made in atomic dynamic calculations using powerful computers. Further, direct structural information on liquids, based on X-ray and neutron diffraction experiments, has been accumulated in the last half century. Thus, the development in liquid state physics would appear to have somewhat reduced the role of the model theories.

The objections to model theories are that, in general, they make drastic assumptions about the structure of a liquid (for example, the assumptions that the spatial arrangement of atoms, or the structure, in a liquid is similar to that of atoms in a crystalline solid, and further, that the mechanism of diffusion is also similar); besides, the model theories contain one or more *a posteriori* parameters which need to be fitted using experimental data.

Unfortunately, none of the equations, described in this section, satisfy the conditions of both accuracy and universality required of any model for predicting diffusivities of metallic liquids.

8.3   Relationship between Viscosity and Self-Diffusivity

8.3.1   The Sutherland–Einstein Equation

Sutherland [19] presented a correction to the Stokes–Einstein equation using hydrodynamic theory:

$$D = \frac{kT(1 + 3\mu / \beta_S r)}{6\pi \mu r (1 + 2\mu / \beta_S r)} \tag{8.22}$$

where r is the radius of the diffusing particle and β_S is the coefficient of sliding friction between the diffusing particle and medium, which depends on the size of the diffusing particle. When the diffusing particle is large compared with the particles of the medium, β_S equals infinity. When the radius of the diffusing particle is approximately equal to that of the medium, β_S becomes zero. For the two extreme cases, Eq. (8.22) may be written

$$D = \frac{kT}{6\pi\mu r}, \quad \text{when} \quad \beta_S = \infty \tag{8.23}$$

$$D = \frac{kT}{4\pi\mu r}, \quad \text{when} \quad \beta_S = 0 \tag{8.24}$$

Both Eqs. (8.23) and (8.24) are generally referred to as the Stokes–Einstein formula. Equation (8.24) is sometimes called the Sutherland–Einstein formula. In these formulae, the friction coefficient is expressed in terms of viscosity and radius of the diffusing particle.

In order to answer the question of whether the diffusing species is in the form of an atom or an ion in metallic liquids, the Goldschmit atomic radii, Pauling ionic radii, and so forth, can be used for r. However, no conclusive answer has yet been presented to indicate which form is better.

8.3.2 Modified Stokes–Einstein Formula

Several investigators have pointed out that the modified Stokes–Einstein formula is in excellent agreement with self-diffusivity data in liquid metals. The formula has the form

$$D = \frac{kT}{\varsigma(V/N_A)^{1/3}\mu} \tag{8.25a}$$

At the melting point,

$$D_m = \frac{kT_m}{\varsigma(V_m/N_A)^{1/3}\mu_m} \tag{8.25b}$$

where the subscript m denotes the melting point and ς is a constant taking a value between 5 and 6.

The radius of a diffusing particle can be obtained on the basis of the hard-sphere model. From the definition of the packing fraction, we have

$$\left(\frac{V}{N_A}\right)^{1/3} = \left(\frac{4\pi}{3\eta}\right)^{1/3} r \tag{8.26}$$

With $\eta = 0.46$, Eq. (8.26) becomes

$$\left(\frac{V}{N_A}\right)^{1/3} = 2.1r \tag{8.27}$$

Substituting this relation into Eq. (8.25), we obtain

$$\frac{kT}{12.6\mu r} \le D \le \frac{kT}{10.5\mu r} \tag{8.28}$$

Equation (8.28) indicates that the modified Stokes–Einstein formula is equivalent to the Stokes–Einstein formula, or the Sutherland–Einstein formula, represent by Eq.(8.24), when the radius of a hard sphere is employed for the diffusing atom. Consequently, the Stokes–Einstein formula, or Sutherland–Einstein formula, using the hard-sphere radius will provide good results for self-diffusivity in liquid metals, even though it does not clarify the atomistic process of diffusion in liquids.

By combining the modified Stokes–Einstein formula with the Andrade formula for melting point viscosity represented by Eq. (7.15), we have an expression for D_m in liquid metals,

$$D_m = 1.20 \times 10^{-9} \left(\frac{T_m}{M}\right)^{1/2} V_m^{1/3}, \text{ (in SI units)} \tag{8.29}$$

in which the numerical factor of $1.20 \times 10^{-9} \text{ kg}^{1/2} \text{ m s}^{-1} \text{ K}^{-1/2} \text{ mol}^{-1/6}$ was determined so as to give the minimum S value for 15 pure liquid metals (see Table 8.4). This type of equation in terms of $(T_m/M)^{1/2} V_m^{1/3}$ is also obtained, on the basis of the corresponding-states principle as well as the hard-sphere model.

Substitution of Eq. (7.41) into Eq. (8.25b) gives an expression for the melting point self-diffusivity in liquid metallic elements in terms of the new dimensionless number $\xi_T^{1/2}$:

$$D_m = 2.19 \times 10^{-16} \left(\frac{\xi_T T_m}{M}\right)^{1/2} \frac{T_m}{C_0 V_m^{1/3} \gamma_m} \tag{8.30}$$

where the numerical factor of $2.19 \times 10^{-16} \text{ kg m}^2 \text{ s}^{-2} \text{ K}^{-1} \text{ mol}^{-1/3}$ was determined so as to give the minimum S value for 18 liquid metallic elements (see Table 8.6). Values for the numerical factor C_0, or $3.771 \times 10^{-2} k_0^{-1}$ (see Section 7.4), can be obtained from experimental viscosity and surface tension data on the basis of Eq. (7.41):

$$C_0 = \frac{\mu_m}{\gamma_m} \left(\frac{\xi_T T_m}{M}\right)^{1/2} \tag{7.41'}$$

Moreover, the relationship of the numerical factor C_0 plotted against atomic number Z (see Figure 15.5), or that of the numerical factor k_0 plotted against atomic number Z (see Figures 6.6 and 15.1), allows us to predict values of $C_0 (= 0.03771/k_0)$ for those not yet available experimentally.

Substituting Eq. (7.41') into Eq. (8.30), the latter can be rewritten in terms of three macroscopic physical quantities:

$$D_m = 2.19 \times 10^{-16} \frac{T_m}{\mu_m V_m^{1/3}} \tag{8.31}$$

(Of course, both Eqs. (8.30) and (8.31) provide the same results of calculations for self-diffusivity in liquid metallic elements.) Comparing Eq. (8.31) with Eq. (8.25b), a ς value of 5.32 for self-diffusivity in liquid metallic elements is obtained.

8.4   Temperature Dependence of Self-Diffusivity

The self-diffusivities of liquid metallic elements are lacking in reliable experimental data because of difficulties encountered in obtaining accurate values at high temperatures. For example, very large discrepancies exist between experimentally derived data for even the common low melting point metals such as tin and lead (see Figures 8.6 and 8.7). Consequently, a detailed comparison of theory or model and experiment for the temperature dependence of self-diffusivities D of liquid metallic elements will not yield fruitful results at the present stage. However, by plotting log D against $1/T$, good straight lines are generally obtained for experimental data on the self-diffusivities of liquid metallic elements. As such, experimental liquid self-diffusivities can be expressed, to a good approximation, by an Arrhenius-type equation:

$$D = D_0 \exp\left(-\frac{H_D}{RT}\right) \tag{8.32}$$

where D_0 and H_D are the Arrhenius parameters. Using Eq. (8.32) and the conditions $T = T_m$, $D = D_m$, in which subscript m denotes the melting point, we have the relation for D_0, i.e. $D_0 = D_m \exp(H_D/RT_m)$. The parameter H_D (sometimes called apparent activation energy) is a measure that represents the temperature dependence of self-diffusivity. Incidentally, even now the Arrhenius-type equation (i.e. Eq. (8.32)) for liquid self-diffusivity has not received a rigorously theoretical explanation. In this book, the authors adopt the Arrhenius-type equation as a mathematical expression for describing the self-diffusivities of liquid metallic elements; however, this does not connote that the authors accept a thermally activated process, like solids, for the mechanism of liquid diffusion, as well as viscosity.

It has long been known that a simple empirical relationship exists between the parameter H_D for liquid metallic elements and their melting point temperatures T_m. Figure 8.3 shows a plot of log H_D against log T_m for various liquid metallic elements. The data for H_D and T_m are given in Tables 17.1 and 17.10. As can be seen, a good straight line is obtained between the two variables; the linear relationship is expressed as

$$\log H_D = 1.10 \log T_m + 1.13 \tag{8.33}$$

or

$$H_D = 13.5 T_m^{1.10} \tag{8.34}$$

From Eqs. (8.32) and (8.34), we have the following equation for describing the self-diffusivities of liquid metallic elements:

$$D = D_m \exp\left(1.62 T_m^{0.10}\right) \exp\left(-\frac{13.5 T_m^{1.10}}{RT}\right) \tag{8.35}$$

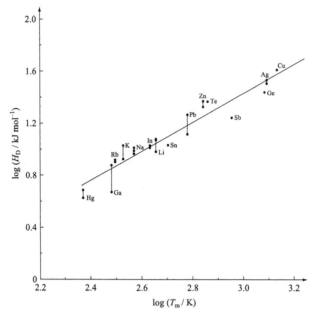

Figure 8.3 *log H_D vs. log T_m. Points linked by a vertical line represent two or three different experimental data for a single metal.*

8.5   Assessment of Self-Diffusivity Equations/Models

Several equations or models for the self-diffusivity of liquid metallic elements are evaluated quantitatively on the basis of the parameters δ_i, Δ and S.

Table 8.1 compares experimental values for the self-diffusivity of liquid metallic elements with those calculated by Protopapas et al. [7] using Eq. (8.13). Equation (8.13) presented by Protopapas et al. performs well for the 15 liquid metallic elements listed in Table 8.1, giving $\Delta(15)$ and $S(15)$ values of 15.3 per cent and 0.268, respectively. However, the agreement between calculation and experiment for antimony (a semimetal) is very poor. With the exception of antimony, Eq. (8.13) works extremely well, with $\Delta(14)$ and $S(14)$ values of 9.6 per cent and 0.114, respectively. These values are within the typical range of uncertainties in the experimental self-diffusivity data for liquid metallic elements.

Chauhan et al. [20] calculated the self-diffusivities of 16 liquid metallic elements at, or near, their melting point temperatures using the Stokes–Einstein formula represented by Eq. (8.24). In their self-diffusivity calculations, the Goldschmit (GS) atomic radii and experimental viscosity data were used for r and μ, respectively. Table 8.2 lists experimental and calculated values for the 16 liquid metallic elements, together with δ_i, Δ and S values. Values of $(D_m)_{exp}$, $(D_m)_{cal}$, and GS diameter are taken from Chauhan et al. [20]. Since two very different experimental values for the self-diffusivity of liquid

Table 8.1 *Comparison of experimental melting point self-diffusivities in liquid metallic elements with those calculated form Eq. (8.13), together with δ_i, Δ, and S values.*

Element		Self-diffusivity $D_m / 10^{-9}$ m^2 s^{-1}		δ_i %
		Experimental	Calculated[†]	
Antimony	Sb	5.18	2.66	94.7
Cadmium	Cd	1.78	2.00	−11.0
Caesium	Cs	2.69	2.31	16.5
Copper	Cu	4.00	3.40	17.6
Gallium	Ga	1.63*	1.73	−5.8
Indium	In	1.67*	1.77	−5.4
Lead	Pb	1.98*	1.67	18.6
Lithium	Li	6.41*	7.01	−8.6
Mercury	Hg	1.03*	1.07	−3.7
Potassium	K	3.68*	3.85	−4.4
Rubidium	Rb	2.68*	2.68	0
Silver	Ag	2.57*	2.77	−7.2
Sodium	Na	4.04*	4.24	−4.7
Tin	Sn	2.19*	1.96	11.7
Zinc	Zn	2.05*	2.55	−19.6
			Δ (15) %	15.3
			S (15)	0.268

[†] Calculated by Protopapas et al. [7].
* The mean values determined from two, three, or five different values of experimental self-diffusivity data for a single metallic element (see Table 17.10), which were used for calculating δ_i, Δ, and S values.

bismuth were reported (as shown in Table 8.2), the δ_i, Δ and S values were determined excluding bismuth. As is obvious from the results of these calculations, with the exception of several metals, calculated values are within the uncertainties of ca. 5–20 per cent, associated with experimental self-diffusivity measurements for liquid metallic elements. Chauhan et al. [20] also computed the temperature dependence of the self-diffusivities of liquid metals using both the Stokes–Einstein formula and the hard-sphere model. According to their conclusions, the Stokes–Einstein formula for self-diffusivity gives reasonable results, when one uses the temperature-dependent diameter as given by Eq.

Table 8.2 *Self-diffusivities in liquid metallic elements at their melting points as predicted by the Stokes–Einstein equation using the Goldschmidt diameter, together with δ_i, Δ, and S values. Values for $(D_m)_{exp}$, $(D_m)_{cal}$, and GS diameter are taken from Chauhan et al. [20].*

Element		Temperature		Diameter	$(D_m)_{exp}$	$(D_m)_{cal}$	δ_i
		K	(°C)	Å $(10^{-10}\,\text{m} = 0.1\,\text{nm})$	$10^{-9}\,\text{m}^2\,\text{s}^{-1}$	$10^{-9}\,\text{m}^2\,\text{s}^{-1}$	%
Bismuth	Bi	544.10	(270.95)	3.64	$(0.8002, 3.78)^\dagger$	2.014	–
Cadmium	Cd	594.05	(320.90)	3.04	1.78	1.8829	−5.5
Caesium	Cs	303	(30)	5.40	2.69	1.956	37.5
Copper	Cu	1356.15	(1083.00)	2.56	3.97	2.6816	48.0
Gallium	Ga	302.93	(29.78)	2.70	1.66*(1.60, 1.71)	1.2705	30.7
Indium	In	429.55	(156.40)	3.14	1.67*(1.68, 1.65)	1.669	0.1
Lead	Pb	623	(350)	3.50	1.94*(1.68, 2.19)	1.770	9.6
Lithium	Li	453	(180)	3.14	5.76	5.37	7.3
Mercury	Hg	273	(0)	3.10	1.15*(1.11, 1.181)	1.1516	−0.1
Potassium	K	343	(70)	4.76	3.64*(3.59, 3.69)	3.103	17.3
Rubidium	Rb	313	(40)	5.04	2.45*(2.22, 2.68)	2.035	20.4
Silver	Ag	1233.65	(960.50)	2.88	2.65	2.198	20.6
Sodium	Na	387	(114)	3.84	3.96*(4.06, 3.85)	3.225	22.8
Thallium	Tl	575.65	(302.50)	3.40	2.009	1.4089	42.6
Tin	Sn	504.99	(231.84)	3.16	2.18*(2.05, 2.31)	1.939	12.4
Zinc	Zn	692.55	(419.9)	2.74	2.05*(2.03, 2.06)	1.5867	29.2
						Δ (15) %	20.3
						S (15)	0.249

† See Table 17. 10.
* The mean values determined from two different values of experimental self-diffusivity data given in parentheses, which are used for calculating δ_i, Δ, and S values.

(8.14), as proposed by Ptotopapas et al. [7]. By contrast, the hard-sphere model, apart from not providing satisfactory numerical accuracy, predicts a much weaker temperature dependence of the self-diffusivity than that observed experimentally.

Yokoyama [21] calculated the self-diffusivities of liquid metallic elements using the hard-sphere model represented by Eq. (8.16). Incidentally, a value of 0.463, through which the hard-sphere structure factor represents experimental structure factor data on liquid metals near their melting point temperatures well, was used for the packing fraction η. Table 8.3 shows the results of Yokoyama's calculations, together with δ_i,

Table 8.3 *Comparison of experimental and Yokoyama's calculated values for the self-diffusivity in liquid metallic elements near their melting point temperatures.*

Element		Temperature		Self-diffusivity $D_m / 10^{-9}$ m^2 s^{-1}		δ_i %
		K	(°C)	Experimental	Calculated	
Antimony	Sb	933	(660)	5.58	2.78	100.7
Caesium	Cs	303	(30)	2.69	2.37	13.5
Copper	Cu	1423	(1150)	4.72	3.57	32.2
Gallium	Ga	323	(50)	1.91*	1.83	4.4
Indium	In	433	(160)	1.71*	1.86	−8.1
Lead	Pb	613	(340)	2.12*	1.72	23.3
Lithium	Li	463	(190)	6.78*	7.33	−7.5
Mercury	Hg	238	(−35)	1.07*	1.00	7.0
Potassium	K	343	(70)	3.92*	4.04	−3.0
Rubidium	Rb	313	(40)	2.69*	2.79	−3.6
Silver	Ag	1273	(1000)	2.83*	2.93	−3.4
Sodium	Na	378	(105)	4.28*	4.45	−3.8
Thallium	Tl	588	(315)	2.01	1.68	19.6
Tin	Sn	523	(250)	2.45*	2.03	20.7
Zinc	Zn	723	(450)	2.41*	2.70	−10.7
					Δ (15) %	17.4
					S (15)	0.295

* The mean values determined from two, three, or five different values of experimental self diffusivity data for a single metallic element (see Table 17. 10), which were used for calculating δ_i, Δ, and S values.

Δ and S. As is evident from Table 8.3, the calculated values for the self-diffusivity of 15 liquid metallic elements are in good agreement with experimental data, except for antimony.

Table 8.4 compares experimental values for the melting point self-diffusivity of liquid metals with those calculated from Eq. (8.29), plus a listing of δ_i, Δ and S values. For the 15 pure metals shown in Table 8.4, the melting point self-diffusivity values calculated from Eq. (8.29) are nearly equal to those obtained experimentally. Table 8.5 gives a comparison of experimental values for the self-diffusivity of 18 liquid metallic elements, apart from bismuth (i.e. 15 pure metals, one semimetal, and two semiconductors) with values calculated from the following equation: $1.92 \times 10^{-9} (T_m / M)^{1/2} V_m^{1/3}$ (this gives the minimum S value for the 18 liquid metallic elements considered), together with δ_i, Δ and S values. As can be seen, the agreement obtained between calculation and experiment is very poor, particularly for germanium.

The results shown in Tables 8.4 and 8.5 indicate that a comparison of experimental and calculated values not only for pure metals but also for various liquid metallic elements (i.e. pure liquid metals, semimetals, and semiconductors) is absolutely essential from the standpoint of materials process science and engineering; we repeat here that in the field of materials process science, both accuracy and universality are required of any model, in particular, for predicting the thermophysical properties of liquid metallic elements.

Table 8.4 *Comparison of experimental values for the self-diffusivities in pure liquid metals with those calculated using Eq. (8.29), together with δ_i, Δ, and S values.*

Element		Self-diffusivity $D_m / 10^{-9}$ m^2 s^{-1}		δ_i %
		Experimental	Calculated	
Cadmium	Cd	1.78	2.10	−15.2
Caesium	Cs	2.69	2.38	13.0
Copper	Cu	4.00	3.50	14.3
Gallium	Ga	1.63*	1.78	−8.4
Indium	In	1.67*	1.86	−10.2
Lead	Pb	1.98*	1.74	13.8
Lithium	Li	6.41*	7.29	−12.1
Mercury	Hg	1.03*	1.00	3.0
Potassium	K	3.68*	4.03	−8.7
Rubidium	Rb	2.68*	2.80	−4.3
Silver	Ag	2.57*	2.91	−11.7
Sodium	Na	4.04*	4.44	−9.0
Thallium	Tl	2.01	1.67	20.4
Tin	Sn	2.19*	2.01	9.0
Zinc	Zn	2.05*	2.66	−22.9
			Δ (15) %	11.7
			S (15)	0.128

* The mean values determined from two, three, or five different values of measured self-diffusivity data for a single metal, which were used for calculating δ_i, Δ, and S values.

Table 8.5 *Comparison of experimental values for the self-diffusivities in liquid metallic elements with those calculated using Eq. (8.29′)†, together with δ_i, Δ, and S values.*

Element		Self-diffusivity D_m / 10^{-9} m^2 s^{-1}		δ_i %
		Experimental	Calculated†	
Antimony	Sb	5.18	4.40	17.7
Cadmium	Cd	1.78	3.36	−47.0
Caesium	Cs	2.69	3.81	−29.4
Copper	Cu	4.00	5.60	−28.6
Gallium	Ga	1.63*	2.85	−42.8
Germanium	Ge	14.2	5.86	142
Indium	In	1.67*	2.98	−44.0
Lead	Pb	1.98*	2.78	−28.8
Lithium	Li	6.41*	11.66	−45.0
Mercury	Hg	1.03*	1.61	−36.0
Potassium	K	3.68*	6.44	−42.9
Rubidium	Rb	2.68*	4.49	−40.3
Silver	Ag	2.57*	4.65	−44.7
Sodium	Na	4.04*	7.11	−43.2
Tellurium	Te	2.72	4.05	−32.8
Thallium	Tl	2.01	2.67	−24.7
Tin	Sn	2.19*	3.22	−32.0
Zinc	Zn	2.05*	4.25	−51.8
			$\Delta(18)\%$	43.0
			S(18)	0.500

† $D_m = 1.92 \times 10^{-9} \left(\frac{T_m}{M}\right)^{1/2} V_m^{1/3}$ (8.29′)

* The mean values determined from two, three, or five different values of measured self-diffusivity data for a single metallic element (see Table 17.10), which were used for calculating δ_i, Δ, and S values.

Figure 8.4 shows the relationship represented by Eq. (8.30), in which various liquid metallic elements, i.e. 15 metals, two semiconductors (germanium and tellurium), and one semimetal (antimony), are plotted. Figure 8.5 shows an enlarged drawing of Figure 8.4, excluding germanium. Table 8.6 compares experimental values for the melting point self-diffusivity of 18 liquid metallic elements (excluding bismuth), with

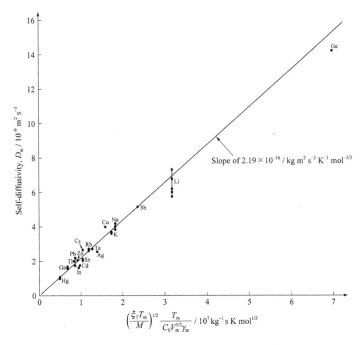

Figure 8.4 *Self-diffusivities of 18 liquid metallic elements at their melting point temperatures as a function of $(\xi_T T_m / M)^{1/2} T_m / C_0 V_m^{1/3} \gamma_m$. Points linked by a vertical line represent two, three, or five different experimental data for a single metal.*

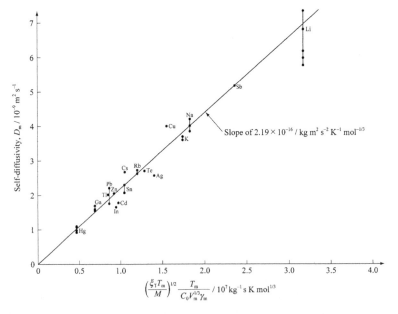

Figure 8.5 *Self-diffusivities of 17 liquid metallic elements at their melting point temperatures as a function of $(\xi_T T_m / M)^{1/2} T_m / C_0 V_m^{1/3} \gamma_m$. Points linked by a vertical line represent two, three, or five different experimental data for a single metal.*

Table 8.6 *Comparison of experimental values for the self-diffusivities in liquid metallic elements with those calculated using Eq. (8.30), or Eq. (8.31), together with δ_i, Δ, and S values.*

Element		Self-diffusivity $D_m / 10^{-9}$ m^2 s^{-1}		δ_i %	C_0† kg$^{-1/2}$ m^{-1} K$^{1/2}$ mol$^{1/2}$
		Experimental	Calculated		
Antimony	Sb	5.18	5.17	0.2	0.280
Cadmium	Cd	1.78	2.16	−17.6	0.331
Caesium	Cs	2.69	2.31	16.5	0.369
Copper	Cu	4.00	3.40	17.6	0.431
Gallium	Ga	1.63*	1.51	7.9	0.292
Germanium	Ge	14.2	15.4	−7.8	0.122
Indium	In	1.67*	2.07	−19.3	0.280
Lead	Pb	1.98*	1.89	4.8	0.389
Lithium	Li	6.41*	6.95	−7.8	0.259
Mercury	Hg	1.03*	1.02	1.0	0.239
Potassium	K	3.68*	3.80	−3.2	0.345
Rubidium	Rb	2.68*	2.63	1.9	0.355
Silver	Ag	2.57*	3.08	−16.6	0.441
Sodium	Na	4.04*	4.01	0.7	0.336
Tellurium	Te	2.72*	2.83	−3.9	0.281
Thallium	Tl	2.01*	1.86	8.1	0.351
Tin	Sn	2.19*	2.30	−4.8	0.313
Zinc	Zn	2.05*	2.02	1.5	0.477
			$\Delta(18)\%$	7.8	
			$S(18)$	0.102	

† A parameter appearing in Eq. (8.30); C_0 values were calculated from Eq. (7.41′).
* The mean values for a single metallic element (see Table 8.5), which were used for calculating δ_i, Δ, and S values.

those calculated from Eq. (8.30), together with C_0 values appearing in Eq. (8.30); the table also lists δ_i, Δ and S values needed for statistical assessment of the model. The data used for calculating self-diffusivities using Eq. (8.30) and experimental data on D_m are given in Chapter 17. As is clear from Table 8.6, Eq. (8.30) performs extremely well with Δ and S values of 7.8 per cent and 0.102, respectively; most importantly, the

Table 8.7 *Comparison of experimental and calculated energies of apparent activations for self-diffusivity in liquid metallic elements, together with δ_i, Δ, and S values.*

Element		H_D / kJ mol^{-1}		δ_i
		Experimental	Calculated	%
Antimony	Sb	17.7	24.1	−26.6
Copper	Cu	40.6	37.7	7.7
Gallium	Ga	6.22*	7.23	−14.0
Germanium	Ge	27.6	33.2	−16.9
Indium	In	10.5*	10.6	−1.0
Lead	Pb	15.8*	15.4	2.6
Lithium	Li	11.1*	11.3	−1.8
Mercury	Hg	4.63*	5.45	−15.0
Potassium	K	9.58*	8.12	18.0
Rubidium	Rb	8.14*	7.49	8.7
Silver	Ag	33.1*	33.9	−2.4
Sodium	Na	9.76*	9.04	8.0
Tellurium	Te	23.2	18.8	23.4
Tin	Sn	10.8	12.7	−15.0
Zinc	Zn	22.4*	18.0	24.4
			$\Delta(15)$%	12.4
			$S(15)$	0.149

* The mean value of two or three different experimental results for a single metal (see Table 17.10).

calculated values using Eq. (8.30) fall within the range of uncertainties associated with the self-diffusivity measurements.

Table 8.7 shows the Arrhenius parameter H_D values (appearing in Eq. (8.32)) measured and calculated from Eq. (8.33) or (8.34), together with corresponding δ_i, Δ and S values. Equation (8.33) or (8.34) performs well for the temperature dependence of self-diffusivity of liquid metallic elements with Δ and S values of 12.4 per cent and 0.149, respectively. Figures 8.6 and 8.7 show comparisons between experimental values for variations in the self-diffusivities of liquid tin and lead with temperature, as determined by several works, and values calculated from Eq. (8.30) or (8.31) in combination with Eq. (8.35). The experimentally derived self-diffusivity data for liquid tin and lead are

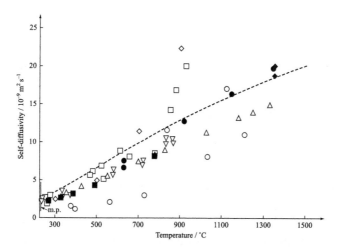

Figure 8.6 *Comparison of experimental self-diffusivity values for liquid tin, as determined by several workers (re-plotted from Masaki et al. [22]), with those calculated (broken curve) from Eq. (8.30) or (8.31) in combination with Eq.(8.35).* ▽, *Davis and Fryzuck;* □, *Ma and Swalin;* ◇, *Kharkov et al.;* △, *Bruson and Gerl;* ○, *Foster and Reynick;* ●, *Itami et al.;* ■, *Frohberg et al.;* ◆, *Yoda et al. Open symbols denote experiments on the ground, closed symbols experiments under microgravity.*

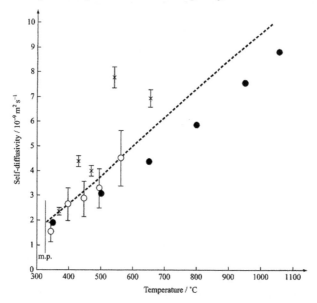

Figure 8.7 *Comparison of experimental self-diffusivity values for liquid lead (re-plotted from Mathiak et al. [23]) with those calculated (broken curve) from Eq. (8.30) or (8.31) in combination with Eq. (8.35).* ×, *Rothman and Hall;* ○, *Döge (×, ○ terrestrial data);* ●, *Frohberg et al. (measured during the Spacelab mission D2).*

taken from Masaki et al.'s [22] and Mathiak et al.'s [23] papers, respectively. As can be seen from these figures, the results of calculations fall well within the range of experimental data.

Lu et al. [24] proposed an expression for the self-diffusivity of liquid metals in terms of T/T_m and two constants for certain materials. Calculated values using the expression were compared with experimental data for several liquid metals; the agreement between calculated and observed values was good.

8.6   Solute Diffusion in Liquid Metals

The phenomenon of solute diffusion, particularly of oxygen and hydrogen, in metallic liquids is an important subject in the field of materials science and technology. Little theoretical work, however, has been done on solute diffusion, because knowledge of the fundamental quantities or states of solute and solvent particles is insufficient. A few investigators have calculated solute diffusivities in liquid metals on the basis of hard-sphere models. Roughly speaking, calculated values for solute diffusivity appear to be reasonable. Unfortunately, a detailed comparison of calculated and experimental values for solute diffusivities in liquid metals is extremely difficult, since the experimental uncertainties are very large as a result of experimental difficulties. Examples of comparisons between calculated and experimental values are given in Figures 8.8 to 8.10.

8.6.1   Equations Based on Model Theories

8.6.1.1   *Swalin's Model Theory*

Swalin's equation for solute diffusivity can be written in the form of the ratio of solute diffusivity D_i to solvent (or base metal) self-diffusivity $D_{S.M}$ [10, 12, 26]:

$$\frac{D_i}{D_{S.M}} = \left\{ 1 - \frac{\varepsilon_D}{\lambda^2 K_f} \left(1 + \frac{2\lambda}{d} + \frac{2\lambda^2}{d^2} \right) \right\}^{-1} \tag{8.36}$$

$$\varepsilon_D = \frac{\beta z^E e^2}{d} \exp\left(-\frac{d}{\lambda}\right)$$

where λ is the screening radius (λ^{-1} is the screening constant), K_f is the force constant of the solvent, d is the distance between solute and solvent ions,[1] z^E is the relative solute to solvent valence, or the excess valence, β is a slowly varying quantity whose magnitude is a function of z^E, and e is the electronic charge.

Solar and Guthrie [27] calculated hydrogen, oxygen, and carbon diffusivities in liquid iron using Swalin's model. In their calculations, the following numerical values

---

[1] The distance between solvent ion centres has usually been used for d, because of a lack of knowledge of that between solute and solvent ions.

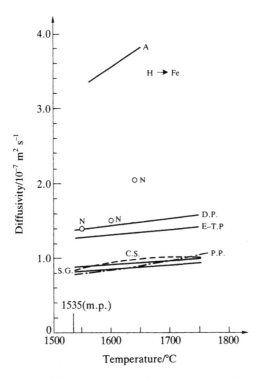

Figure 8.8 *Interdiffusion of hydrogen in liquid iron. A, Arkharov et al.; DP, Depuydt and Parlee; E - TP El-Tayeb and Parlee; SG, Solar and Guthrie; N, Nyguist; CS, calculation through the proposed theory implemented with Carnahan–Starling pair correlation functions; PP, calculation through the proposed theory implemented with Protopapas–Parlee pair correlation functions. (after Protopapas and Parlee [25]).*

for the parameters of Swalin's equations were employed. The solutes were assumed to be in the form H^{1+}, O^{2-}, C^{4+}, giving z^E values of -1, -4, and $+2$ for hydrogen, oxygen, and carbon, respectively, where the charge of the iron particles is taken as $+2$. The corresponding β values were estimated as 1, 1.5, and 1 after Alfred and March, and the screening radius for liquid iron was calculated as 0.503 Å ($= 0.503 \times 10^{-10}$ m $= 0.0503$ nm) after Mott. The liquid iron structural data used were those of Waseda et al., i.e. $Z = 9.5$ and $d = 2.58$ Å. A force constant of 6×10^4 dyne cm^{-1} ($= 60$ N m^{-1}) is extrapolated from the temperature dependence of K_f reported by Waser and Pauling [11]. In Table 8.8, the calculated values are listed together with experimental data; the agreement with the data is not satisfactory.

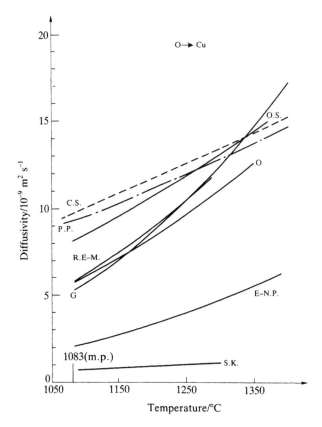

Figure 8.9 *Interdiffusion of oxygen in liquid copper. OS,*
Osterwald and Schwarzlose; RE-M, Rickert and El-Miligy;
O, Oberg et al.; G, Gerlach et al.; E-NP, El-Naggar and
Parlee; SK, Shurygin and Kryuk; CS, same as in Figure 8.8;
PP, same as in Figure 8.8 (after Protopapas and Parlee [25]).

8.6.1.2 *Reynik's Model Theory (a Semi-Empirical Small Fluctuation Model)*

According to Reynik [29], the solute diffusivity can be expressed as

$$D_i = 2.08 \times 10^9 Z x_0^2 T - 1.72 \times 10^{24} Z x_0^4 K_f \tag{8.37}$$

$$x_0 = d - (r + r_i)$$

where r and r_i are the effective atomic radii of solvent and solute, respectively.

Solar and Guthrie [27] illustrated the solute diffusivities of liquid iron as a function of x_0. Figure 8.11 demonstrates this relationship for three liquids having force constants of 4×10^4, 5×10^4, and 6×10^4 dyne cm^{-1}, respectively (the other parameters being

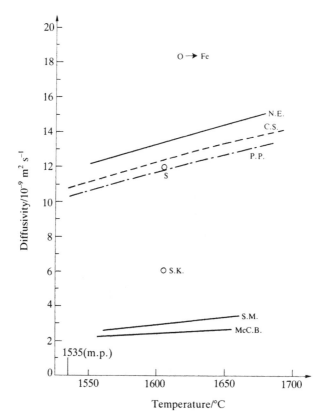

Figure 8.10 *Interdiffusion of oxygen in liquid iron. NE, Novokhatskii and Ershov; S, Scwerdtferger; SK, Shurygin and Kryuk; SM, Suzuki and Mori; McCB, MaCarron and Belton; CS, same as in Figure 8.8; PP, same as in Figure 8.8 (after Protopapas and Parlee [25]).*

Table 8.8 *Diffusivities of dissolved hydrogen, oxygen, and carbon in pure liquid iron.*

	Solute diffusion coefficients 10^{-9} m^2 s^{-1}									
	Theories calling (a) for ionic radii of diffusing particle								(b) for atomic radii	
Diffusing species	Stokes–Einstein	Suther-land	Li and Chang	Eyring		Walls and Upthegrove	Cohen–Turnbull	Swalin	Reynik	Expt. data
				b.c.c. liq.	f.c.c. liq.					
Hydrogen	~300 000	~450 000	~720 000	~720 000	~480 000	~10^7	47.04	12.19	−44.31	92
Oxygen	2.31	3.46	5.44	5.44	3.62	3.43	15.34	9.79	+4.40	14
Carbon	19.05	28.58	44.88	44.88	29.92	50.84	22.72	14.33	+50.45	10

* After Solar and Guthrie [27], and Gourtsoyannis [28].

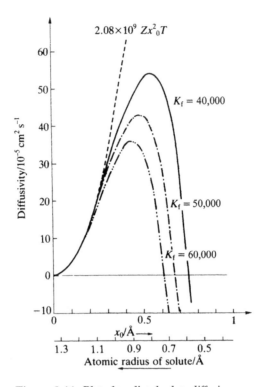

Figure 8.11 *Plot of predicted solute diffusion coefficients (10^{-5} cm^2 s^{-1}) in liquid iron vs. corresponding values of diffusion displacement x_0 for various values of force constant K_f appearing in Eq. (8.37). Solute radii (Å) corresponding to those values of x_0 are also provided to illustrate the unsuitability of Eq. (8.37) for small diffusing particles (after Solar and Guthrie [27]).*

taken equal to the structural data of liquid iron). As can be seen, for low values of x_0 the D curves are asymptotic to the temperature term of Eq. (8.37) and the magnitude of the force constants have little influence on the predicted values. Most of the common ions always fall in this range and it can be verified that the predicted diffusivities are usually within an acceptable range of measured values. However, the behaviour of the D curve at large x_0 values (i.e. smaller solutes) is startling: due to the increasing influence of the force constant term (proportional to $x_0^4 K_f$), the curve drops abruptly and even becomes negative. The elements oxygen, nitrogen, and hydrogen fall in this range. Thus the values predicted for hydrogen ($r_i = 0.46$ Å, $x_0 = 0.88$ Å) are all negative, which is, of course, incorrect. Even granted that the theory does not apply to hydrogen, the shape of the D curves at high x_0 values ($x_0 > 0.5$ Å), together with the uncertainty on atomic radii and lack of information on force constants, make useful predictions for other small solutes impossible.

Other impossibilities also invalidate the use of this semi-empirical small fluctuation theory. For example, the model does not allow for diffusion of solutes in liquid iron if their atomic radii are in excess of $1.34 \text{ Å}(d_{Fe} - r_{Fe})$ since x_0 then becomes negative.[2]

One further discrepancy is the fact that Reynik's model (apparently applicable to all liquids) predicts that atomic movements stop at temperatures above 0 K, while all other approaches, including Swalin's, predict that such movements stop at exactly 0 K, in accord with common belief.

All model theories which have been tested by Solar and Guthrie (see Table 8.8) appear to be incapable of predicting liquid phase diffusivities for the important systems carbon, oxygen, nitrogen, hydrogen, etc. in liquid iron or sulphur, and oxygen in liquid nickel or copper. We must conclude that the major discrepancies between various theoretical predictions and experimental values for those systems may well be due to a lack of knowledge of the physical state of solutes in metallic liquids.

8.6.2 Empirical Relations

Several empirical relations between solute diffusivity and thermodynamic quantities for various elements in liquid metals have been presented. Examples are given in Figures 8.12 to 8.14. Nevertheless, we have already mentioned that transport coefficients (i.e. viscosity, diffusivity) cannot be expressed in terms of thermodynamic quantities alone.

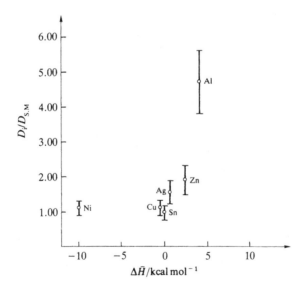

Figure 8.12 $D_i / D_{S,M}$ *vs. relative partial molar enthalpy of solutes in liquid tin (after Ma and Swalin [30]).*

[2] The atomic radii used were 1.24, 0.46, 0.60, and 0.77 Å for iron, hydrogen, oxygen, and carbon, respectively.

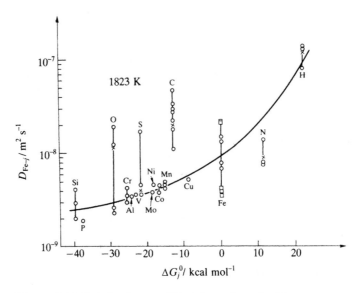

Figure 8.13 *Relation between diffusivity and standard free energy*
ΔG_j^0 *of solution for various solute elements j in liquid iron at 1550 °C*
(after Ono [31]): ○, observed values; □, predicted values; ×,
recommended values of Bester and Lange.

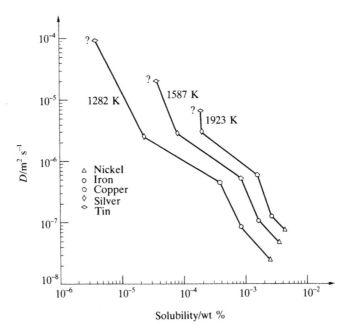

Figure 8.14 *Correlation between diffusion coefficients and*
hydrogen solubilities in liquid nickel, iron, copper, silver, and tin
(after Sacris and Parlee [32]).

8.7   Methods of Self-Diffusivity Measurement

Diffusion occurs as a result of concentration gradient or a density gradient. The flux of matter $\mathcal{J}$, i.e. the amount of diffusing particles which passes per unit time through unit area of a plane perpendicular to the direction of diffusion, is generally proportional to a concentration gradient $\partial c / \partial x$ of the diffusing particles:

$$\mathcal{J} = -D\frac{\partial c}{\partial x} \tag{8.38}$$

where c is the concentration, x is the distance in the direction in which diffusion occurs, and D is the diffusivity or diffusion coefficient (D is introduced as a proportionality factor). This equation is known as Fick's first law. The diffusing particles in binary systems always flow in the direction of decreasing concentration gradient, so a negative sign is therefore introduced in order to make the diffusivity D positive. Diffusivity has the dimensions of $(\text{length})^2/(\text{time})$, i.e. m^2/s in SI units.

On combining Eq. (8.38) and the equation of continuity,[3] we have

$$\frac{\partial c}{\partial t} = \frac{\partial}{\partial x}\left(D\frac{\partial c}{\partial x}\right) \tag{8.39}$$

If the diffusivity is independent of concentration, Eq. (8.39) can be simplified to (i.e. D can be taken outside the differential):

$$\frac{\partial c}{\partial t} = D\frac{\partial^2 c}{\partial x^2} \tag{8.40}$$

Equations (8.39) and (8.40) are known as Fick's second law.

Particular solutions to Fick's second law depend on the initial and the boundary conditions employed in experiments. These solutions provide information on concentration with respect to the time of contact, or diffusion, and system geometry.

The average value of the square of the linear displacement of particles after a time t is given by

$$\overline{x^2} = 2Dt \tag{8.41}$$

This is the diffusion law of Einstein and Smoluchkowsli. The relation can be obtained by solving the diffusion equation, and also by considering a random walk process.

In the liquid state, the regular arrangement of atoms is destroyed, and atoms undergo Brownian motion according to the Langevin equation of motion. In this treatment, the

---

[3] The equation of continuity is given by

$$\frac{\partial c}{\partial t} = -\frac{\partial \mathcal{J}}{\partial x}$$

This equation corresponds to the law of conservation of particles.

diffusing atoms are impeded by the frictional forces of their neighbours. The diffusivity can be formulated in terms of the friction coefficient ς_f or the mobility U_D

$$D = \frac{kT}{\varsigma_f} = U_D kT \tag{8.42}$$

This is called the Einstein relation. According to the Einstein relation, a calculation of diffusivity devolves, in essence, to a calculation of friction coefficient ς_f.

There are a number of methods for measuring liquid diffusivity; of these the following methods have been used to measure liquid metal diffusivities: (a) capillary reservoir, (b) diffusion couple, (c) shear cell, (d) plane source, (e) electrochemical concentration-cell, (f) slow neutron scattering, and (g) nuclear magnetic resonance. Details of these methods have been reviewed [33–36] and only outlines will be described here.

8.7.1   Capillary Reservoir Method

In the capillary reservoir method, the metallic sample containing the solute to be investigated is contained in a capillary tube of uniform diameter with one end sealed. This capillary is then immersed in a large bath, i.e. reservoir, of liquid solvent metal at a chosen temperature, as illustrated in Figure 8.15. In the case of self-diffusivity measurements, the radioisotope can be added to the liquid either in the capillary tube or in the bath. The capillary tubes used are usually a few millimetres or less in diameter, and from one to a few centimetres in length. After a measured time, the capillary is taken out of the reservoir, and the concentration of the specimen in the capillary is determined. The diffusivity can be calculated from this information and the length of the capillary, and so forth. In experiments to determine the diffusivity of metallic liquids, convection currents are generally a critical problem. The capillary reservoir technique largely allows us to avoid convection, thanks to the small diameter of the capillary. Another merit of this method is its relative insensitivity to slight vibrations or disturbances in the vicinity of

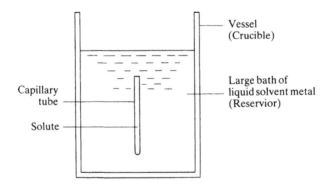

Figure 8.15 *Schematic drawing of the capillary reservoir method.*

the diffusion apparatus. Because of these advantages, the capillary reservoir technique has frequently been adopted for measurements with metallic liquids.

In order to obtain data of high accuracy, the use of a long capillary tube of fine diameter is desirable, even though attendant experimental difficulties increase.

8.7.2 Diffusion Couple Method

A capillary of uniform cross-section is half filled with solute or radioactive metal and half with solvent metal (see Figure 8.16). The lengths of the two specimens are much longer than the anticipated interdiffusion distances, so that the system can be considered as being infinitely long on both sides of the interface. The specimens are rapidly heated to the desired temperature and the solute or tracer element is allowed to diffuse into the solvent metal. After a given time lapse, the capillary tube containing the specimen is cooled rapidly. Diffusivities can then be calculated from measurements of solute concentrations, or of tracer activities, at various points along the rods. A disadvantage of this technique is that the two specimens are in contact during the heating and cooling stages, as well as during the diffusion period proper. As such, some diffusion will occur during heating and cooling. The method is useful for cases where two metallic samples melt at similar temperatures and where diffusivities are to be determined relatively close to their melting points.

8.7.3 Shear Cell Method

So far, the shear cell method has only been used for diffusivity measurements at low temperatures (e.g. gallium, tin, lead). The shear cell consists of two optically flat cylindrical disks mounted coaxially. Off-centre holes are drilled in these disks, which may be aligned to form the diffusion path. The solvent and solute or radioactive metal are kept separated until the run is begun by aligning the holes that make up the diffusion path.

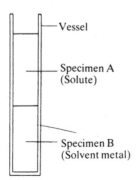

Figure 8.16 *Schematic drawing of the diffusion couple method.*

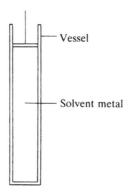

Figure 8.17 *Schematic drawing of the plane source method.*

At the end of the run, the disks are again rotated to misalign the compartments. An advantage of this technique is its usefulness for diffusivity measurements at high pressure. The principal disadvantages of the method are that stirring can occur while the disks are being rotated at the start or end of an experiment, and that the equipment needed is relatively complex.

8.7.4   Plane Source Method

In this technique, one part of a diffusion cell comprises a thin planar section of about one millimetre or less (the 'plane source'), while the other is made up of a longer section of a few centimetres deep, but much longer than diffusion distances (see Figure 8.17). At the start of a diffusion run, this thin piece disperses into the larger piece. The experimental procedure is similar to that of the diffusion couple method.

8.7.5   Electrochemical Concentration-Cell Method

Diffusivities can also be obtained from a liquid-metal concentration cell. An advantage of this method is that it can substitute measurements of electrical quantities (i.e. current and voltage) for tedious chemical or radiochemical analyses. In other words, the technique allows direct measurements of diffusivity in metallic liquids, without solidifying the liquid metal specimen. A disadvantage of the technique is that only a restricted number of metal systems and concentration intervals are available for study.

Details of the electrochemical concentration-cell method are described by Edward et al. [34].

Diffusivity can also be determined using the **neutron scattering** and **nuclear magnetic resonance (NMR) techniques** [35]. In the neutron scattering method, diffusivities are obtained by measuring, as a function of position and time, the intensity of emissions from a radioactive isotope of the element being investigated. In the NMR method, diffusivities are determined by observing the diffusion rate for nuclear spins, i.e. the amplitude of the magnetization as a function of time (for a high diffusion rate, the amplitude decays rapidly).

The neutron method uses measurements over distances of a few angstroms, the NMR method is over distances of several hundred angstroms, and the tracer method over distances of millimetres. Nevertheless, all these methods give, within experimental uncertainties, equivalent diffusivities for liquid metals.

The absence of accurate data for the diffusivity of metallic liquids is mainly due to the existence of convective flow in the liquid samples. In order to avoid natural convection, caused by gravity, in liquid diffusion experiments, measurements of liquid metallic diffusivity under microgravity have been carried out (using mostly the shear cell method) since the 1970s [22, 23, 36–38].

8.8 Experimental Data for the Self-Diffusivity in Liquid Metallic Elements

As already mentioned, it would appear, at present, that experimentally derived self-diffusivity data are available for only some 20 metallic elements, and moreover, even in the case of the common low melting point metals such as gallium, tin, and lead (see Figures 8.6 and 8.7 and Table 17.10), fairly large discrepancies in the self-diffusivities at, or near, their melting point temperatures exist between the experimental results of several investigators. Experimental uncertainties in the self-diffusivity values for liquid metallic elements at, or near, their melting point temperatures may be estimated to be ca. 5 to 20 per cent or more.

Experimental data for the self-diffusivity of liquid metallic elements are listed in Table 17.10.

....

REFERENCES

1. S.A. Rice and N.H. Nachtrieb, *Adv. Phys.*, **16** (1967), 351.
2. E. Helfand, *Phys. Fluids.*, **4** (1961), 681.
3. H.T. Davis and J.A. Palyvos, *J. Chem. Phys.*, **46** (1967), 4043.
4. S.A. Rice and J.G. Kirkwood, *J. Chem. Phys.*, **31** (1959), 901.
5. C.T. Vadovic and C.P. Colver, *Phil. Mag.*, **21** (1970), 971.
6. T.E. Faber, *Introduction to the Theory of Liquid Metals*, Cambridge University Press, 1972, p.174.
7. P. Protopapas, H.C. Andersen and N.A.D. Parlee, *J. Chem. Phys.*, **59** (1973), 15.
8. R.J. Speedy, *Mol. Phys.*, **62** (1987), 509.
9. A.D. Pasternak, *Phys. Chem. Liq.*, **3** (1972), 41.
10. R.A. Swalin, *Acta Metall.*, **7** (1959), 736.

11. J. Waser and L. Pauling, *J. Chem. Phys.*, **18** (1950), 747.
12. V.G. Leak and R.A. Swalin, *Trans. Met. Soc. AIME*, **230** (1964), 426.
13. S.M. Breitling and H. Eyring, in *Liquid Metals, Chemistry and Physics*, edited by S.Z. Beer, Marcel Dekker, New York, 1972, Chap. 5.
14. S. Glasstone, K.J. Laidler, and H. Eyring, *The Theory of Rate Processes*, McGraw-Hill, New York, 1941, Chap. IX.
15. G. Careri and A. Paoletti, *Nuovo Cimento*, **2** (1955), 574.
16. M.H. Cohen and D. Turnbull, *J. Chem. Phys.*, **31** (1959), 1164.
17. N.H. Nachtrieb, *Adv. Phys.*, **16** (1967), 309; see also N.H. Nachtrieb, *Ber. Bunsenges Phys. Chem.*, **80** (1976), 678.
18. J.R. Cahoon, *Metall. Mater. Trans. A*, **34** (2003), 882.
19. W. Sutherland, *Phil. Mag.*, **9** (1905), 781.
20. A.S. Chauhan, R. Ravi, and R.P. Chhabra, *Chem. Phys.*, **252** (2000), 227.
21. I. Yokoyama, *Physica B*, **291** (2000), 145.
22. T. Masaki, T. Fukazawa, S. Matsumoto, T. Itami, and S. Yoda, *Meas. Sci. Technol.*, **16** (2005), 327.
23. G. Mathiak, A. Griesche, K.H. Kraatz, and G. Frohberg, *J. Non-Cryst. Solids*, **205–207** (1996), 412.
24. H.M. Lu, G. Li, Y.F. Zhu, and Q. Jiang, *J. Non-Cryst. Solids*, **352** (2006), 2797.
25. P. Protopapas and N.A.D. Parlee, *High Temp. Sci.*, **8** (1976), 141.
26. R.A. Swalin and V.G. Leak, *Acta Metall.*, **13** (1965), 471.
27. M.Y. Solar and R.I.L. Guthrie, *Met. Trans.*, **3** (1972), 2007.
28. L. Gourtsoyannis, *Kinetics of Compound Gas Absorption by Liquid Iron and Nickel*, Ph.D. Thesis, McGill University, 1978.
29. R.J. Reynik, *Trans. Met. Soc. AIME*, **245** (1969), 75.
30. C.H. Ma and R.A. Swalin, *Acta Metall.*, **8** (1960), 388.
31. Y. Ono, *Tetsu-to-Hagané*, **63** (1977), 1350.
32. E.M. Sacris and N.A.D. Parlee, *Met. Trans.*, **1** (1970), 3377.
33. J.O'M. Bockris, J.L. White, and J.D. Mackenzie, *Physicochemical Measurements at High Temperatures*, Butterworths, London, 1959.
34. J.B. Edwards, E.E. Hucke, and J.J. Martin, *Met. Rev.*, **120**(Part II) (1968), 13.
35. N.H. Nachtrieb, in *Liquid Metals, Chemistry and Physics*, edited by S.Z. Beer, Marcel Dekker, New York, 1972, p.509.
36. H. Fukuyama and Y. Waseda (eds.), *High-Temperature Measurements of Materials*, Springer, Berlin, 2009.
37. T. Hibiya and I. Egry, *Meas. Sci. Technol.*, **16** (2005), 317.
38. G. Mathiak, E. Plescher, and R. Willnecker, *Meas. Sci. Technol.*, **16** (2005), 336.

9
Electrical and Thermal Conductivity

9.1 Introduction

In this chapter, we review some of the more important electronic transport properties of metallic liquids: their electrical conductivities or electrical resistivities, and their thermal conductivities. As is well known, solid metals are uniquely characterized by their high electrical and thermal conductivities. Similarly, metallic liquids are also reasonably good conductors of electricity and heat. The high electrical and thermal conductivities of metals in their condensed states can be attributed almost entirely to freely moving electrons, or conduction electrons.

During the last half century, theoretical considerations of electronic transport properties of liquid metals, particularly their electrical resistivities, have made considerable progress thanks to the advent of pseudopotential theory. Notwithstanding this, only a few theoretical studies have been made on the subject of liquid metal thermal conductivities. In order to understand clearly the behaviour of electrons, which are the carriers of the electric and heat currents in metals, some knowledge of quantum mechanics and quantum statistical mechanics is needed.

From the viewpoint of materials process science and technology, relatively few studies have been carried out on the electronic transport properties of metallic liquids. However, electrical conductivities of metallic liquids are of obvious importance to many liquid metal processing operations, e.g. electric furnace steelmaking and refining operations, electromagnetic stirring for melt cleanliness and microstructural control, and the electrowinning of aluminium from alumina in the Hall–Héroult cell. Thermal conductivities are also of significance in liquid metal processing operations. For instance, the generation of thermal natural convection phenomena in bath or furnaces can have important technological implications for the cast structure of that metal. Similarly, the melting rates of furnace or ladle additions, or the formation of protective thermal accretions around submerged nozzles, are all linked, in a fundamental sense, to the value of a metal's thermal conductivity. Furthermore, to give another important example, thermal conductivity is a main factor exerting an influence on the properties of semiconductor crystals, e.g. the crystal pulling method such as Czochoralski crystal growth. As a final example, the high thermal conductivities of liquid metals make them good media for removing heat from nuclear reactors. Thus, in the last 30 years or so, experimental investigations have been

The Thermophysical Properties of Metallic Liquids: Volume 1 – Fundamentals. First Edition.
Takamichi Iida and Roderick I. L. Guthrie. © Takamichi Iida and Roderick I. L. Guthrie 2015.
Published in 2015 by Oxford University Press.

intensively made on the thermal conductivities of metallic liquids, from the standpoint of materials process science and technology. Nevertheless, experimental investigations dealing with the thermal conductivities of metallic liquids are still not satisfactory, and this is largely due to the extreme difficulty in making measurements for metallic liquids in the absence of any convection effects.

9.2 Theoretical Equations for the Electrical and Thermal Conductivities of Metallic Liquids

9.2.1 Electrical Conductivity or Electrical Resistivity of Metals

9.2.1.1 Electrical Conductivity Based on the Free Electron Model

The electric current is characterized as the motion of electrically charged particles within conductors.

The electric current density j at any point in space is proportional to the electric filed E at that point, namely

$$j = \sigma_e E \tag{9.1}$$

or

$$E = \rho_e j \tag{9.2}$$

where σ_e is the electrical conductivity and ρ_e is the electrical resistivity, i.e. the resistivity is defined as the reciprocal of the conductivity.

The electric current density is defined as the electric charge transported through unit area in unit time. As is generally known, liquid metals as well as solid metals are good conductors; the freely movable charges are the free electrons, or the conduction electrons, with each electron of charge, $-e$. The net number of electrons passing through a unit area in unit time is nv, where n is the number of electrons per unit volume, i.e. the electron number density, and v the average electron velocity. The electric current density is therefore given by

$$j = -nev \tag{9.3}$$

In the absence of an applied electric field, the average velocity $v = 0$, i.e. $j = 0$. Thus, under conditions of equilibrium, all directions of moving-electrons may be regarded as equally probable.

Now consider the motion of a free electron gas in an electric field. We may write an equation of motion for the drift velocity in a constant electric field, as [1]:

$$m_e \frac{dv}{dt} = -eE - \frac{m_e}{\tau} v \tag{9.4}$$

where m_e is the electron mass and τ is the mean free time, or the relaxation time. Equation (9.4) has the particular solution

$$v = -\frac{eE}{m_e}\tau \tag{9.5}$$

From Eqs. (9.1), (9.3), and (9.5), we have the important result

$$\sigma_e = \frac{ne^2\tau}{m_e} \tag{9.6}$$

or

$$\rho_e = \frac{m_e}{ne^2\tau} \tag{9.7}$$

Incidentally, quantum theory does not alter the result of Eq. (9.6) in any essential way.

9.2.1.2 Ratio of Liquid / Solid Conductivity

Mott [2] presented an equation for the ratio of liquid/solid conductivity, $\sigma_{e,l}/\sigma_{e,s}$, at a metal's melting point which is based on the simple assumption that the atoms in a liquid metal vibrate about slowly-varying mean positions with a frequency ν_l. This equation is expressed in the form

$$\frac{\sigma_{e,l}}{\sigma_{e,s}} = \left(\frac{\nu_l}{\nu_s}\right)^2 = \exp\left(-\frac{80\Delta_s^l H_m}{T_m}\right) \tag{9.8}$$

where $\Delta_s^l H_m$ is the enthalpy of melting, or the latent heat of fusion, in kJ mol^{-1}, and ν_s is the frequency of atomic vibrations in the solid.

In spite of his simple treatment, the agreement obtained between theory and experiment for normal metals is surprisingly good.

Ziman [3,4] has provided a reasonable interpretation for the ratio $\rho_{e,l}/\rho_{e,s}$ $(=\sigma_{e,s}/\sigma_{e,l})$, which is based on the modern theory of electronic transport.

9.2.1.3 Electrical Resistivity Based on the Nearly Free Electron Model

The nearly free electron model gives the following expression for the electrical resistivity of metals:

$$\rho_e = \frac{12\pi V}{e^2 \hbar v_F^2}\int_0^1 |U(Q)|^2 S(Q)\left(\frac{Q}{2k_F}\right)^3 d\left(\frac{Q}{2k_F}\right) \tag{9.9}$$

where e is the electron charge, v_F and k_F are the Fermi velocity and the Fermi wave vector, respectively, V is the molar volume, $U(Q)$ is the pseudopotential, $S(Q)$ is the (static) structure factor, and $\hbar$ $(= h/2\pi,\ h = $ Planck constant) is the Dirac constant. Equation (9.9) is known as Ziman's formula.

Evans et al. [5,6] extended Ziman's formula to make it suitable for strong-scattering liquids.

$$\rho_e = \frac{12\pi V}{e^2 \hbar v_F^2}\int_0^1 |t|^2 S(Q)\left(\frac{Q}{2k_F}\right)^3 d\left(\frac{Q}{2k_F}\right) \tag{9.10}$$

The t matrix can be expressed in terms of the phase shifts η_l of the various partial waves, as

$$t(q) = -\frac{2\pi\hbar^3}{m_e(2m_eE_F)^{1/2}V}\sum_l (2l+1)\sin\eta_l\exp(i\eta_l)P_l(\cos\theta)$$

where $q = |k_F' - k_F|$, E_F is the Fermi energy, and P is the Legendre polynomial.

Resistivities calculated through the use of Eq. (9.10) are in good agreement with experimental data [7].

9.2.1.4 Electrical Resistivity Based on the Fluctuation Scattering Model

Takeuchi and Endo [8,9] proposed an expression for the electrical resistivity of metallic liquids which makes use of the fluctuation scattering model. Subsequently, Takeuchi and Endo's approach was extended by Tomlinson and Lichter [10].

9.2.2 Thermal Conductivity of Metals

9.2.2.1 Thermal Conductivity Based on the Free Electron Model [1]

From the kinetic theory of the thermal conductivity of a classical electron gas, we have

$$\lambda = \frac{1}{3}C_{el}vl = \frac{1}{3}C_{el}v^2\tau \tag{9.11}$$

where C_{el} is the heat capacity of the electron gas per unit volume,[1] v is the average electron velocity, and l is the mean free path of an electron between collisions ($l = v\tau$). Using the model for a free electron Fermi gas and Fermi–Dirac statistics, Eq. (9.11) can be rewritten as

$$\lambda = \frac{\pi^2 k^2}{3m_e}n\tau T \tag{9.12}$$

where k is the Boltzmann constant and T is the absolute temperature.

9.2.2.2 A Hard-Sphere Model of Theoretical Conductivity [4]

An equation has been proposed for the thermal conductivity of a hard-sphere fluid, which can be expressed in the form

$$\frac{\mu}{\lambda} = \frac{2m}{5k} \tag{9.13}$$

where μ is the viscosity and m is the atomic mass. This relationship agrees rather well with experimental data for liquid argon and alkali halides (e.g. NaCl, LiBr).

In Chapters 6 to 8, we found that the hard-sphere models give good agreement with experimental data for the surface tension, viscosity, and diffusivity of pure liquid metals. Equation (9.13), however, cannot be applied to metallic liquids because the atomic (or ionic) contribution to the thermal conductivity, which Eq. (9.13) should describe, would be negligibly small compared with electronic contribution. Incidentally, a quarter of a century ago, the present authors [11] proposed a similar expression for dielectric liquids (i.e. at the melting point, $\mu/\lambda = m/3k$).

---

[1] C_{el} is given by $C_{el} = \frac{\pi^2 nk^2 T}{2E_F}$.

9.3 Relationship between Electrical Conductivity and Thermal Conductivity of Metallic Liquids: the Wiedemann–Franz–Lorenz Law

A simple theoretical relationship exists between the electrical conductivity and the thermal conductivity of metals [1,4]:

$$\frac{\lambda}{\sigma_e T} = \frac{\pi^2}{3}\left(\frac{k}{e}\right)^2 = 2.45 \times 10^{-8} \ W\,\Omega\,K^{-2} \tag{9.14}$$

This inter-relationship, namely the fact that $\lambda / \sigma_e T$ $(= \lambda \rho_e / T)$ is approximately the same for all metals, is well known as the Wiedemann–Franz–Lorenz (WFL) law.[2] The right-hand side of Eq. (9.14), i.e. $\pi^2 k^2 / 3 \ e^2 \equiv L_0$, is called the Lorenz number.

The Lorenz number of 2.45×10^{-8} W Ω K^{-2} appearing in Eq. (9.14) can be derived on the basis of the free electron model and the Fermi–Dirac quantum statistics (based on the theoretical treatment of Sommerfeld): combination of Eqs. (9.6) and (9.12) leads to the WFL law represented by Eq. (9.14) (i.e. the value for the Lorenz number 2.45×10^{-8} W Ω K^{-2}). This is a remarkable and useful result, particularly for researchers in the area of materials process science and engineering, because it involves neither n, m_e, nor τ. Nevertheless, the physical meaning of both Eq. (9.6) and Eq. (9.12) may be of great importance for further studies in materials process science in the future.

Busch et al. [12,13] and Haller et al. [14] developed a new apparatus which allowed them to measure, with high accuracy, the Lorenz number of liquid metals up to 500 °C. Their experimental results for the Lorenz number of liquid gallium, mercury, and tin show good agreement with the WFL law, contrary to considerably larger deviations (about 30–100 per cent) from the law in earlier measurements.[3] An article on the thermal conductivities of pure liquid metals by Mills et al. [15] indicates that the results of calculations on the WFL law are in good agreement with experiment for a number of pure metals (i.e. $L_m / L_{0,m} = 1.0$, where the subscript m denotes the melting point), as shown in Table 9.1. Furthermore, Busch et al. found that liquid mercury–indium alloys also obey the WFL law. Yamasue et al. [16] measured the thermal conductivities of liquid silicon and germanium using the non-stationary hot wire method. According to their experimental results, the thermal conductivity values for liquid silicon and germanium are close to those calculated from electrical conductivity values based on the WFL law, respectively. Thus, on combining the WFL law and experimentally derived electrical resistivity data, the prediction of thermal conductivities of liquid metals is possible because comparatively accurate data on electrical conductivities of various liquid metals are available. The WFL law is, nevertheless, an approximate relationship; therefore, direct measurements of accurate thermal conductivities for metallic liquids are required from both the standpoints of practical operations and materials process science.

---

[2] Wiedemann and Franz observed, in 1853, that λ / σ_e is approximately the same for metals at the same temperature. Later work by Lorenz, who observed that the ratio is proportional to the absolute temperature, has established the more general form of the law over a wide range of temperatures.

[3] In earlier experiments, most determinations of the Lorenz number had been made by separate measurements of the electrical conductivity and the thermal conductivity.

Table 9.1 Summary of thermal conductivities and Lorenz ratios at the melting point for metallic elements.

Element			T_m		$\lambda_{s,m}$ W m^{-1} K^{-1}	$\lambda_{l,m}$ W m^{-1} K^{-1}	$b/10^{-2}$ W m^{-1} K^{-2}	$L_m/L_{0,m}$	
			K	(°C)				Solid	Liquid
Aluminium	Al		933	(660)	211	91	3.4	1.0	1.0
Antimony	Sb		904	(637)	17	26	1.0	–	1.5
Bismuth	Bi		545	(272)	7.6	12	1.0	1.0	1.2
Cadmium	Cd		594	(321)	93	41	3.3	1.07	1.0
Caesium	Cs		302	(29)	35.9	20	0.27	1.05	1.0
Cerium	Ce		1072	(799)	–	22	1.25	–	1.0
Chromium	Cr		2180	(1907)	45	35	–	–	–
Cobalt	Co		1768	(1495)	45	36	–	1.07	–
Copper	Cu		1358	(1085)	330	163	2.67	1.0	1.0
Gadolinium	Gd		1585	(1312)	19	19	1.9	–	–
Gallium	Ga		303	(30)	–	28	6.2	–	1.0
Germanium	Ge		1211	(938)	15	39	–	–	–
Gold	Au		1337	(1064)	247	105	3.0	1.0	1.0
Hafnium	Hf		2506	(2233)	39	–	–	–	–
Indium	In		430	(157)	76	38	0.9	1.1	1.1
Iridium	Ir		2719	(2446)	95	76	–	–	–
Iron	Fe		1811	(1538)	34	33	–	0.95	1.06

Lanthanum	La	1193	(920)	–	17	0.5	–	–
Lead	Pb	601	(328)	30	15	0.75	1.0	1.0
Lithium	Li	454	(181)	71	43	2	–	–
Magnesium	Mg	923	(650)	145	79	7	–	–
Manganese	Mn	1519	(1246)	24	22	–	1.0	–
Mercury	Hg	234	(−39)	–	–	–	–	–
Molybdenum	Mo	2896	(2623)	87	72	–	1.0	1.0
Neodymium	Nd	1289	(1016)	–	18.4	0.53	–	–
Nickel	Ni	1728	(1455)	70	60	–	1.0	1.09
Niobium	Nb	2750	(2477)	78	62	–	–	–
Palladium	Pd	1828	(1555)	99	87	–	–	–
Platinum	Pt	2042	(1769)	80	53	–	–	–
Potassium	K	337	(64)	98.5	56	−3.8	1	0.9
Praseodymium	Pr	1204	(931)	–	22	1.4	0.91	–
Rhenium	Re	3459	(3186)	65	55	–	–	–
Rhodium	Rh	2237	(1964)	110	69	–	–	–
Rubidium	Rb	312	(39)	58	33.2	−1.9	–	0.96
Scandium	Sc	1814	(1541)	24.5	225	–	–	–
Silicon	Si	1687	(1414)	25	56	–	–	–
Silver	Ag	1235	(962)	363	175	4.3	1.0	1.0

continued

Table 9.1 (continued)

Element		T_m		$\lambda_{s,m}$ W m^{-1} K^{-1}	$\lambda_{l,m}$ W m^{-1} K^{-1}	$b/10^{-2}$ W m^{-1} K^{-2}	$L_m/L_{0,m}$	
		K	(°C)				Solid	Liquid
Sodium	Na	371	(98)	120	88	−5	–	–
Tantalum	Ta	3290	(3017)	70	58	–	–	–
Tin	Sn	505	(232)	–	27	2	–	1.05
Titanium	Ti	1941	(1688)	31	31	–	–	1.1
Tungsten	W	3695	(3422)	95	63	–	–	0.98
Vanadium	V	2183	(1910)	51	43.5	–	–	1.0
Zinc	Zn	693	(420)	90	50	6	0.90	1.04
Zirconium	Zr	2128	(1855)	38	36.5	–	–	–

List of symbols: T temperature, λ thermal conductivity, b temperature coefficient of thermal conductivity, L experimental value of Lorenz number, L_0 theoretical value of Lorenz number.

Subscripts: l liquid, m melting temperature, s solid.

Data are taken from Mills et al. [15]; $\lambda_1 = \lambda_{1,m} + b(T - T_m)$, W m^{-1} K^{-1}.

From the practical point of view, for example, ever-increasing demands for the manufacture of high quality (multicomponent) metallic materials are in need of direct measurements for thermal conductivity (e.g. the alloy casting industries).

From the standpoint of materials process science, both the accumulation of experimental data and its systematic analysis based on the WFL law as a yardstick, or the accurate determination of the extent of deviation from the WFL law, are needed for progress in understanding the thermal conductivity of metallic liquids.

9.4 Methods of Electrical Conductivity/Resistivity and Thermal Conductivity Measurement

9.4.1 Methods of Electrical Resistivity (or Electrical Conductivity) Measurement

Electrical resistivity is a measure of a material's ability to oppose the flow of an electric current. The resistivity ρ_e is the electrical quantity defined by the equation: $\rho_e = RA/l$, where R is the resistance of a uniform specimen (or wire) of length l and cross-sectional area A. It is measured in ohm metres in SI units (Ω m). Incidentally, the electrical conductivity $\sigma_e (= 1/\rho_e)$ is measured in siemens per metre in SI units (S m^{-1}; 1 S = $1/\Omega$).

The d.c. four-probe method, the rotating magnetic field method, and the noncontact inductive method have been used for measuring electrical resistivities of metallic liquids.

9.4.1.1 D.C. Four-Probe Method

This technique is based on Ohm's law. Resistivity measurements are carried out by determining the potential drop across the liquid in a capillary tube of a known length and cross-section, while maintaining a constant current density. In general, a cell constant of a given capillary is determined experimentally using a liquid of known electrical resistivity (usually mercury). A schematic diagram of resistivity cells is shown in Figure 9.1. The principle and construction of the method are both quite simple. As a result, the d.c. four-probe (potentiometric) technique has been widely used for measuring electrical resistivities of metallic liquids.

A serious problem with this technique, however, is one of materials for the cell or capillary tube, and especially the electrodes. Although molybdenum, tungsten, or graphite can be used for electrodes, dissolution of electrodes in the liquid metal sample, or chemical reaction between the two, may take place. In order to eliminate these problems, a method (or an improved four-probe method), which consists of four solid electrodes made of identical material to the liquid metal sample and providing for a temperature gradient in the lower part of the cell, has been developed, as indicated in Figure 9.2. However, the improved four probe method is limited to simple liquid metallic substances and alloys having a narrow liquid–solid range.

In resistivity measurements by these methods, degassing of the liquid metal specimen in the fine capillary tube is of great importance.

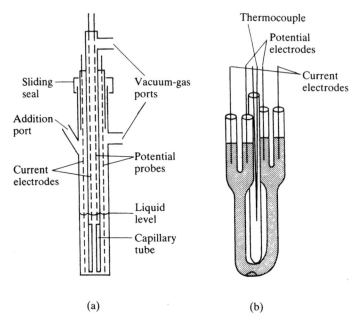

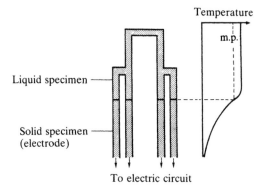

Figure 9.1 *Schematic diagrams of the liquid-metal resistivity cell (after (a) Adams and Leach [17] and (b) Mera et al. [18]).*

Figure 9.2 *Schematic diagram of cell and specimen of an improved four-probe method (after Kita et al. [19]).*

Careful measurements using the four-probe method can provide accurate values for the electrical resistivity of simple liquid metallic substances.

9.4.1.2 *Rotating Magnetic Field Method*

In measuring electrical resistivities of metallic liquids at high temperatures, the rotating magnetic field method without electrodes has sometimes been used.

This technique is based on the phenomenon that a cylindrical sample in a rotating magnetic field shows a torque depending linearly on the conductivity (i.e. the torque is inversely proportional to the electrical resistivity). A schematic diagram of the apparatus is shown in Figure 9.3. The constant for the apparatus is determined experimentally by the use of a standard sample (e.g. mercury).

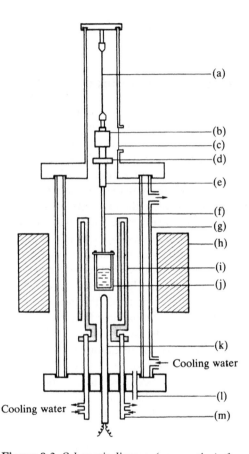

Figure 9.3 *Schematic diagram (cross section) of the apparatus for measuring electrical resistivity of liquid metals by the rotating magnetic field method (after Ono and Yagi [20]): (a) tungsten suspension wire, (b) mirror, (c) window, (d) brass disk, (e) brass rod, (f) graphite rod, (g) silica jacket,*
(h) electromagnet, (i) graphite heating element,
(j) alumina crucible with metal specimen,
(k) thermocouple (Pt 30% Rh–Pt 6% Rh),
(l) vacuum line, (m) power supply.

Disadvantages are that the method requires density values of the liquid metallic sample, and some corrections for calculating resistivities from experimental data.

9.4.1.3 *Noncontact Inductive Method*

In the last 15 years or so, the electrical resistivities of metallic liquids including undercooled states have been measured using noncontact (or containerless) inductive techniques.

Measurements of the electrical resistivity (or conductivity) of a levitated liquid drop are possible using noncontact techniques. Experimental details on these techniques are described in several articles [21,22].

9.4.2 Methods of Thermal Conductivity Measurement

Thermal conductivity λ is a measure of the ability of a substance to conduct heat: λ = heat flow conducted across unit area per second per unit temperature gradient. Units of measurement for λ in SI system are watt metre^{-1} kelvin^{-1} (W m^{-1} K^{-1}).

Methods used for measuring thermal conductivities of metallic liquids can be classified into steady state, non-steady state, and transient techniques.

The main point in thermal conductivity experiments is to suppress convection taking place in a liquid sample, or to complete the experiments before the onset of convection.

In the past 30 years or so, non-steady state and transient techniques have been widely used for measuring thermal conductivities of metallic liquids, e.g. the laser flash (or pulse) method, and the hot wire method. A characteristic feature of these methods is that the measurements are carried out rapidly. The recent developments and experimental measurements using these methods are described in several articles [15,23,24].

Here, we will outline a steady state method (the longitudinal heat flow method). The longitudinal heat flow method (or the axial heat flow method) has been most frequently used for measuring the thermal conductivities of low melting point metallic liquids (e.g. mercury, tin, lead). An apparatus for the technique is shown schematically in Figure 9.4. The liquid metal sample is contained in a cell (see Figure 9.4(b)) for which, usually, length/breadth $\approx$ 10, through which heat flows in a vertical direction. Heat losses in the horizontal direction are prevented by thermal shields.

The measurement of thermal conductivity is primarily concerned with the difference in temperature as related to heat flow. Thermal conductivity can be computed from the measured heat flow and longitudinal temperature gradient. For example, Duggin [25] calculated the thermal conductivity of liquid mercury in a Pyrex cell from the following equation:[4]

$$\lambda = \frac{1}{A}\left(\frac{\dot{q}}{dT/dx} - \lambda_c A_c\right) \tag{9.15}$$

where $\dot{q}$ is the measured power input to the lower heater, λ_c is the thermal conductivity of the cell, A is the average cross-sectional area of the cavity in the cell containing the

---

[4]Pyrex cells were used because of ease of fabrication.

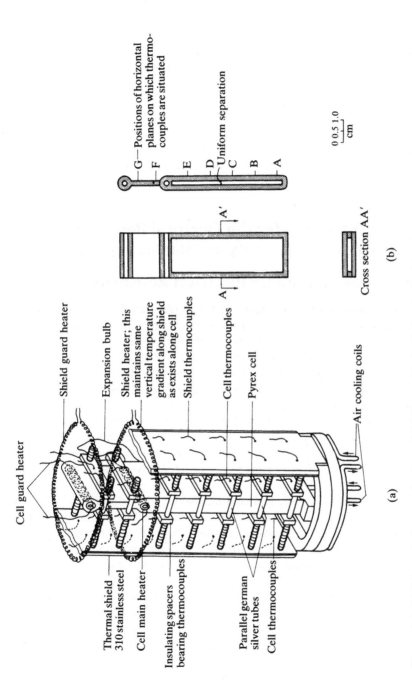

Figure 9.4 (a) Schematic diagram of a thermal shield assembly for the longitudinal heat flow method of thermal conductivity measurement and (b) (Pyrex) specimen cell (after Duggin [25]).

metallic liquid, A_c is the average cross-sectional area of the (Pyrex) walls of the cell, and dT/dx is the longitudinal temperature gradient over the measured portion (B–E in Figure 9.4(b)) of the cell.

In the early years of this century, Sklyarchuk and Plevachuk [26] measured the thermal conductivities of liquid lead, tin, and a semiconductor alloy, TlAsTe$_2$, using a modified apparatus based on the steady state method, over a wide temperature range (up to 1427 °C). According to their conclusions, the construction of the apparatus ensures the maximum reduction in heat leakage and minimizes the heat flux convection. The instrument developed in their investigation is capable of measuring the thermal conductivity of melts to an accuracy of 7 per cent. The results obtained for liquid lead, tin, and TlAsTe$_2$ in their investigation are in good agreement with the values obtained in other investigations.

9.5   Experimental Data for the Electrical Resistivity and Thermal Conductivity of Metallic Liquids

9.5.1   Experimental Data for Electrical Resistivity

The electrical resistivities of most metals in the liquid state just above their melting points are about 1.5–2.3 times as great as those of solid metals just below their melting points. This is caused by the fact that liquid metals with a relatively disordered arrangement of ions have higher resistivities, or lower conductivities, as compared to crystalline solid metals with their regular arrangement. Because the electron mean free path is shorter when the electrons are moving through the disordered liquids, these decreases in electrical conductivity are to be expected.

Tables 9.2 to 9.4 list experimental data for the electrical resistivities of liquid metallic elements. The accuracy of the electrical resistivity data at their melting points is comparatively good (about 2–5 per cent). For example, electrical resistivity data for liquid tin, copper, and iron group metals, as measured by different investigators, are summarized in Table 9.5

As can be seen from Table 9.2, germanium, silicon, and tellurium, which are semiconductors in the solid state, show a remarkable decrease of their resistivities on melting, as they become metallic liquids. Similarly, antimony, bismuth, and manganese also show considerable decreases in resistivities on melting. This phenomenon of increasing conductivity can be attributed to an increase in the number of conduction electrons on melting.

The electrical resistivities of liquid metals increase linearly with increasing temperature, except for cadmium and zinc which both have negative temperature coefficients just above their melting points. Figure 9.5 indicates the electrical resistivity of several liquid metallic elements as a function of temperature. Incidentally, we should note that fairly large discrepancies exist between experimentally obtained temperature coefficient $d\rho_e/dT$ values for high melting point metallic liquids.

Table 9.2 *Changes of electrical resistivity associated with the melting of metallic elements.*

Element		T_m °C	$\rho_{e,s}$ μ Ω cm	$\rho_{e,1}$ μ Ω cm	$\rho_{e,1}/\rho_{e,s}$
Aluminium	Al	660.3	10.9	24.2	2.20
Antimony	Sb	630.6	183	113.5	0.62
Barium	Ba	727	~82	~134	1.6
Beryllium	Be	1287	No data		
Bismuth	Bi	271.4	290.8	130.2	0.45
Cadmium	Cd	321.1	17.1	33.7	1.97
Caesium	Cs	28.4	21.7	36.0	1.66
Calcium	Ca	842	No data		
Cobalt	Co	1495	97	102	1.05
Copper	Cu	1084.6	10.3(5)	21.1	2.04
			9.4(5)	20.0	2.1
Gallium	Ga	29.8	~56 c axis		0.46 c
			~18 a	25.8	1.4 a
			~8 b		3 b
Germanium	Ge	937	900	60	0.067
			1200	63	0.053
			1100	85	0.077
Gold	Au	1064.2	13.68	31.2(5)	2.28
Indium	In	156.6	15.2	33.1	2.18
Iron	Fe	1538	127.5	138.6	1.09
Lead	Pb	327.5	49.0	95.0	1.94
Lithium	Li	180.5	Insufficient data		
Magnesium	Mg	650	15.4	27.4	1.78
Manganese	Mn	1246	66	40	0.61
Mercury	Hg	−38.8	18.4 (par.)	90.96	4.94 (par)
			24.3 (perp.)		3.74 (perp)
Nickel	Ni	1455	65.4	85.0	1.30
Niobium[a]	Nb	2468*	61	85.5	1.4
Palladium[a]	Pd	1554.9*	–	83±2	–
Potassium	K	63.4	8.32	12.97	1.56

continued

Table 9.2 *(continued)*

Element		T_m °C	$\rho_{e,s}$ $\mu\,\Omega\,cm$	$\rho_{e,1}$ $\mu\,\Omega\,cm$	$\rho_{e,1}/\rho_{e,s}$
Rhodium[a]	Rh	1960*	95.2	108.5	1.14
Rubidium	Rb	39.3	13.7	22.0	1.61
Selenium	Se	220.5	$\sim 10^{3.5}$	$\sim 10^7$	~ 1000
Silicon	Si	1412	~ 2400	~ 81	~ 0.034
Silver	Ag	961.8	8.2(5)	17.2(5)	2.1
Sodium	Na	97.7	6.598	9.573	1.451
Strontium	Sr	777	No data		
Tellurium	Te	450	~ 5500 (par.)	~ 500	~ 0.091 (par.)
			~ 11000 (perp.)		~ 0.045 (perp.)
Thallium	Tl	304	35.5	73.1	2.06
Tin	Sn	231.9	22.8	48.0	2.11
Zinc	Zn	419.5	16.7	37.4	2.24

$\Omega\,cm = 10^{-2}\,\Omega\,m\,(\mu\Omega\,cm = 10^{-8}\,\Omega\,m)$
Data, except for those bearing the superscript a, are taken from Cusack and Enderby [27].
[a] Data from Kozuka and Morinaga [28].
* Melting point (from Kozuka and Morinaga [28]).

Table 9.3 *Temperature dependence of electrical resistivity for liquid metallic elements.* $\rho_{e,1}$ *(in* $\mu\Omega\,cm$*) is given by* $\alpha T + \beta$ *from the melting point to the temperature in column 5.*

Element		T_m K	T_m (°C)	$\alpha/10^{-2}$ $\mu\Omega\,cm\,K^{-1}$	β $\mu\Omega\,cm$	Temp. range, T_m to K
Aluminium	Al	933.5	(660.3)	1.45	10.7	1473
				$24.19 + 1.306 \times 10^{-2}\,(T-T_m)^a$		1160
Antimony	Sb	903.8	(630.6)	2.70	87.9	1273
Barium	Ba	1000	(727)	Insufficient data		
Bismuth	Bi	544.2	(271.4)	5.70	99.2	1273
Cadmium	Cd	594.2	(321.1)	Not linear		
Cobalt	Co	1768	(1495)	6.12	−6	1973

Table 9.3 *(continued)*

Element		T_m		$\alpha/10^{-2}$ $\mu\Omega\,\mathrm{cm\,K^{-1}}$	β $\mu\Omega\,\mathrm{cm}$	Temp. range, T_m to K
		K	(°C)			
Copper	Cu	1357.8	(1084.6)	0.89	9.1	1473
				1.02	6.2	1873
Gallium	Ga	302.9	(29.8)	1.95	19.9	670
Germanium	Ge	1210	(937)	Insufficient data		
Gold	Au	1337.3	(1064.2)	1.4	12.5	1473
Indium	In	429.7	(156.6)	2.55	22.2	1273
				$28.43 + 2.48 \times 10^{-2}\,T(°C)^b$		673
Iron	Fe	1811	(1538)	3.3	50	1973
Lead	Pb	600.6	(327.5)	4.79	66.6	1273
				$79.25 + 4.82 \times 10^{-2}\,T(°C)^b$		666 *to*
Magnesium	Mg	923	(650)	0.50	22.9	1173
Manganese	Mn	1519	(1246)	Insufficient data		
Mercury	Hg	234.3	(−38.8)	Not linear		
Nickel	Ni	1728	(1455)	1.27	63	1973
Potassium	K	336.5	(63.4)	6.4	−8.5	573
Rubidium	Rb	312.5	(39.3)	8.6	−4.8	373
Selenium	Se	493.7	(220.5)	Not linear		−
Silicon	Si	1685	(1412)	11.3	−113	1820
Silver	Ag	1234.9	(961.8)	0.90	6.2	1473
Sodium	Na	370.9	(97.7)	3.8	−4.5	573
Tellurium	Te	723	(450)	Not linear		
Thallium	Tl	577	(304)	2.92	56.3	1073
Tin	Sn	505.1	(231.9)	2.49	35.4	1473
				$41.16 + 2.64 \times 10^{-2}\,T(°C)^b$		803
				$40.64 + 2.68 \times 10^{-2}\,T(°C)^b$		
Zinc	Zn	692.7	(419.5)	Not linear		

Data, except for those bearing the superscript a or b, are taken from Cusack and Enderby [27].

[a] Data from Rhim and Ishikawa [21].

[b] Data from Kozuka and Morinaga [28].

Table 9.4 *Electrical resistivities of the periodic Group IIA metals in the liquid state.*

Group IIA metal		T_m °C	$\rho_{e,1}$* μΩ cm	$\partial\rho_{e,1}/\partial T$ μΩ cm K^{-1}	Temp. range, T_m to °C
Barium	Ba	729	306 ± 11	−0.04 ± 0.05	780
Calcium	Ca	839	33.0 ± 1.4	0 ± 0.01	880
Magnesium	Mg	648	26.1 ± 0.7	0 ± 0.03	750
Strontium	Sr	768	84.08 ± 2.0	−0.003 ± 0.007	820

Data are taken from Kozuka and Morinaga [28]. (See also Van Zytveld et al. [29]).
* At the melting point.

Table 9.5 *Electrical resistivities of several pure liquid metals measured by different investigators.*

Metal	Investigator	$\rho_{e,1}$† μΩ cm	$(\rho_{e,1}/\rho_{e,s})$†	$d\rho_{e,1}/dT/10^{-2}$ μΩ cm K^{-1}
Sn	Roll and Motz	48.0	2.10	2.50
	Kita et al.	47.42		2.673
Cu	Roll and Motz	21.1	2.04	0.866
	Kita et al.	21.50	2.03	0.837
Fe	Powell	138.6	1.09	
	Arsentiev et al.	135.1	1.06	3.89
	Ono and Yagi	137.8	1.05	4.34
	Kita et al.	135.9	1.06	1.54
Co	Regeli et al.	102	1.05	6.1
	Ono and Yagi	126.5	1.21	3.84
	Kita et al.	120.5	1.14	1.92
Ni	Regeli et al.	85.0	1.3	1.3
	Ono and Yagi	85.0	1.47	2.80
	Kita et al.	87.0	1.40	1.16

† At the melting point.   Ω cm = 10^{-2} Ω m
(After Kita et al. [19]).

In general, a liquid metal's electrical resistivity is raised when foreign atoms are introduced into the melt. In other words, a liquid alloy's electrical resistivity generally shows positive deviations from additivity of its components. Some examples for the resistivities of liquid alloys are given in Figures 9.6 to 9.9. However, in the case of a liquid alloy system which is composed of polyvalent components, the resistivity can sometimes

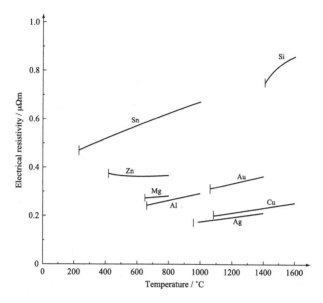

Figure 9.5 *Electrical resistivity of several liquid metallic elements as a function of temperature. Data from Gale and Tolemeier [30].*

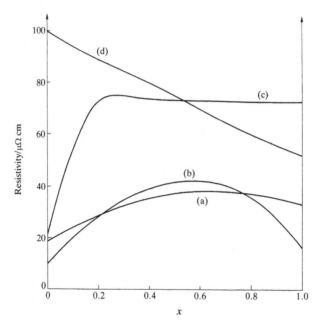

Figure 9.6 *Resistivity ρ_e vs. atomic concentration x for four typical binary liquid alloys: (a) Ag–Au at 1200 °C; (b) Na–K at 100 °C; (c) Cu–Sn at 1200 °C; (d) Pb–Sn at 400 °C (after Faber and Ziman [31]).*

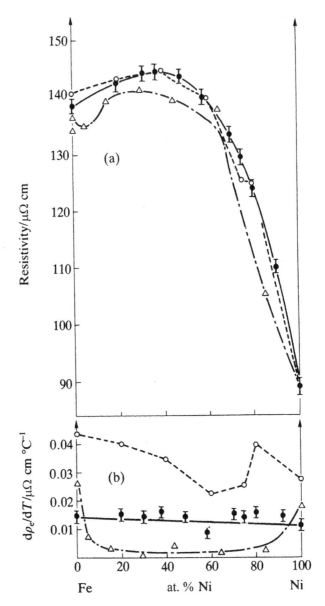

Figure 9.7 *(a) Electrical resistivity ρ_e at 1600 °C and (b) temperature coefficient $d\rho_e/dT$ for liquid iron–nickel alloys: •, Kita and Morita; ○, Ono and Yagi; △, Baum et al. (after Kita and Morita [32]).*

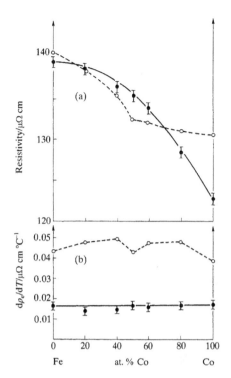

Figure 9.8 *(a) Electrical resistivity ρ_e at 1600 °C and (b) temperature coefficient $d\rho_e/dT$ for liquid iron–cobalt alloys: •, Kita and Morita; ○, Ono and Yagi (after Kita and Morita [32]).*

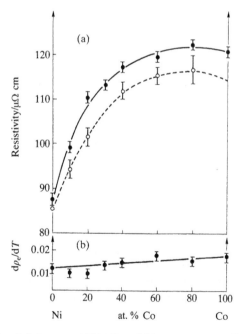

Figure 9.9 *(a) Electrical resistivity ρ_e at 1500 °C and (b) temperature coefficient $d\rho_e/dT$ for liquid nickel–cobalt alloys: •, Kita and Morita; ○, Dupree et al. (after Kita and Morita [32]).*

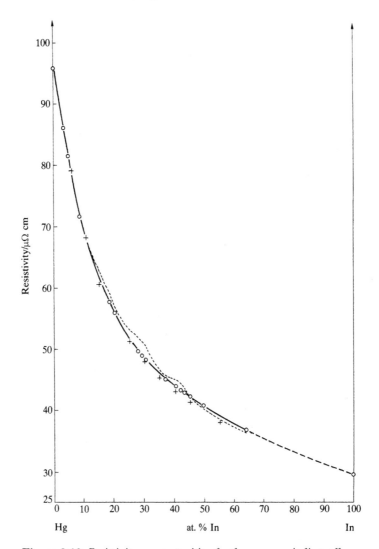

Figure 9.10 *Resistivity vs. composition for the mercury–indium alloy system at 20 °C: ○, Cusack et al.; +, Schulz; – – –, Roll and Swamy (after Cusack et al. [33]).*

show a negative deviation from additivity of component resistivities. As indicated in Figure 9.10, the isothermal electrical resistivity of liquid mercury–indium alloy system has a negative deviation from that based on additivity of resistivities.

The electrical resistivities of a large number of liquid metallic elements are given in Table 17.11.

9.5.2 Experimental Data for Thermal Conductivity

It is not easy to establish accurate and reliable experimental data as well as rigorous theories for the thermal conductivities of metallic liquids. Fairly large discrepancies exist between experimental data for the thermal conductivity of metallic liquids; reported thermal conductivities of metallic liquids vary by about ±7 to ±25 per cent around the mean.

The thermal conductivity of a metal, as well as its electrical conductivity, normally decreases on melting because the disordered arrangement of ions in the liquid state. However, metallic liquids are still much better heat conductors than non-metallic liquids. This is shown in Figure 9.11 (see also Figures 9.12 and 9.13).

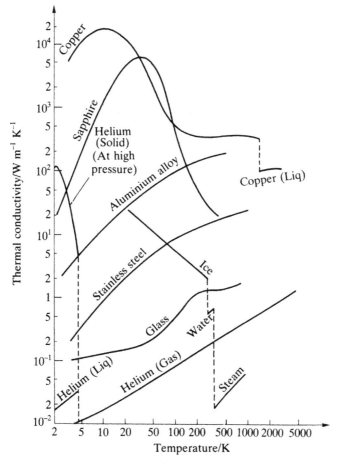

Figure 9.11 *Typical curves showing temperature dependence of thermal conductivity (after Powell and Childs [34]).*

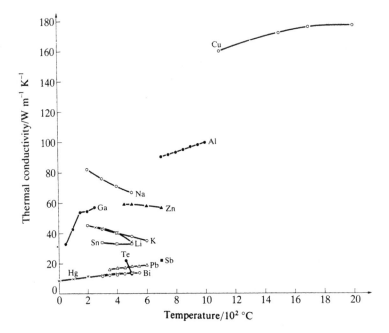

Figure 9.12 *Thermal conductivity of some liquid metallic elements as a function of temperature. Data from Powell and Childs [34].*

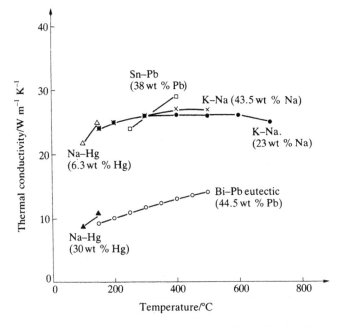

Figure 9.13 *Thermal conductivity of several binary liquid alloys as a function of temperature. Data from Powell and Childs [34].*

Figures 9.12 and 9.13 illustrate some experimental vales of λ for liquid metallic elements and binary liquid alloys, respectively. As is obvious from these figures, the temperature dependence of thermal conductivities of metallic liquids is quite complex and variable.

Experimental values for the thermal conductivity of various liquid metallic elements are also shown in Table 17.11.

....

REFERENCES

1. C. Kittel, *Introduction to Solid State Physics*, 7th ed., John Wiley & Sons, Inc., 1996, Chaps. 5 and 6 (see also 2nd ed., 1956, Chaps. 6 and 10).
2. N.F. Mott, *Proc. R. Soc. Lond.*, *A*, **146** (1934), 465.
3. J.M. Ziman, *Phil. Mag.*, **6** (1961), 1013.
4. T.E. Faber, *Introduction to the Theory of Liquid Metals*, Cambridge University Press, 1972, Chaps. 3 and 5.
5. R. Evans, D.A. Greenwood, and P. Lloyd, *Phys. Lett. A*, **35** (1971), 57.
6. R. Evans, B.L. Gyorffy, N. Szabo, and J.M. Ziman, in *Proceeding of the 2nd International Conference on Liquid Metals*, Tokyo, edited by S. Takeuchi, Taylor and Francis, London, 1973, p. 319.
7. Y. Waseda, *The Structure of Non-Crystalline Materials, Liquids and Amorphous Solids*, McGraw-Hill, New York, 1980, p.203.
8. S. Takeuchi and H. Endo, *J. Japan Inst. Met.*, **26** (1962), 498.
9. S. Takeuchi and H. Endo, *Trans. JIM.*, **3** (1962), 30; S. Takeuchi and H. Endo, *Trans. JIM.*, **3** (1962), 35.
10. J.L. Tomlinson and B.D. Lichter, *Trans. Met. Soc. AIME*, **245** (1969), 2261.
11. T. Iida and R.I.L. Guthrie, *The Physical Properties of Liquid Metals*, Clarendon Press, Oxford, 1988, p.250.
12. G. Busch, H.-J. Güntherodt, and P. Wyssmann, *Phys. Lett. A*, **39** (1972), 89.
13. G. Busch, H.-J. Güntherodt, W. Haller, and P. Wyssmann, *Phys. Lett. A*, **41** (1972), 29; G. Busch, H.-J. Güntherodt, W. Haller, and P. Wyssmann, *Phys. Lett. A*, **43** (1973), 225.
14. W. Haller, H.-J. Güntherodt, and G. Busch, *Liquid Metals 1976*, Institute of Physics Conference Series No. 30, 1977, p.207.
15. K.C. Mills, B.J. Monaghan, and B.J. Keene, *Int. Mater. Rev.*, **41** (1996), 209.
16. E. Yamasue, M. Susa, H. Fukuyama, and K. Nagata, *J. Cryst. Growth*, **234** (2002), 121.
17. P.D. Adams and J.S.Ll. Leach, *Phys. Rev.*, **156** (1967), 178.
18. Y. Mera, Y. Kita, and A. Adachi, *Technol. Rep. Osaka Univ.*, **22** (1972), 445.
19. Y. Kita, S. Oguchi, and Z. Morita, *Tetsu-to Hagané*, **64** (1978), 711.
20. Y. Ono and T. Yagi, *Trans. ISIJ.*, **12** (1972), 314.
21. W.-K. Rhim and T. Ishikawa, *Rev. Sci. Instrum.*, **69** (1998), 3628.
22. G. Lohöfer, *Meas. Sci. Technol.*, **16** (2005), 417.
23. K.C. Mills, in *Fundamentals of Metallurgy*, edited by S. Seetharaman, Woodhead Publishing, Cambridge, 2005, p.109.
24. H. Fukuyama and Y. Waseda (eds.), *High-temperature Measurements of Materials*, Springer, Berlin, 2009.

25. M.J. Duggin, in *Thermal Conductivity*, edited by C.Y. Ho and R.E. Taylor, Plenum Press, 1969, p.727.
26. V. Sklyarchuk and Yu Plevachuk, *Meas. Sci. Technol.*, **16** (2005), 467.
27. N. Cusack and J.E. Enderby, *Proc. Phys. Soc.*, 75 (1960), 395.
28. Z. Kozuka and K. Morinaga, in *Handbook of Physico-Chemical Properties at High Temperatures*, edited by Y. Kawai and Y. Shiraishi, Iron and Steel Institute of Japan, Tokyo, 1988, Chap. 8.
29. J.B. Van Zytveld, J.E. Enderby, and E.W. Collings, *J. Phys. F, Met. Phys.*, 2 (1972), 73.
30. W.F. Gale and T.C. Tolemeier (eds.), *Smithels Metals Reference Book*, 8th ed., Elsevier Butterworth-Heinemann, Oxford, 2004, 14–9.
31. T.E. Faber and J.M. Ziman, *Phil. Mag.*, **11** (1965), 153.
32. Y. Kita and Z. Morita, in *Liquid and Amorphous Metals, Part II*, edited by C.N.J. Wagner and W.L. Johnson, North-Holland Physics Publishing, Amsterdam, 1984, p.1079.
33. N. Cusack, P. Kendall, and M. Fielder, *Phil. Mag.*, **10** (1964), 871.
34. R.L. Powell and G.E. Childs, *American Institute of Physics Handbook*, McGraw-Hill, New York, 1972, 4–142.

Index

A

absolute
  density determinations
    97–98
  surface tension 212
  temperature 7, 23, 188,
    236, 283, 314
  viscosities 264, 265
absorption coefficient per unit
  mass 102
accuracy 19, 24–25, 101, 103,
    135, 142, 167, 174,
    221, 255, 258, 268,
    278, 284, 292
accurate
  density 78, 97
  determination 210, 319
  experimental
    determinations 22, 250
  (property) predictions
    22, 135
  self-diffusivities 277
  sound velocities 167
  surface tensions 186, 213
  viscosities 236, 264, 268
acoustic technique 101
actinides (or actinoids) 33
actinium, Ac 8, 88, 115
actinoid metals/elements
    74, 76
activation energy 234, 235,
    237, 284
activation-state theory (or
    model) 284
activity 198
  coefficient 251
  components 205
  solute 197
additional binding forces 36
additivity
  component molar
    volumes 105
  components (electrical
    resistivity) 328–332
  densities 107

adiabatic (or isentropic)
    compressibility 141
adsorbing species 202
adsorption 197
  behaviour of solute 197
  coefficient 202
  of monolayer 202
aerodynamic (technique) 101
affinity for oxygen 212
Ag-Au 206, 207, 252, 329
Ag-Cu 252
Ag-O 203
Ag-Sb 252, 253
Ag-Sn 252
A groups (periodic table) 33
Al-Cu 97, 105, 252
Alder transition 77
Alder-Wainwright correction
    C_{AW} 281
algebraic expression 129
alkali metals 33, 36, 37,
    39, 79, 82, 112, 146,
    175–177, 212, 226, 264
alloying elements 200
Al-Mg 252
alumina 96, 98, 311, 321
Aluminium, Al 8, 33, 41,
    60–62, 75, 88, 97, 103,
    115, 119, 124, 144,
    145, 149, 154, 156,
    159, 166, 183, 190,
    194, 196, 212, 220,
    239, 241, 243, 244–246,
    248, 270, 272, 278,
    311, 316, 325, 326
aluminium-copper alloys 97,
    105, 252
Al-Zn 252
amalgam 199, 201, 250
americium, Am 8, 74, 76,
    88, 115
amorphous
  semiconductor silicon 68
  solids 45, 78
amplitude
  magnetization 309
  plate oscillations 261

analytical formula (viscosity
    calculations) 258
Andrade
  coefficient 25, 185, 186,
    226, 227, 233, 246
  correlation 240
  derivation 226
  formula/equation 22, 23,
    25, 223, 226–228, 235,
    237, 240–244, 286
angular
  displacement (suspension
    fibre) 260
  velocity 260
anharmonic
  effects 141, 163
  motions 22, 86
anomalous (behaviour) 37, 74,
    145, 227
  discontinuities 107
  metals/metallic elements
    167, 177, 186, 189,
    246, 249
  sound velocity 145,
    167–170, 186–187
  viscosity 186–187, 227,
    240, 272
antimony, Sb 1, 8, 12, 36, 42,
    60, 62, 76, 88, 103,
    115, 126, 144, 145,
    149, 154, 156, 166,
    177, 183, 186, 189,
    190, 194, 196, 226,
    241, 243–246, 248, 258,
    288, 289, 291, 293,
    295, 296, 316, 324–326
Apparent
  activation energy (viscous
    flow, diffusion) 236,
    248, 287, 296
  frequency 228, 230
  loss of weight 95
  mass transfer
    coefficients 171
applied electric field 312